I0044455

Animal Biology and Diversity

Animal Biology and Diversity

Editor: Adalina Woodbury

www.callistoreference.com

Callisto Reference,
118-35 Queens Blvd., Suite 400,
Forest Hills, NY 11375, USA

Visit us on the World Wide Web at:
www.callistoreference.com

ISBN: 978-1-64116-125-1 (Hardback)

Cataloging-in-Publication Data

Animal biology and diversity / edited by Adalina Woodbury.
p. cm.
Includes bibliographical references and index.
ISBN 978-1-64116-125-1
1. Zoology. 2. Animal diversity. 3. Animals. 4. Biology.
I. Woodbury, Adalina.
QL45.2 .A55 2019
590--dc23

Table of Contents

Preface

Every book is a source of knowledge and this one is no exception. The idea that led to the conceptualization of this book was the fact that the world is advancing rapidly; which makes it crucial to document the progress in every field. I am aware that a lot of data is already available, yet, there is a lot more to learn. Hence, I accepted the responsibility of editing this book and contributing my knowledge to the community.

Animal biology or zoology is the scientific study of all structural, embryological, evolutionary and taxonomical aspects of animals. The study of animal behavior as well as distribution is also encompassed in this domain. Zoology has been divided into a number of sub-disciplines and related fields. Some of these are comparative anatomy, animal physiology, vertebrate zoology, invertebrate zoology, behavioral ecology, paleontology, etc. This book aims to serve as a resource guide in the fields of animal biology and diversity by presenting upcoming theories and concepts relevant to these fields. Different approaches, evaluations, methodologies and advanced studies in the area of zoology have been included herein. This book is a complete source of knowledge on the present status of this important field.

While editing this book, I had multiple visions for it. Then I finally narrowed down to make every chapter a sole standing text explaining a particular topic, so that they can be used independently. However, the umbrella subject sinews them into a common theme. This makes the book a unique platform of knowledge.

I would like to give the major credit of this book to the experts from every corner of the world, who took the time to share their expertise with us. Also, I owe the completion of this book to the never-ending support of my family, who supported me throughout the project.

Editor

1

Syzygium aromaticum (L.) elicits lifespan extension and attenuates age-related Aβ-induced proteotoxicity in *Caenorhabditis elegans*

Aakanksha Pant[1*], Prem Prakash[1], Rakesh Pandey[2] and Rishendra Kumar[3]

*Corresponding author: Aakanksha Pant, Department of Botany, Government P.G. College, Kumaun University, Dwarahat, Almora 263653, India
E-mail: pant.aakanksha@yahoo.com

Reviewing editor: Tsai-Ching Hsu, Chung Shan Medical University, Taiwan

Additional information is available at the end of the article

Abstract: Clove (*Syzygium aromaticum*) is a popular medicinal plant which has been traditionally used in India as spice and medicine to counter various ailments. *However, the stress modulatory and antiaging potential of this plant is yet to be characterized.* Therefore, the present study evaluates the effect of clove oil (CO) on oxidative stress, lifespan, mobility, and the expression of aging-related proteins using *Caenorhabditis elegans* model system. The CO (10 ppm) was found to extend mean lifespan in worms by 21.4% ($p < 0.001$) under normal and by 63% ($p < 0.0001$) under juglone-induced oxidative stress conditions. The extension of mean lifespan in *mev-1* mutant and elevated expression of *gst*-4 and *sod*-3 confirmed stress modulatory effects of CO. Additionally, the CO reduced intracellular ROS and $A\beta_{1-42}$ proteotoxicity. Altogether, the present study unravels the anti-aging and stress modulatory potential of CO and suggests CO as a potential pharmaceutical entity in modulating aging process.

Subjects: Bioscience; Microbiology; Neuroscience; Pharmacology; Toxicology

Keywords: clove oil; *Caenorhabditis elegans*; lifespan; Alzheimer's disease

1. Introduction

Ayurveda, the word derived from Sanskrit itself means "the scripture of longevity". The practice finds its foundation in holistic view of treatment which cures human diseases all the way through the establishment of equilibrium between elements of human life, the body, the mind, and the soul. The scientific investigation concerning the best known of these traditionally practiced herbs is need of the hour and therefore, we investigated anti-aging and stress modulating potential of widely used

ABOUT THE AUTHOR

Aakanksha Pant has published peer-reviewed articles in the area of aging, stress biology, dietary interventions regulating aging and age associated disorder. She has studied human homologous *Caenorhabditis elegans* model system for deciphering role of various signaling pathway in regulating aging phenomenon. The author has keen interest in identifying plant-based natural bioactive molecule which will be able to elicit anti-aging effects and can be utilized for formulating therapies for various stress and age disorders.

PUBLIC INTEREST STATEMENT

Spices and herbs have been part of various food preparations since ancient time till today. These spices contribute to various beneficial therapeutic effects in human. Clove is one of these spices/herbs which has proven efficacy in treating dental problems, inflammation, and neurodegenerative disorders. Therefore, present study evaluates effect of clove oil on aging and age-related neurodegenerative Alzheimer's disease using a free living soil nematode *Caenorhabditis elegans* organism model which shares genetic similarity with human. Clove oil was found to improve health span and lifespan of *C. elegans* in dose-dependent manner. The study unravels potential of clove oil in delaying aging and age-related pathologies.

popular medicinal and aromatic plant clove (*Syzygium Aromaticum*). This popular spice has proven efficacy in treating dental problems, inflammation, and neurodegenerative disorders (Adams, Gmünder, & Hamburger, 2007; Chaieb et al., 2007). Despite its well-known therapeutic potential, the plant is yet not fully explored against aging disorders. Aging is a multifaceted phenomenon and a time-dependent decline in physiological functions which is universal to every living organism (Kenyon, 2010). The hallmark of cell senescence is the onset of various age-related afflictions including neurodegeneration, cardiovascular disease, and cancer (Guarente & Kenyon, 2000). The main goal of gerontological research is to identify pharmacological molecules that can delay age-related diseases and maintain vitality of later life in humans. A number of earlier reports have established that active constituents of different plant extracts are efficient in reversing aging, extending lifespan, and improving stress tolerance (Arya, Dwivedi, & Subramaniam, 2009; Asthana et al., 2015; Kampkötter et al., 2008). Thus, a proper understanding of clove oil (CO) on aging and longevity modulation is still a unfolded mystery and should be investigated. As an organism, soil nematode *Caenorhabditis elegans* ages and shares similar aspects of aging like human (Kenyon, 2010). Further, the advantageous characteristics of this model organism are attributed to its genetic pliability, invariant and fully described developmental program, well characterized genome, ease of maintenance, short and fertile life cycle, and small body size (Kenyon, 2010). *C. elegans* shares more than 80% gene homology with human and protein network regulating aging is conserved between worm and human which makes this nematode model a powerful tool for the screening of lifespan modulators (Kenyon, 2010). The present investigation attempts to determine the effects of CO on oxidative stress, lifespan, mobility, neurotoxicity, and the expression of aging-related proteins.

2. Experimental

2.1. Plant material and GC/MS analysis

The *Syzygium aromaticum* buds were procured from a local market. The volatile oil was obtained by conventional hydrodistillation of the buds of *S. aromaticum* in a clevenger-type apparatus. The GC and GC/MS analyses were performed using previously described standard methods (Adams, 2012; Kollmannsberger & Nitz, 1994; Srivastava, Srivastava, & Syamsundar, 2005). The GC and GC/MS analyses resulted in the identification of major constituent as eugenol (65%, Figure S1).

2.2. Caenorhabditis elegans maintenance and egg preparation

The wild-type *C. elegans* Bristol strain N_2, TK22, *mev-1* (kn1), CL4176 (dvIs27 [myo3::Aβ let 3′UTR (pAF29); pRF4 (rol6 (su1006)]), CL 2006 (dvIs2 [pCL12 (unc-54/human Aβ peptide 1–42 minigene) + pRF4]) was used in this experiment. Animals were maintained on nematode growth medium (NGM) and fed with OP-50 strain *Escherichia coli* at 20°C (unless otherwise indicated). A synchronized culture was obtained by sodium hypochlorite treatment (50% sodium hypochlorite [12% Cl]; 2.5 M sodium hydroxide), which kills adult worms, but not their eggs (Brenner, 1974). The strain used in this experiment was obtained from the *Caenorhabditis* Genetics Centre, University of Minnesota, and Minneapolis, USA.

2.3. Lifespan analysis

Age-synchronized N2 worms were used for lifespan assay (Pant et al., 2014). Worms were synchronized by alkaline hypochlorite treatment. The isolated eggs were allowed to hatch on NGM plates previously spotted with or without different concentrations of CO different doses (1, 10, 100 ppm) till L4 stage. About 25–30 L4 molts were then transferred to NGM plates previously spotted with corresponding test concentration and 50-μM FUdR (Sigma-Aldrich) to block progeny development. Worms were then observed daily for survival and transferred to fresh plates after every 3–4 days to avoid contamination and to assure the presence of the compound throughout experiment. The experiment was terminated when all worms were scored as dead or censored. Three independent trials were performed for all treatments, and the data shown represent three replicates with similar effects on longevity.

2.4. Stress assays

The effect of CO on stress response of worm was assessed by exposing treated worms to juglone--induced oxidative stress and heat-induced thermal stress (Pant et al., 2014). For assessing thermo-tolerance, age-synchronized N2 worms were raised on treatment (CO) and control plates at 20°C and shifted to 35°C as day-2 adults. The survival of worms was scored by touch provoke method (Lithgow, White, Melov, & Johnson, 1995). Furthermore, oxidative stress resistance was assessed by exposing age-synchronized CO and EU pre-treated and control day-2 adult worms to lethal dose (250 µM) of juglone (5-Hydroxy-1,4-napthoquinone, Sigma-Aldrich). The survival was scored every hour after the treatment. The experiments were done in three replicates independent of each other.

2.5. Measurement of body size

The fourth larval stage (L4) worms exposed onto a bacterial lawn spot of an NGM plate containing different doses (1, 10, 100 ppm) of CO and control plates without the phytomolecule which were incubated for 24 h were directly picked for body size measurement (Pant et al., 2014). The body sizes of more than 20 animals in the photo pictures with a scale were randomly measured using the Leica Application Suite V3 software (version 3.4.0). The experiment was performed independently thrice.

2.6. Brood size assay

Synchronous wild-type hermaphrodites were cultured at 20°C for 3 days until they reached the L4 stage. One worm each (three replicates of each treatment) was then transferred and maintained on a separate NGM plate with (10, 100 ppm) of CO or without treatment and incubated at 20 ± 0.5°C in incubator. Individual worms were transferred daily; the progeny were left to develop for 3 days before counting (Pant et al., 2014). After three days, the progeny were counted under stereoscopic microscope.

2.7. Measurement of aging phenotype

2.7.1. Assessment of feeding behavior

The progression in age is correlated to decline in pharyngeal pumping rate. The effect of CO on feeding behavior was evaluated by counting the pharyngeal contractions and relaxation. Pharyngeal pumping is defined as number of contractions (i.e. backward grinder movements in the terminal bulbs). This assay was performed with day-2 and day-4 adult worms on approximately 40 worms at regular interval of 20 s at room temperature (Pant et al., 2014). The *p*-value was calculated using Assistat 7.7 beta statistical assistance software. The assay was performed three times independently.

2.7.2. Lipofuscin assay

The effect of CO treatment on age pigment lipofuscin was evaluated using previously described method (Pant et al., 2014). The day-4 adult worms ($n = 20$) were randomly selected from each treatment group (CO and control) and observed for autoflorescence. The lipofuscin levels were quantified by determining the average pixel intensity in each worm using Image-J software (NIH). The *p*-value was calculated using Assistat 7.7 beta statistical assistance software. The assay was performed thrice independently.

2.8. Measurement of intracellular ROS in C. elegans

The intracellular ROS levels were determined according to Smith and Luo, 2003 with minor modifications. Adult day-4 worms treated with or without CO were used for intracellular ROS determination. Worms were collected in 300 µl of 0.1% PBST and equally timed homogenization and sonication. The homogenized samples were transferred to 96 well plate and prior to reading, 15 µl of 10 mM H_2DCF-DA was added to each well. Fluorescent readings were measured using Spectra Max M2 multimode microplate reader, (Molecular Devices) at 485 nm excitation and 530 nm emission. Observations were recorded at every 20 min for 2 h and 30 min at 37°C. The test was performed three times independently.

2.9. Worm paralysis assays

The CL4176 strain [dvIs2 [pCL12 (unc54/human Aβ minigene) + pRF4] containing a heat-sensitive mutation developed to express human amyloid β_{1-42} ($A\beta_{1-42}$) present in the muscle tissue was maintained on NGM plates at 16°C (Dostal, Roberts, & Link, 2010). Previous to the beginning of the experiment, *C. elegans* were age synchronized at 16°C on CO pre-treated NGM plates. L1 worms from F2 generation were transferred to control or treatment plates and allowed to mature gravid adult stage to lay eggs. After reaching the L3 stage, the incubation temperature of the plates was increased from 16°C to 25°C, in order to induce the expression of $A\beta_{1-42}$. The evaluation of the mobility of worms was started 18–20 h after increasing the incubation temperature in 2 h increments until all worms were paralyzed. The worms are considered paralyzed if they failed to respond to prodding and demonstrated "halos" of cleared bacteria around their heads (indicative of insufficient body movement to access food), eggs accumulation close to the body. The experiment was performed in three independent trials.

2.10. In vivo gene expression studies in transgenic C. elegans

The *in vivo* expression of stress response genes *sod*-3 and *gst*-4 was quantified with the aid of transgenic GFP reporters CF1553 (muIs84) and CL2166 (dvIs19) using fluorescence microscopy. The treated/untreated worms were photographed individually by piptetting onto 2% agarose pad on glass slide. The worms were anaesthetized with 1% sodium azide prior to microscopy. The assay was performed according to previously described methods (Pant et al., 2014) using a GFP filter (with excitation at 365 nm and emission at 420 nm) using a fluorescence microscope DMI 3000 B (Leica, Wetzlar, Germany) at 20X. The fluorescence levels were quantified using Image-J software (NIH).

2.11. Statistical analysis

Experiments were performed at least in triplicate. Significant differences between the lifespan of treated and control worms under normal/stressed conditions were determined using Kaplan-Meir survival assay in Med Calc software version 12.7.7.0. Data are presented as mean ± SD, and student's *t*-test or one-way ANOVA analysis was used to determine the statistical significance between experimental groups. Statistical significance was defined as $^{*}p < 0.05$, $^{**}p < 0.01$, and $^{***}p < 0.001$.

3. Results and discussion

3.1. Clove oil promotes healthy lifespan in C. elegans

An increment in aging world population and age-related ailments have sidetracked the aging research towards dietary interventions promoting longevity with maintenance of vitality of later life. Aging is a global phenomena accompanied with decline in ability of an organism to maintain cellular homeostasis with time (Kenyon, 2010). The significant progress in field of gerontological sciences has focused research towards aging and factors modulating lifespan, but mechanisms extending lifespan of an organism still remain elusive. Therefore, we evaluated the age defying effects of CO which is commonly used in traditional Ayurvedic medicine, dentistry, and food preparations for its potential health benefits in humans. The N2 wild-type worms were exposed to different doses of CO (1, 10 and 100 ppm) from early stages of lifespan and maximal lifespan extension of 21.4% ($p < 0.0001$) was recorded in 10 ppm followed by 14.35% ($p < 0.0001$) in 100 ppm and 6.99% ($p < 0.0001$) in 1 ppm (Figure 1(A), Table 1). The extension in lifespan by different interventions is oftenly associated with decline in feeding behavior, growth, and fertility of an organism (Gruber, Tang, & Halliwell, 2007; Kenyon, 2010). The extension in mean lifespan in exchange of health span and vitality is not the prime goal of aging research. Therefore, we evaluated effect of CO on pharynx pumping rate, growth and fecundity of worm. The non-significant difference between treated (CO) and untreated (control) was observed suggesting lifespan extension with maintenance of health span (Figure S2(A)–(C)). The present findings are supported by previous studies where lifespan extension is mediated by various natural herbs and molecules.

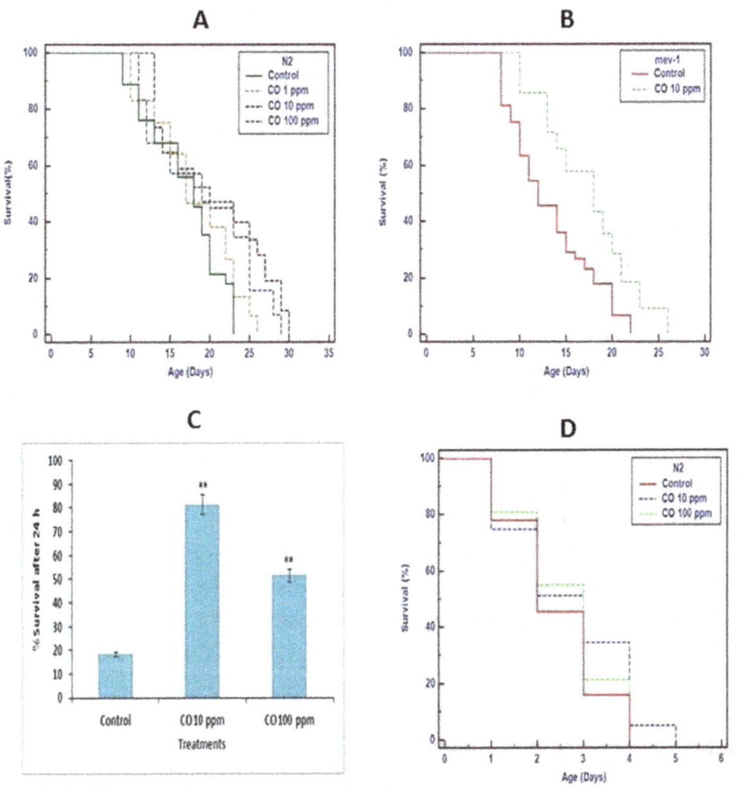

Figure 1. Effect of CO on lifespan in wild type and mutant strain of *C. elegans* at 20°C. (A) The maximum extension of 21.4% (*p* < 0.0001) in lifespan was recorded in 10 ppm CO followed by 14.35% (*p* < 0.0001) in 100 ppm and 6.99% (*p* < 0.0001) in 1 ppm; (B) the percentage mean survival in 10 ppm CO-treated *mev-1 (kn1)* mutants is enhanced by 30.94% (*p* < 0.0001) in comparison to control. The data were analyzed using the Kaplan–Meir survival analysis in Medcalc 12.7.7.0 software. Effect of CO on stress level; (C) CO (10 ppm) treatment significantly reduced the juglone sensitivity in pre-treated worms compared to untreated control worms mediating improved survival under oxidative stress conditions; (D) CO (10 ppm) treatment significantly enhanced mean lifespan of worms in comparison to control under thermal stress conditions. The data are statistically analyzed using ANOVA in ASSISTAT 7.7 beta statistical assistance software. Differences between the data were considered significant at *p* ≤ 0.05.

Note: Error bars represent means ± S.E.M. **p* ≤ 0.05. ***p* ≤ 0.01.

3.2. CO reduces intestinal lipofuscin autoflorescence

The lifespan progression is marked with increment in oxidative stress and aggregation of byproducts of macromolecular damage which causes stress in an organism (Garigan et al., 2002). The accumulation of auto fluorescent intestinal age pigment lipofuscin is a biomarker of aging in worms. The aggregation of lipofuscin protein is an aging biomarker. The age-related stress is directly linked with lipofuscin aggregation. The significant decline in intestinal lipofuscin was observed in CO-treated worm by

Table 1. Lifespan analysis of wild-type N2 and *mev-1* mutant at 20°C

Genotypes	Treatments	Mean lifespan	±SD	±SE	Sample size (N)	% Change	*p*-value
N2 wild type	Control	16.86	4.73	0.386	150		
N2 wild type	CO 1 (ppm)	18.04	5.21	0.406	165	6.99	<0.0001
N2 wild type	CO 10 (ppm)	20.51	6.43	0.480	178	21.64	<0.0001
N2 wild type	CO 100 (ppm)	19.11	6.24	0.45	185	13.34	<0.0001
mev-1 (kn1)	Control	13.28	4.52	0.348	169		
mev-1 (kn1)	CO 10 (ppm)	17.39	4.85	0.391	152	30.94	<0.0001

15.78% ($p = 0.0019$). Whereas, significant increase was recorded in control worms (Figure S2(D)–(F)). The significant decline in lipofuscin aggregation suggests that CO can maintain function of primary-targeted organ with progression in aging process.

3.3. CO confers stress tolerance and attenuates ROS level in C. elegans

The increment in stress and intracellular ROS levels leads to decline in mean survival in organisms (Abdollahi, Moridani, Aruoma, & Mostafalou, 2014; Epel & Lithgow, 2014). The progression in lifespan is accompanied by cellular metabolic decline and disruption of normal cellular redox function which hampers cellular macromolecular organization (Kenyon, 2010). Therefore, the present study evaluates effect of CO on stress level and intracellular ROS level in worms. The CO pre-treated worms demonstrated enhanced mean survival by 63% ($p < 0.0001$) in 10 ppm under intracellular ROS generator juglone-induced oxidative stress (Figure 1(C)). Additionally, enhanced survival was observed in CO (10 ppm)-treated worms by 10.87% ($p = 0.0119$) under thermal stress condition (Figure 1(D)). To investigate whether the increment in mean survival under normal and oxidative stress conditions is associated with reduction in ROS level, we employed DCF-DA method using whole live nematode for quantification of *in vivo* ROS. The decline of 26% ($p = 0.047$) in ROS level was observed in CO (10 ppm)-treated worms in comparison to untreated control worms (Figure S3). Furthermore, CO-treated *mev-1* mutant worm demonstrated prolonged lifespan by 30.94 ($p < 0.0001$) in comparison to untreated control suggesting stress alleviating potential of CO (Figure 1(B), Table 1). Altogether, the results suggest CO exposure mediates increment in stress tolerance level compared to untreated control. The results are supported by previous findings where enhanced stress tolerance was found to extend mean lifespan in various organisms.

3.4 CO-delayed β-amyloid proteotoxicity in C. elegans

The extension in lifespan is of little benefit if it doesn't revive overall health of living organism. The progression in age is accompanied by an increment in oxidative stress which has implication in age-related pathologies like neurodegeneration (Shaw, Werstuck, & Chen, 2014). The age-related pathologies elicit protein aggregation and misfolding (Dostal & Link, 2010). The disruption of protein

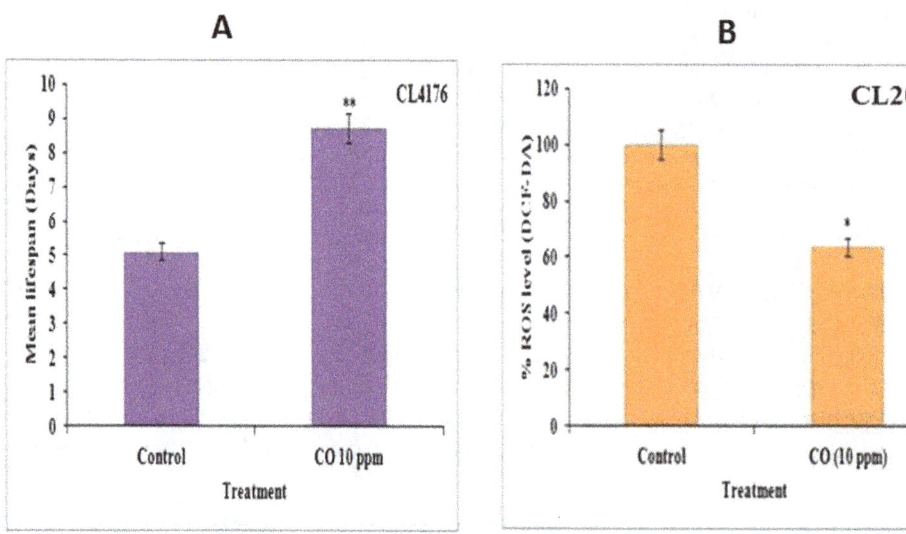

Figure 2. CO treatment-attenuated Aβ-induced paralysis in *C. elegans* CL4176 and intracellular ROS mediated oxidative stress due to Aβ proteotoxicity in CL2006 transgenic strains. (A) CO (10 ppm) treatment was able to rescue worms from Aβ-induced paralysis thereby enhancing mean lifespan in worms in comparison to control; (B) CO (10 ppm) treatment reduced Aβ proteotoxicity in CL2006 transgenic worms by reducing intracellular ROS mediated oxidative stress. The data are statistically analyzed using ANOVA in ASSISTAT 7.7 beta statistical assistance software. Differences between the data were considered significant at $p \leq 0.05$.

Note: Error bars represent means ± S.E.M $^*p \leq 0.05$. $^{**}p \leq 0.01$.

homeostasis leads to accumulation of insoluble proteins aggregate such as β-amyloids which is associated with Alzheimer's associated neurotoxicity and cell death (Dostal & Link, 2010). The β-amyloids aggregation is target of various therapeutic studi es as it is a marker of AD progression (Gutierrez-Zepeda, Santell, Wu, Brown, & Wu, 2005). Therefore, we exploited *C. elegans* transgenic model of human proteotoxic disease CL4176 (dvIs27 [myo3::Aβ let 3′UTR (pAF29); pRF4 (rol6(su1006))]), which expresses an aggregating amyloid β_{1-42} peptide in muscle tissue (Dostal et al., 2010; Link, 1995, 2006). When this transgenic strain is subjected to temperature up shift from 16 to 26°C, worms express β-amyloid protein aggregates in body muscles and followed by paralysis. We observed delayed paralysis in CO-treated worms in comparison to untreated control worms (Figure 2(A)). The decline in percentage of paralyzed worms was observed in CO-treated worms as compared to control. The CO treatment decreased the proportion of paralyzed worms due to β-amyloid aggregates followed by increment in mean survival in CL4176 worms. In addition to that CO treatment demonstrated decline in ROS level in β-amyloid expressing transgenic strain CL2006 in comparison to untreated control CL2006 worms (Figure 2(B)). In *C. elegans*, increment in oxidative stress level strongly correlates with Aβ toxicity, and it has been shown that a number of natural products that reduce ROS are neuroprotective (Dostal & Link, 2010). Altogether, CO rescued paralysis phenotype and reduced ROS level in Alzheimer's worm model suggesting CO can ameliorate detrimental effects of β-amyloid induced proteoxicity.

3.5. CO alters expression of stress response genes

The aging phenomenon is followed by elevation in oxidants production and decline in antioxidant enzymes which leads to compromised cellular redox homeostasis (Kenyon, 2010). The CO treatment lead to decline in oxidative stress, ROS level and age-related β-amyloid with extension in lifespan. Furthermore, the transgenic strains stably expressing SOD-3 and GST-4 were evaluated for change in expression on CO treatment. CO exposure up regulated the expression of *sod-3* and *gst-4* (Figure 3). The CO-treated worms demonstrated elevation in SOD-3 expression by 29.7% ($p = 0.0222$) in comparison to control. Additionally, up regulation in GST-4 expression was observed on CO treatment by 19.38% ($p = 0.001$) in comparison to untreated control worms. Altogether, extension in mean lifespan and decline in cellular stress can be attributed to up regulation of antioxidant genes *sod-3* and *gst-4*.

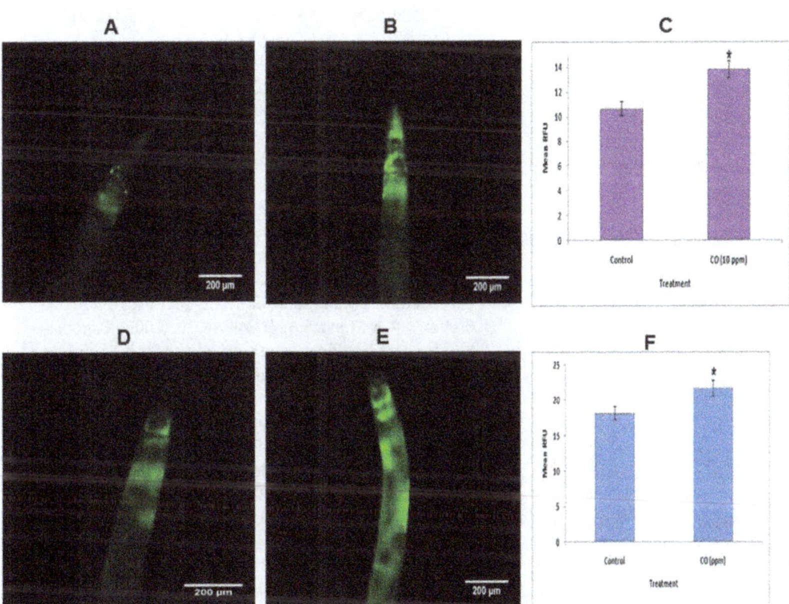

Figure 3. Effect of CO on stress response gene *sod-3*. (A) Control *sod-3*::gfp transgenic strain of *C. elegans* (*n* = 89); (B) 10 ppm CO-treated *sod- 3*::gfp transgenic strain; (C) Quantification of SOD-3::GFP expression; (D) Control *gst-* 4::gfp; (E) 10 ppm CO-treated *gst-4*::gfp transgenic strain of *C. elegans* ; (F) Quantification of GST-4::GFP expression. Scale bar = 200 μm. Scale bar = 200 μm. The data are statistically analyzed using ANOVA in ASSISTAT 7.7 beta statistical assistance software. Differences between the data were considered significant at $p \le 0.05$.

Note: Error bars represent means ± S.E.M. $^{*}p \le 0.05$. $^{**}p \le 0.001$.

4. Conclusion

The present study suggests that CO modulates stress level, which leads to lifespan extension and alleviation of Aβ proteotoxicity in *C. elegans.* The CO exposure prolonged lifespan with alleviation of stress and age-related Aβ-proteotoxicity which can be attributed to up regulation of stress response gene *gst*-4 and *sod*-3. The neuroprotective and lifespan prolonging effects demonstrated by CO can be subjected to future investigations. The neuromodulatory and longevity promoting effects are of medical interest because of its impact on age-related pathologies like AD. Thus, EU might facilitate the development of targeted therapy for AD and several age-related disorders.

Acknowledgments

The authors are also thankful to Director, CSIR-CIMAP, Lucknow and the Head of Department of Biotechnology and Botany, Kumaun University for their valuable support.

Funding

The authors are highly grateful to the Caenorhabditis Genetics Centre (CGC), Minneapolis, MN, USA, which is funded by the NIH Office of Research infrastructure Programs [grant number P40 OD010440], National Centre for Research Resources (USA), for providing the *C. elegans* strains.

Competing Interests

The authors declare no competing interest.

Author details

Aakanksha Pant[1]

E-mail: pant.aakanksha@yahoo.com

Prem Prakash[1]

E-mail: drpp_bot@yahoo.co.in

Rakesh Pandey[2]

E-mail: r.pandeycimap@gmail.com

Rishendra Kumar[3]

E-mail: kumar.rishendra@gmail.com

[1] Department of Botany, Government P.G. College, Kumaun University, Dwarahat, Almora 263653, India.

[2] Department of Microbial Technology and Nematology, Central Institute of Medicinal and Aromatic Plants, P.O. CIMAP, Almora, Lucknow 226 015, India.

[3] Department of Biotechnology, Kumaun University, Nainital 263136, India.

Cover image

Source: Author.

References

Abdollahi, M., Moridani, M. Y., Aruoma, O. I., & Mostafalou, S. (2014). Oxidative stress in aging. *Oxidative Medicine Cellular Longevity, 2014*, 1–2. http://dx.doi.org/10.1155/2014/876834

Adams, M., Gmünder, F., & Hamburger, M. (2007). Plants traditionally used in age related brain disorders—A survey of ethnobotanical literature. *Journal of Ethnopharmacology, 113*, 363–381. http://dx.doi.org/10.1016/j.jep.2007.07.016

Adams, R. P. (2012). *Identification of essential oils by ion trap mass spectroscopy.* San Diego, CA: Academic Press.

Arya, U., Dwivedi, H., & Subramaniam, J. R. (2009). Reserpine ameliorates Aβ toxicity in the Alzheimer's disease model in *Caenorhabditis elegans. Experimental Gerontology, 44*, 462–466. http://dx.doi.org/10.1016/j.exger.2009.02.010

Asthana, J., Pant, A., Yadav, D., Lal, R. K., Gupta, M. M., & Pandey, R. (2015). *Ocimum basilicum* (L.) and *Premna integrifolia* (L.) modulate stress response and lifespan in *Caenorhabditis elegans. Industrial Crops and Products, 76*, 1086–1093. http://dx.doi.org/10.1016/j.indcrop.2015.08.032

Brenner, S. (1974). The genetics of *Caenorhabditis elegans. Genetics, 77*, 71–94.

Chaieb, K., Hajlaoui, H., Zmantar, T., Kahla-Nakbi, A. B., Rouabhia, M., Mahdouani, K., & Bakhrouf, A. (2007). The chemical composition and biological activity of clove essential oil, *Eugenia caryophyllata (Syzigium aromaticum* L. Myrtaceae): A short review. *Phytotherapy Research, 21*, 501–506. http://dx.doi.org/10.1002/(ISSN)1099-1573

Dostal, V., & Link, C. D. (2010). Assaying β-amyloid toxicity using a transgenic *C* elegans model. *Journal of visualized experiments*, (44), e2252–e2252.

Dostal, V., Roberts, C. M., & Link, C. D. (2010). Genetic mechanisms of coffee extract protection in a *Caenorhabditis elegans* Model of β-amyloid peptide toxicity. *Genetics, 186*, 857–866. http://dx.doi.org/10.1534/genetics.110.120436

Epel, E. S., & Lithgow, G. J. (2014). Stress biology and aging mechanisms: Toward understanding the deep connection between adaptation to stress and longevity. *The Journals of Gerontology Series A: Biological Sciences and Medical Sciences, 69*, S10–S16. http://dx.doi.org/10.1093/gerona/glu055

Garigan, D., Hsu, A. L., Fraser, A. G., Kamath, R. S., Ahringer, J., & Kenyon, C. (2002). Genetic analysis of tissue aging in *Caenorhabditis elegans*: A role for heat-shock factor and bacterial proliferation. *Genetics, 161*, 1101–1112.

Gruber, J. A. N., Tang, S. Y., & Halliwell, B. (2007). Evidence for a trade-off between survival and fitness caused by resveratrol treatment of *Caenorhabditis elegans. Annals of the New York Academy of Sciences, 1100*, 530–542. http://dx.doi.org/10.1196/annals.1395.059

Guarente, L., & Kenyon, C. (2000). Genetic pathways that regulate ageing in model organisms. *Nature*, 255–262. http://dx.doi.org/10.1038/35041700

Gutierrez-Zepeda, A., Santell, R., Wu, Z., Brown, M., & Wu, Y. (2005). Soy isoflavone glycitein protects against beta-amyloid-induced toxicity and oxidative stress in transgenic *Caenorhabditis elegans. BMC Neuroscience*, 6–54.

Kampkötter, A., Timpel, C., Zurawski, R. F., Ruhl, S., Chovolou, Y., Proksch, P., & Wätjen, W. (2008). Increase of stress resistance and lifespan of *Caenorhabditis elegans* by quercetin. *Comparative Biochemistry and Physiology Part B: Biochemistry and Molecular Biology, 149*, 314–323. http://dx.doi.org/10.1016/j.cbpb.2007.10.004

Kenyon, C. J. (2010). The genetics of ageing. *Nature, 464*, 504–512. http://dx.doi.org/10.1038/nature08980

Kollmannsberger, H., & Nitz, S. (1994). The flavour-composition of supercritical gas extracts: III. Clove (*Syzygium aromaticum*) [Ueber die aromastoffzusammensetzung von hochdruck-extrakten: III. Gewuerznelken (*Syzygium aromaticurn*)]. *Chemie, Mikrobiologie, Technologie der Lebensmittel, 16*, 112–123.

Link, C. D. (1995). Expression of human beta-amyloid peptide in transgenic *Caenorhabditis elegans*. *Proceedings of the National Academy of Sciences, 92*, 9368–9372. http://dx.doi.org/10.1073/pnas.92.20.9368

Link, C. D. (2006). *C. elegans* models of age-associated neurodegenerative diseases: Lessons from transgenic worm models of Alzheimer's disease. *Experimental Gerontology, 41*, 1007–1013. http://dx.doi.org/10.1016/j.exger.2006.06.059

Lithgow, G. J., White, T. M., Melov, S., & Johnson, T. E. (1995). Thermotolerance and extended life-span conferred by single-gene mutations and induced by thermal stress. *Proceedings of the National Academy of Sciences, 92*, 7540–7544. http://dx.doi.org/10.1073/pnas.92.16.7540

Pant, A., Saikia, S. K., Shukla, V., Asthana, J., Akhoon, B. A., & Pandey, R. (2014). Beta-caryophyllene modulates expression of stress response genes and mediates longevity in *Caenorhabditis elegans*. *Experimental Gerontology, 57*, 81–95. http://dx.doi.org/10.1016/j.exger.2014.05.007

Shaw, P. X., Werstuck, G., & Chen, Y. (2014). Oxidative stress and aging diseases. *Oxidative Medicine and Cellular Longevity, 2014*, 569146.

Smith, J. V., & Luo, Y. (2003). Elevation of oxidative free radicals in Alzheimer's disease models can be attenuated by Ginkgo biloba extract EGb 761. *Journal of Alzheimer's Disease, 5*, 287–300.

Srivastava, A. K., Srivastava, S. K., & Syamsundar, K. V. (2005). Bud and leaf essential oil composition of *Syzygium aromaticum* from India and Madagascar. *Flavour and Fragrance Journal, 20*, 51–53. http://dx.doi.org/10.1002/(ISSN)1099-1026

2

Weather constraints on Burmese python survival in the Florida Everglades, USA based on mechanistic bioenergetics estimates of core body temperature

Randal S. Stahl[1*], Richard M. Engeman[1], Michael L. Avery[2] and Richard E. Mauldin[1]

*Corresponding author: Randal S. Stahl, United States Department of Agriculture/ Animal and Plant Health Inspection Service/Wildlife Services/National Wildlife Research Center, 4101 LaPorte Ave., Fort Collins, CO, USA

E-mail: randal.s.stahl@aphis.usda.gov

Reviewing editor: Inga Zeisset, University of Brighton, UK

Additional information is available at the end of the article

Abstract: Invasive Burmese pythons (*Python bivittatus*) succumbed to weather-induced mortality in the Florida Everglades in January 2010. We use a mechanistic bioenergetics model to calculate body temperature in various-sized pythons for the successive months of December and January from 2009 to 2014 using daily weather data for the Everglades area. We incorporate python thermal behaviors judged to mitigate weather effects on body temperature. These models suggest that for at least one month in every year except 2013 pythons experienced body temperatures that would subject them to significant physiological stress. However, estimated body temperatures as low as those reported for pythons that succumbed to exposure occurred only in January and December 2010. Results demonstrate the importance of weather variability on survival.

Subjects: Animal Ecology; Biodiversity & Conservation; Vertebrates

Keywords: *Python bivittatus*; climate; air temperature; site occupancy; thermal behavior; solar radiation

1. Introduction

The invasive Burmese python (*Python bivittatus*) has been established in Southern Florida, USA, in the vicinity of Everglades National Park for around a quarter century (Meshaka, Loftus, & Steiner, 2000; Snow, Krysko, Enge, & Oberhoffer, 2007). The python population in Florida is attributed to illegal pet releases, although the highly destructive Hurricane Andrew in 1992 may also have released many from captive breeding and holding facilities (Bilger, 2009; Engeman, Jacobson, Avery, & Meshaka, 2011; Snow et al., 2007; Willson, Dorcas, & Snow, 2011). The origin of the individuals

ABOUT THE AUTHORS

Current research addresses issues focused on understanding the potential for the spread of vertebrate invasive species and developing tools to prevent or mitigate the spread of vertebrate invasive species through human activities. Recent work has examined climate constraints on the survival of the Burmese python in Florida and developing repellents that can be applied to cargo to prevent the transport of the brown treesnake from Guam.

PUBLIC INTEREST STATEMENT

It has been reported in the popular media that the climate along the Eastern seaboard and in many southern states of the United States would allow for the potential expansion of the invasive Burmese python population from the Florida Everglades into these regions. We demonstrate that extremes in weather using data from the Florida Everglades area would likely preclude expansion outside the current range using a more detailed evaluation of the temperature effect on the core body temperature of pythons. In general, temperature extremes are more likely to constrain invasive species expansion and are better estimates for establishment of reptiles than average temperature, which is commonly used.

available in the pet trade at the inception and early growth of the population came from a subset of the native range, initially Thailand near Bangkok and subsequently Vietnam near Ho Chi Minh City after 1994 (Barker & Barker, 2008a, 2008b). The numbers of pythons that founded the population and the ability of these pythons to expand into the temperate regions of the US are largely unknown but attributed to a small founder population (Dorcas, Willson, & Gibbons, 2011; Snow et al., 2007; Willson et al., 2011).

Rodda, Jarnevich, and Reed (2009, 2011) and Pyron, Burbrink, and Guiher (2008) applied climate data from the native ranges of the Burmese python and the closely related Indian python (*Python molurus*) to evaluate possible range expansion of the Burmese python in the contiguous US as the combined native range of these species covers both tropical and temperate regions in Asia (Groombridge & Luxmoore, 1991; Whitaker & Captain, 2004; Zhao & Alder, 1993). Climate matching has been used extensively to predict establishment of an invasive population by contrasting the climate in the species' home range with the local climate of the site of invasion (e.g. Bomford, Kraus, Barry, & Lawrence, 2009). This has often been the only consistent predictor of invasion success across biological groups (Hayes & Barry, 2008) when combined with previous history of invasion success and the numbers of released individuals. The approach is sensitive to both climate parameterization and the statistical methods used in evaluating the likelihood of the climates being comparable between the home range and the new range being evaluated for expansion (Rodda et al., 2011). For example, Rodda et al. (2009) predicted a possible range expansion encompassing the southern tier of the US from the Delmarva Peninsula in the east to portions of California in the west, while Pyron et al. (2008) identified only southern portions of Florida and Texas as suitable for expansion. Both approaches used mean monthly minimum and maximum air temperatures at large spatial scales as part of the parameterization of the climate-matching modeling.

Climate matching does not account for the effect of weather on the body temperature of an organism nor does it directly couple the energy environment resulting from weather to the organism as modified by thermal behavior. As a result, there is no way to mechanistically determine what aspects of a climate preclude site occupancy in a landscape. This can be accomplished by calculating an energy budget for an organism applying physical mechanisms of energy exchange between an organism and its environment.

There are four fundamental physical mechanisms, derived from first principles, used to estimate the rates of energy exchange (fluxes) between an organism and its environment: radiation, latent heat, conduction, and sensible heat transfer (Montieth & Unsworth, 2013). To elucidate; radiation is energy transmitted as light and is associated with sunlight but has a thermal component in the infrared and longer wavelengths; sensible heat exchange occurs as a result of a state change for example when water evaporates; conduction occurs at an interface between two surfaces, and is moderated by a boundary layer resistance; sensible heat transfer accounts for the energy flows in a medium, as in air or soil, diurnally.

As an example, the magnitude of these four mechanisms, calculated for the average global mean energy budget under contemporary climate conditions results in a net positive energy imbalance of 0.6 W m^{-2}, with solar radiation accounting for 161 W m^{-2}, sensible heat loss corresponding to 20 W m^{-2}, latent heat loss contributing 84 W m^{-2}, and conduction having a net loss of 56 W m^{-2} (Hartmann et al., 2013). It is this energy imbalance that results in the observed increase in surface temperature globally. These same mechanisms are used internationally to predict freeze damage when coupled to organism physiology (Snyder & De Melo-Abreu, 2005) as the mechanisms allow for the calculation of an organism's temperature. Reviews of the application of these four mechanisms to individual animals can be found in Campbell (1977), Campbell and Norman (1998), and Montieth and Unsworth (2013). For poikilotherms, organisms that do not use metabolism to maintain body temperature, it is possible to use this approach to predict when an organism's body temperature falls outside some empirically determined lethal limit and can be used to identify periods when local weather precludes site occupancy. In contrast, climate match, as a result of the parameters used, can never identify when an organism's body temperature falls outside a lethal limit and thus can never identify weather events that result in mortality. Additionally, these mechanisms can be coupled to organism behavior on short time scales to evaluate behavioral effects on body temperature.

Thermoregulatory behaviors involve changing body position (Johnson, 1972, 1973) and the use of particular microclimates to improve heat balance (Pearson, Shine and Williams, 2003; Shine and Fitzgerald, 1996; Shine & Madsen, 1996; Slip & Shine, 1988) and to maintain body temperature (Shine, 1981; Storey & Storey, 1992). The use of burrows by captive pythons during cold spells was reported by Dorcas et al. (2011), and the practice is widely reported in related species (Dorcas & Willson, 2011). This phenotypic attribute is set in early developmental stages and may not be plastic, preventing the snake from adapting to extremes in climate variation (Aubret & Shine, 2010; Shine, Madsen, Elphick, & Harlow, 1997). The sum of thermoregulatory tactics, physiological adaptation to the cold, and behavioral strategies are mechanisms by which pythons may survive exposure to weather extremes.

Within Everglades National Park, pythons prefer saline glades and mangroves (Meshaka et al., 2000; Snow et al., 2007) and broadleaf and coniferous forest with bordering marsh (Walters, Mazzotti, & Fitz, 2016) but avoid marsh and open water. Pythons maintained in semi-natural enclosures in the Upper Coastal Plain of South Carolina containing aquatic, terrestrial, arboreal, and underground refuges exhibited strong seasonal shifts in habitat use, moving from aquatic habitat use in the late summer to terrestrial and underground refuges in the fall and winter (Dorcas et al., 2011).

Many species of pythons are reported to shift use of habitat diurnally to moderate body temperature (Avery et al., 2010; Dorcas et al., 2011; Mazzotti et al., 2011; Pearson, Shine, & Williams, 2003; Shine & Fitzgerald, 1996; Shine & Madsen, 1996; Slip & Shine, 1988). Pythons are also reported to shift body posture to increase heat exchange with their environment, adopting positions that are outstretched, loosely coiled, tightly coiled, and tightly coiled with the head positioned under the coils (Johnson, 1972, 1973; Pearson et al., 2003). In carpet pythons, Pearson et al. (2003) measured significant changes in body temperature ranging from 24.0°C stretched out, 25.1°C tightly coiled, to 26.1°C loosely coiled. During cold periods in Florida, pythons were often observed coiled along river banks or canal banks (Dorcas & Willson, 2011).

We calculated an energy budget to estimate body temperatures of a python basking tightly coiled, and either exposed to ambient climatic conditions on the ground or in a burrow or refuge 30 cm below the soil surface. This modeling exercise allowed us to investigate the utility of adopting a fixed-heat conserving body position as a thermoregulatory behavior in one of two microclimates on body temperature. We chose these scenarios as unusually cold, overcast weather in January 2010 contributed to weather-induced mortality among Burmese pythons in captive and free-ranging populations (Avery et al., 2010; Dorcas et al., 2011; Mazzotti et al., 2011). Avery et al. (2010) and Dorcas et al. (2011) observed captive pythons basking in a coiled body position, a thermal regulatory behavior, at ambient air conditions determined to be lethal near or below 0°C. Additionally, Avery et al. (2010) reported that captive pythons left heated refuges to bask during this cold spell but that behavioral change proved maladaptive.

In calculating the radiation input into the energy budget for a python, we followed the approach of Campbell (1977) and defined short-wave radiation as energy having wavelengths between 300 and 4,000 nm and long-wave radiation having wavelengths between 4,000 and 80,000 nm. Sunlight produces a radiant flux density of 1.36 kW/m^2 at the edge of the Earth's atmosphere (Campbell, 1977). During the daylight hours short-wave radiation has a large impact on body temperature due to efficient absorption of this energy associated with the dark coloration of a python above ground. The use of a refuge shifts the energy balance from radiative and conductive fluxes at the surface to sensible heat transfer from the surface to the refuge and then long-wave radiation fluxes from the refuge walls coupled with conduction from the air in the refuge.

We model body temperatures for pythons using temperature data available for Everglades National Park for each day in the months of December 2009 and January 2010 and for the next four December–January periods. We examine ambient air temperature in the context of being a conductive mechanism for energy transfer and demonstrate its impact on body temperature. We summarize this approach over the sequential two month periods to assess the frequency with which low body temperatures occur. Duration of body temperature falling outside a lethal limit and the frequency of occurrence of these events underlie the importance of weather in restricting site occupancy and determine the limitations for python range expansion.

2. Methods

Diurnal fluctuations in the body temperature of a python are calculated using a biophysical energy-flux model from Campbell (1977) and Campbell and Norman (1998). The body temperature of a python results from the energy flux between its body and its environment. The energy budget for the python that allows for the calculation of body temperature is described as follows (Equation (7.19) in Campbell, 1977):

$$M - \lambda E = \rho c_p (T_b - T_e)/(r_b + r_e) \tag{1}$$

where λE is evaporative heat loss, M is energy produced as a result of metabolism, ρc_p is the product of air density and the specific heat capacity of air, T_b is core body temperature, T_e is equivalent environmental temperature (corresponding to the energy radiating from a black body cavity at this temperature), r_b is whole-body thermal resistance, and r_e is the sum of parallel resistances at the body surface to radiative heat loss (r_r) and convective heat loss (r_a). For a poikilotherm, it is assumed that $M - \lambda E = 0$. The equation predicting a python's core body temperature mathematically simplifies to $T_b = T_{e.}$

The environmental temperature is calculated as (Equation (7.17) in Campbell, 1977):

$$T_e = T_a + r_e(R_{\text{abs}} - \varepsilon\sigma T_a^4)/\rho c_p \tag{2}$$

where $\varepsilon\sigma T_a^4$ is the long-wave radiation energy loss, ε is the surface emissivity, σ is the Stephan-Boltzman constant, and T_a refers to ambient temperature. T_a was calculated from Equations (2.2) and (2.3) in Campbell and Norman (1998) for air temperature. For the soil temperature at 30 cm depth, T_a was calculated from Equation (2.8) in Campbell (1977). R_{abs} is total energy absorbed from both long-wave and short-wave radiation.

The equation that describes R_{abs} at night is (Equation (7.11) in Campbell, 1977):

$$R_{\text{abs}} = a_L \varepsilon\sigma T_a^4 \tag{3}$$

where a_L is the long-wave absorptivity of the python, and ε is the average emissivity of the surroundings. This equation includes a short-wave radiation component during the day which has the following form (Equation (7.12) in Campbell, 1977):

$$\text{Total } R_{\text{abs}} - \text{night } R_{abs} + a_s S_p (A_p/A) + S_d. \tag{4}$$

The terms S_p and S_d are projected and diffuse short-wave irradiance incident to the snake. The projected short-wave irradiance is calculated as a ratio of the projected surface area of the python normal to the incident radiation divided by the total surface area of the python as a product of the short-wave absorptivity of the python, a_s. S_p varies throughout the day in a sinusoidal manner following Equations (5.8)–(5.10) in Campbell (1977). Diffuse radiation is direct solar radiation attenuated by clouds, if present, and calculated from S_p using Equation (5.11) in Campbell (1977). As the short-wave radiation term is impacted by the size of an animal, simulations with snakes coiled with circular diameters of 0.092, 0.127, 0.376, and 0.81 m were evaluated.

Table 1. The monthly mean air and soil temperatures for December or January 2009 to 2014 in the data sets used in the models

Year	Month	Air temp mean daily max (°C)	Air temp mean daily min (°C)	Soil temp daily mean at 10 cm (°C)	Soil temp mean daily amplitude (°C)
2009	December	26.5	19.1	21.1	1.7
2010	January	22.4	13.2	16.2	2.1
	December	21.7	10.6	17.0	2.4
2011	January	24.3	14.7	18.5	2.1
	December	26.4	19.4	20.3	2.9
2012	January	24.7	15.9	18.4	3.2
	December	26.3	18.1	18.6	2.7
2013	January	26.5	19.3	19.5	2.8
	December	27.2	20.6	18.8	2.2
2014	January	24.3	15.6	15.8	2.8

Note: For January 2014 soil temperatures were recorded from 1/1 to 1/25.

Monthly summaries of daily weather data (Monthly Climate Data F6 Product) collected by the National Weather Service at the Miami Weather Station (http://w2.weather.gov/climate/index.php?wfo=mfl, Accessed 4/30/2014) were used to parameterize the biophysical model for the scenario where the pythons are lying in a coil, above ground (exposed) for the months of December and January in the years spanning 2009–2014. The daily maximum and minimum air temperatures (Table 1), as well as the average daily wind speed and cloud cover, are used to parameterize the model.

Data from the National Resource Conservation Service soil-monitoring network are used to parameterize the second scenario in which pythons are in a burrow 30 cm below the soil surface and is based on depth to ground water at the Everglades station (www.NRCS.USDA.gov; SCAN site FL Everglades). The soil temperature data are measured at a soil depth of 10 cm. The daily maximum and minimum soil temperatures from each day (Table 1) and the average difference between these values are used to calculate the temperature amplitude in the soil temperature calculation. Soil moisture, measured at the same location, is used to calculate damping depth for the soil temperature calculation. The soil texture at the monitoring site is loam, the damping depth (the depth at which surface temperature fluctuations will be reduced by a fixed amount) based on soil moisture is set to 0.105 m, following Campbell (1977).

The model calculates snake body temperature at hourly time steps over a 24 h interval for both the above-ground scenario and the burrow scenario for each day in a month. We examine model output for each scenario to record the number of hours in a given 24 h interval that a python's body temperature is either <5°C or >5°C but <10°C. The average body temperature for each temperature range is also determined. The average number of hours in a month where a python's body temperature meets one of these two criteria is also determined. The models are programmed in MATLAB (Mathworks, Natick, MA, USA).

3. Results

January (13.2°C) and December (10.6°C) 2010 had the lowest mean daily temperatures over the five-year period examined. To illustrate the effect of extreme cold on predicted body temperature, we present data from January 2010 when mortality in captive pythons was reported by Avery et al. (2010) and in telemetered pythons by Mazzotti et al. (2011). For 9 January 2010, when the recorded high temperature was 17.2°C, the low was 3.3°C, and the sky was cloudy, model-simulation results reveal that air temperature exceeds python body temperature before sunrise (1:00–7:00 h) and after

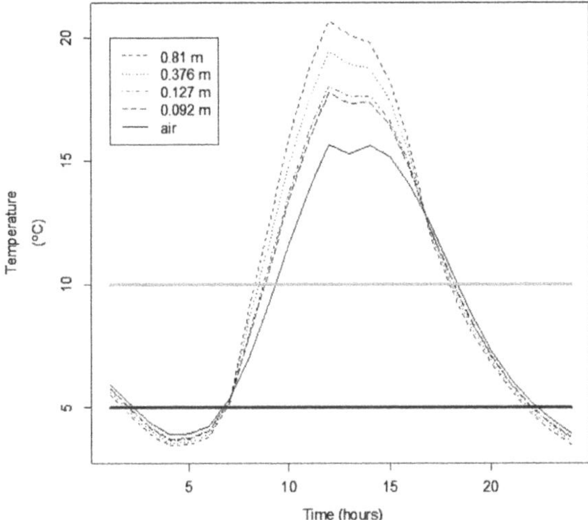

Figure 1. Hourly air temperature for 9 January 2010 compared to the estimated body temperature of Burmese pythons arranged in a coil of varying dimension exposed above ground to ambient conditions on the same date. Body temperature is dependent on size as the energy balance is dominated by solar short-wave radiation.

Notes: Significant mortality has been reported in captive snakes with body temperatures below the horizontal grey line at 10°C. Body temperatures below the horizontal black line at 5°C are perceived as lethal.

sunset (17:00–24:00 h; Figure 1). The air temperatures provide a positive energy flux from conduction. However, the python's energy balance is dominated by long-wave radiation in the dark, and the snakes are radiating more heat into the environment than they are receiving from it, due in part to the cold night sky. Python body temperature rises above air temperature from 7:00 to 17:00 h due to radiation fluxes and the efficient absorption of short-wave radiation with larger snakes having higher body temperatures because their larger surface area allows for greater absorption of short-wave radiation. On 9 January 2010, there is a 14 h period (body temperatures falling below the horizontal grey line in Figure 1) when a python's body temperature would have been <10°C and at least a 7 h period (body temperatures falling below the horizontal black line in Figure 1) when the body temperature would have been close to the lower lethal limit for a python (<5°C; Jacobson et al., 2012). This is consistent with Dorcas et al. (2011) reporting mortality at body temperatures between 5 and 10°C in captive pythons.

In both January and December 2010, there were four or more days when pythons would have experienced predicted body temperatures <5°C for an average interval greater than four hours/day across all size classes evaluated by us (Table 2). Pythons are projected to experience body temperatures >5 and <10°C for one or more days for one or more size classes every month evaluated except for January and December 2013 (Table 2). January and December 2010 were the harshest months, with pythons experiencing body temperatures between 5 and 10°C from 12 to 19 days, depending on body size. Pythons with largest body size were projected to experience the largest number of days with these low body temperatures (19 days in December 2010).

For pythons in a burrow 30 cm below the soil surface, our modeling predicts that body temperature is unaffected by either body size or air temperature on 9 January 2010. The diel fluctuation in body temperature, irrespective of python body size, ranges from a minimum of 10.3°C to a maximum of 10.9°C while air temperature ranges from a minimum of 3.9°C to a maximum of 15.6°C on this date (data not shown). The bioenergetics of an organism in a burrow are dominated by long-wave radiation from the surrounding soil and are not sensitive to surface area. For this particular day, the body temperature predictions of the pythons remain above 10°C. Soils moderate surface-ambient temperature fluctuations, via sensible heat transfer, as illustrated by the daily average air temperature for January 2010 and the corresponding average soil temperature at 30 cm (Figure 2).

Table 2. The number days per month, the average number of hours per day, and the average body temperature of Burmese pythons were estimated to fall between 5 and 10°C for December or January, 2009 to 2014 when the pythons were exposed above ground to ambient climatic conditions

Year	Month	Air temp mean daily max (°C)	Air temp mean daily min (°C)	Body size (m)	Number of days where body temp is >5°C <10°C	Average # of hours >5°C <10°C	Mean body temp >5°C <10°C
2009	December	26.5	19.1	0.81	1	6	8.4 ± 0.3
				0.376	1	5	8.7 ± 0.2
				0.127	1	5	9.3 ± 0.2
				0.091	1	4	9.2+0.1
2010	January	22.4	13.2	0.81	11	8.2	7.3 ± 0.1
				0.376	12	7.7	7.5 ± 0.1
				0.127	12	7.7	7.7 ± 0.1
				0.091	12	7.5	7.7 ± 0.1
	December	21.7	10.6	0.81	19	5.1	7.8 ± 0.2
				0.376	15	5.9	7.8 ± 0.2
				0.127	14	5.8	7.8 ± 0.1
				0.091	13	6.4	7.7 ± 0.2
2011	January	24.3	14.7	0.81	8	6.4	8.2 ± 0.2
				0.376	8	5.2	8.2 ± 0.2
				0.127	7	5.1	8.4 ± 0.2
				0.091	7	5	8.4 ± 0.2
	December	26.4	19.4	0.81	1	2	9.8
				0.376	-		
				0.127	-		
				0.091	-		
2012	January	24.7	15.9	0.81	7	6.6	7.8 ± 0.2
				0.376	7	6.3	8.0 ± 0.2
				0.127	6	5.8	8.1 ± 0.2
				0.091	6	5.5	8.1 ± 0.2
	December	26.3	18.1	0.81	3	4	8.7 ± 0.3
				0.376	3	2.7	8.7 ± 0.3
				0.127	1	5	8.7 ± 0.2
				0.091	1	5	8.8 ± 0.2
2013	January	26.5	19.3	0.81	-		
				0.376	-		
				0.127	-		
				0.091	-		
	December	27.2	20.6	0.81	-		
				0.376	-		
				0.127	-		
				0.091	-		
2014	January	24.3	15.6	0.81	9	6.3	8.3 ± 0.2
				0.376	8	6	8.5 ± 0.2
				0.127	7	5.5	8.6 ± 0.1
				0.091	7	5.5	8.6 ± 0.1

Note: Dashes denote no estimated body temperature in this range for that month.

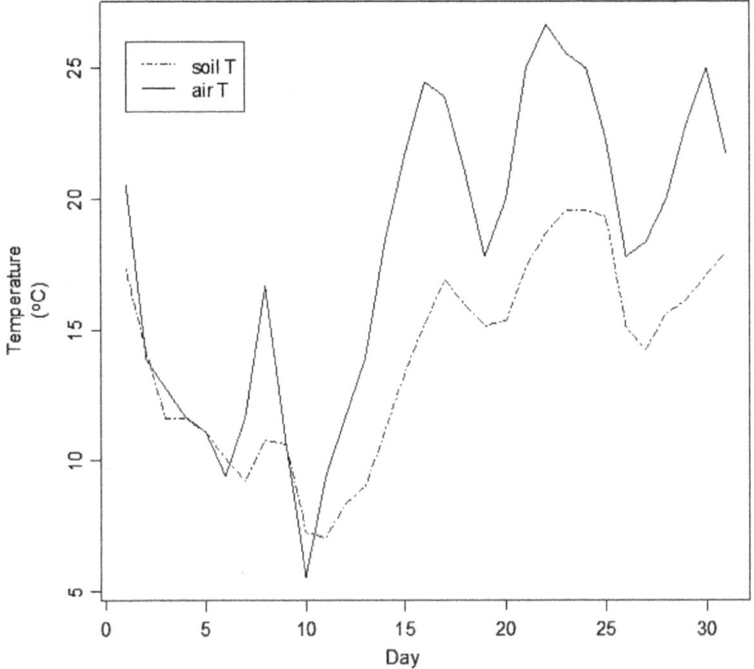

Figure 2. Daily average air temperature for January 2010 compared to daily soil temperature on the same date at a soil depth of 30 cm.

January 2014 was the only month in which python body temperatures are predicted to fall below 5°C in the burrow scenario. It is predicted to have occurred on a single day for a period of 11 h with a mean body temperature of 0.5 ± 0.7°C, irrespective of body size. The month of January 2014 had only 25 days data collected at a soil depth of 10 cm by the monitoring station. Times which predicted python temperatures were >5 and <10°C in the years of 2010, 2013, and 2014 ranged from two to six days/month.

4. Discussion

For pythons above ground, in all size classes, there are extended periods of time when body temperature is predicted to fall below 5 or 10°C. Dorcas et al. (2011) reported minimum body temperatures for 8 out of 10 snakes succumbing at temperatures higher than 4.1°C, predicted by our model to have occurred for 9 January 2010. Mazzotti et al. (2011) also reported mortality in 9 of 10 telemetered pythons in the Everglades with body temperatures falling below 10°C prior to 9 January 2010. Pearson et al. (2003) reported body temperatures as low as 10°C in female carpet pythons in a region at the limit of their known thermal range for portions of the day. The body temperatures calculated by the model are not at or below freezing, and mortality was not likely associated with the formation of ice crystals in inter or intracellular spaces (Storey & Storey, 1992). More likely, the inability to generate energy at low temperatures and a corresponding decreased respiration rate and associated hypoxia (Davies & Bennett, 1981) are contributing factors to mortality at the body temperatures predicted by the model. Burmese pythons are reported to maintain a low energy metabolism (Dorcas & Willson, 2011), and cold may exacerbate this low energy state.

Although a burrow or refuge at a depth of 30 cm is expected to have a more stable temperature, the soil temperatures predicted by the model would result in body temperatures fatal in pythons held in captivity in South Carolina during this time period (Dorcas et al., 2011). Moreover, considerable evidence suggests that Burmese pythons are not behaviorally programmed to seek shelter in cooler temperatures, but rather to bask, even in lethally cold air temperatures (Avery et al., 2010; Barker, 2008; Jacobson et al., 2012). This behavioral pattern of leaving a refuge to bask is reported in other python species that experience winter temperatures similar to those observed in Florida (Pearson et al., 2003; Shine et al., 1997). Mazzotti et al. (2011) reported maximum body temperatures in telemetered pythons above 30°C which they attributed to basking during this cold period.

Basking under extreme cold may reflect the thermal history of the pythons in FL and the current climatic conditions they are experiencing whereby phylogenetic inertia leads to maladaptive behavior in severe cold. Use of a refuge moderates thermal heat loss but does not prevent it. If the objective is to regulate body temperature, then remaining in a refuge as it cools will not be a successful strategy. In the absence of a metabolic means to increase body temperature, increasing exposure to solar radiation by basking is the only option, and adopting a tightly coiled body posture helps maintain body temperature.

Monthly mean maximum and minimum air temperatures are poor predictors of an ectotherm's body temperature (Tables 1 and 2), as the use of microclimates and modifications to body position allows snakes to achieve body temperatures above ambient (Avery et al., 2010; Engeman, Avery, & Jacobson, 2014; Engeman et al., 2011; Jacobson et al., 2012). Furthermore, the mean monthly minimum air temperature is 7 to 10°C higher than the lowest modeled average body temperature for pythons in January and December 2010. The importance of this is that weather observed in 2010 is unusual, climatically, in that it is the only year in which body temperatures predicted by the models for snakes above ground were associated with death from exposure, with significant mortality reported during January 2010 in both captive and wild populations (Avery et al., 2010; Dorcas et al., 2011; Mazzotti et al., 2011).

The average temperatures predicted by the models for snakes 30 cm below ground in the Everglades area were not cold enough to kill pythons, in contrast to the conditions reported by Dorcas et al. (2011) and this may be attributed to the difference in depth to ground water, a change in soil texture and soil moisture content. Soils provide a high degree of thermal insulation, and considerable thermal moderation is predicted by our model to occur over the month of January 2010 (Figure 2). Soils temperatures lag air temperatures temporally, and short-term extremes in air temperature are buffered with increasing soil depth. Soil temperatures lag air temperature and Dorcas et al. (2011, Figure 4) present trends in air and refuge temperatures illustrating this on a longer time scale in a semi-natural refuge.

Snakes with larger body sizes are predicted to have experienced a larger number of days with low body temperatures in 2010 could have implications for population dynamics. Python size is directly related to breeding potential, with females averaging larger than males (De Vosjoli, 1991; Shine et al., 1997). Thus, exposed breeding females would be most at risk of cold-induced mortality. As such, a cold spell might not only suppress populations, but could also immediately diminish breeding potential among the surviving population. However, body-size effects may be mitigated by use of other microclimates potentially available in the landscape (Pearson et al., 2003).

The role of microclimates as moderate weather may be reflected in the results of aerial surveys conducted by Mazzotti et al. (2011) between 2 January and 4 February 2010, whereby they located a total of 104 non-telemetered pythons, of which 60 were found alive. This result does not represent a survival rate as the surveys were conducted over a month's time and there is no information available on the total population this subsample represents (Engeman et al., 2014). However, in the context of the two scenarios we modeled, the majority (84%) of these pythons were located in artificial habitats identified as levees, canals, and roads (Mazzotti et al., 2011). Anthropogenic structures have distinct microclimates that may increase survival, particularly as they are often elevated, drier, and warmer sites in the landscape. Insufficient detail is presented in Mazzotti et al. (2011) to assess how the behaviors associated with use of these anthropogenic structures impacted the energy budgets of the pythons that were detected over this month-long time frame. Based on the results of the model, we would anticipate that access to refuges at depths greater than 30 cm, as might exist in canal banks or levees, would provide a more stable thermal environment and may not have elicited behavioral changes such as basking to increase body temperature during the coldest periods. Extensive use of anthropogenic structures by carpet pythons was reported by Shine and Fitzgerald (1996), with an associated increase in population in a mosaic rural landscape, contrasting with a general population decline in other parts of Australia.

Changes in the energetic physiology of pythons under the temperature extremes observed in Florida in 2010 would likely be required to ensure survival. The generally moderate climate in Florida may preclude opportunities for these attributes to develop (Aubret & Shine, 2010). Rapid weather changes, as observed in 2010, may preclude the development of metabolic adaptations, as this requires several weeks in reptiles (Davies & Bennett, 1981). Additionally, these temperature changes occurred rapidly enough that changes in body position or microclimate use would not preclude the onset of potentially lethal physiological responses.

An important limitation to using a mechanistic energy budget for the calculation of body temperature in a poikilotherm is that body temperature does not mechanistically determine mortality unless temperatures correspond to those at which freezing of tissues occurs. The body temperatures calculated from a mechanistic energy budget must be compared with those empirically measured in pythons. The critical thermal minimum is especially important in this context and has not been definitively determined for pythons (Avery, 1982; Huey, 1982; Jacobson et al., 2012). The role of frequency of exposure must also be empirically determined as the effects of repeated exposure have not been accessed as compared to continuous exposure. The weather events evaluated over the five-year period by applying a mechanistic bioenergetics model to predict body temperature in pythons demonstrate the rarity of the severely cold weather observed in 2010 in the Everglades and the importance of climate variability as contrasted with an average representation of climate used to predict invasion risk for this species.

5. Conclusion

We use a mechanistic bioenergetics modeling approach to investigate the effect of cold weather conditions during the successive months of December and January from 2009 to 2014 on Burmese python body temperature. The results predict body temperatures for pythons that would result in significant physiological stress for all size classes investigated in at least one month for all years we evaluated except 2013 (Jacobson et al., 2012). The model results predict critically low body temperatures across all size classes in 2010 for periods long enough to result in likely death based on minimum body temperatures reported for pythons (Dorcas et al., 2011; Mazzotti et al., 2011). Our approach is unique in that it is mechanistic and couples the environment to the body temperature of a python and allows for the inclusion of thermoregulatory behaviors exhibited by pythons in the wild. Over the period examined in this study, the bioenergetics approach predicts body temperatures low enough to result in mortality as compared to body temperatures empirically determined in pythons that were reported to have died as a result of exposure. This approach could be applied to other locations to evaluate the possible range expansion of the Burmese python through Florida and into other locations in the contiguous US.

Acknowledgments
We would like to thank Fred Kraus and two anonymous reviewers for their comments and suggestions on earlier versions of this manuscript.

Funding
The authors received no direct funding for this research.

Competing Interests
The authors declare no competing interest.

Author details
Randal S. Stahl[1]
E-mail: randal.s.stahl@aphis.usda.gov
Richard M. Engeman[1]
E-mail: Richard.m.engeman@aphis.usda.gov
Michael L. Avery[2]
E-mail: michael.l.avery@aphis.usda.gov
Richard E. Mauldin[1]
E-mail: raichard.e.mauldin@aphis.usda.qov
[1] United States Department of Agriculture/Animal and Plant Health Inspection Service/Wildlife Services/National Wildlife Research Center, 4101 LaPorte Ave., Fort Collins, CO, USA.
[2] United States Department of Agriculture/Animal and Plant Health Inspection Service/Wildlife Services/National Wildlife Research Center Florida Field Station, 2820 East University Ave., Gainesville, FL, USA.

Cover image
Source: Eric Tilman an Employee of the USDA.

References

Aubret, F., & Shine, R. (2010). Thermal plasticity in young snakes: how will climate change affect the thermoregulatory tactics of ectotherms? *Journal of Experimental Biology, 213*, 242–248. http://dx.doi.org/10.1242/jeb.035931

Avery, M. L., Engeman, R. M., Keacher, K. L., Humphrey, J. S., Bruce, W. E., Mathies, T. C., & Mauldin, R. E. (2010). Cold weather and the potential range of invasive Burmese pythons. *Biological Invasions, 12*, 3649–3652. http://dx.doi.org/10.1007/s10530-010-9761-4

Avery, R. A. (1982). Field studies of body temperatures and thermoregulation. In C. Gans & F. H. Poug (Eds.), *Biology of the Reptilia*, (Vol. 12, pp. 93–166). New York, NY: Academic Press.

Barker, D. G. (2008). Will they come in out of the cold? Observations of large constrictors in cool and cold conditions. *Bulletin of the Chicago Herpetological Society, 43*, 93–97.

Barker, D. G., & Barker, T. M. (2008a). The distribution of the Burmese python, *Python molurus bivittatus. Bulletin of the Chicago Herpetological Society, 43*, 33–38.

Barker, D. G., & Barker, T. M. (2008b). Comments on a flawed herpetological paper and an improper and damaging news release from a government agency. *Bulletin of the Chicago Herpetological Society, 43*, 45–47.

Bilger, B. (2009, April 20). Swamp things. *The New Yorker*, pp. 80–89.

Bomford, M., Kraus, F., Barry, S. C., & Lawrence, E. (2009). Predicting establishment success for alien reptiles and amphibians: A role for climate matching. *Biological Invasions, 11*, 713–724. http://dx.doi.org/10.1007/s10530-008-9285-3

Campbell, G. S. (1977). *An introduction to environmental biophysics*. New York, NY: Springer-Verlag. http://dx.doi.org/10.1007/978-1-4684-9917-9

Campbell, G. S., & Norman, J. M. (1998). *An introduction to environmental biophysics* (2nd ed.). New York, NY: Springer. http://dx.doi.org/10.1007/978-1-4612-1626-1

Davies, P. M. C., & Bennett, E. L. (1981). Non-acclimatory latitude-dependent metabolic adaptation to temperature in juvenile natricine snakes. *Journal of Comparative Physiology ? B, 142*, 489–494. http://dx.doi.org/10.1007/BF00688980

De Vosjoli, P. (1991). *The general care and maintenance of burmese pythons*. Lakeside, CA: Advanced Vivarium Systems.

Dorcas, M. E., & Willson, J. D. (2011). *Invasive pythons in the United States ecology of an introduced predator*. Athens GA: The Univerity of Georgia Press.

Dorcas, M. E., Willson, J. D., & Gibbons, J. W. (2011). Can invasive Burmese pythons inhabit temperate regions of the southeastern United States? *Biological Invasions, 13*, 793–802. http://dx.doi.org/10.1007/s10530-010-9869-6

Engeman, R. M., Avery, M. L., & Jacobson, E. (2014). Weighing empirical and hypothetical evidence for assessing potential invasive species range limits: A review of the case of Burmese pythons in the USA. *Environmental Science and Pollution Research, 21*, 11973–11978. http://dx.doi.org/10.1007/s11356-014-3173-4

Engeman, R. M., Jacobson, E., Avery, M. L., & Meshaka, Jr, W. E. (2011). The aggressive invasion of exotic reptiles in Florida with a focus on prominent species: A review. *Current Zoology, 57*, 599–612. http://dx.doi.org/10.1093/czoolo/57.5.599

Groombridge, B., & Luxmoore, R. (1991) Pythons in south-east Asia. A review of distribution, status, and trade in three selected species. *Secretariat of the Convention on International Trade in Endangered Species of Wild Fauna and Flora*, Lausanne.

Hartmann, D. L., Klein Tank, A. M. G., Rusticucci, M., Alexander, L. V., Brönnimann, S., Charabi, Y., ... Zhai, P. M. (2013). Observations: Atmosphere and surface. In T. F. Stocker, D. Qin, G.-K. Plattner, M.Tignor, S. K. Allen, J. Boschung, ... P. M. Midgley (Eds.), *Climate Change 2013: The Physical Science Basis. Contribution of Working Group I to the Fifth Assessment Report of the Intergovernmental Panel on Climate Change* (pp. 159–254). Cambridge and New York, NY: Cambridge University Press.

Hayes, K. R., & Barry, S. (2008). Are there any consistent predictors of invasion success? *Biological Invasions, 10*, 483–506. http://dx.doi.org/10.1007/s10530-007-9146-5

Huey, R. B. (1982). Temperature, physiology, and the ecology of reptiles. In C. Gans & F. H. Pough (Eds.), *Biology of the reptilia* (Vol. 12, pp. 25–91). Academic Press.

Jacobson, E. R., Barker, D. G., Barker, T. M., Mauldin, R., Avery, M. L., Engeman, R., & Secor, S. (2012). Environmental temperatures, physiology and behavior limit the range expansion of invasive Burmese pythons in southeastern USA. *Integrative Zoology, 7*, 271–285. http://dx.doi.org/10.1111/inz.2012.7.issue-3

Johnson, C. R. (1972). Thermoregulation in pythons—I. Effect of shelter, substrate type and posture on body temperature of the australian carpet python, Morelia spilotes variegata. *Comparative Biochemistry and Physiology Part A: Physiology, 43*, 271–278. http://dx.doi.org/10.1016/0300-9629(72)90185-5

Johnson, C. R. (1973). Thermoregulation in pythons—II. Head-body temperature differences and thermal preferenda in australian pythons. *Comparative Biochemistry and Physiology Part A: Physiology, 45*, 1065–1087. http://dx.doi.org/10.1016/0300-9629(73)90343-5

Mazzotti, F. J., Cherkiss, M. S., Hart, K. M., Snow, R. W., Rochford, M. R., Dorcas, M. E., & Reed, R. N. (2011). Cold-induced mortality of invasive Burmese pythons in south Florida. *Biological Invasions, 13*, 143–151. http://dx.doi.org/10.1007/s10530-010-9797-5

Meshaka, Jr, W. E. Loftus, W., & Steiner, T. (2000). The herptofauna of Everglades National Park. *Florida Scientist, 63*, 84–103.

Montieth, J. L., & Unsworth, M. (2013). *Principles of environmental physics. Plants, animals and the atmosphere* (4th ed.). New York, NY: Academic Press.

Pearson, D., Shine, R., & Williams, A. (2003). Thermal biology of large snakes in cool climates: a radio-telemetric study of carpet pythons (Morelia spilota imbricata) in south-western Australia. *Journal of Thermal Biology, 28*, 117–131. http://dx.doi.org/10.1016/S0306-4565(02)00048-7

Pyron, R. A., Burbrink, F. T., & Guiher, T. J. (2008). Claims of Potential Expansion throughout the U.S. by Invasive Python Species Are Contradicted by Ecological Niche Models. *PLoS ONE, 3*, e2931. doi:10.1371/journal.pone.002931

Rodda, G. H., Jarnevich, C. S., & Reed, R. N. (2009). What parts of the US mainland are climatically suitable for invasive alien pythons spreading from Everglades National Park? *Biological Invasions, 11*, 241–252. http://dx.doi.org/10.1007/s10530-008-9228-z

Rodda, G. H., Jarnevich, C. S., & Reed, R. N. (2011). Challenges in identifying sites climatically matched to the native ranges of animal invaders. *PLoS ONE, 6*, e14670. doi:10.1371/journal.pone.0014670

Shine, R. (1981). Venomous snakes in cold climates: Ecology of the Australian genus Drysdalia (*Serpentes: Elapidae*). *Copeia, 1981*, 14–25. http://dx.doi.org/10.2307/1444037

Shine, R., & Fitzgerald, M. (1996). Large snakes in a mosaic rural landscape: The ecology of carpet pythons Morelia spilota (serpentes: Pythonidae) in coastal eastern Australia. *Biological Conservation, 76*, 113–122. http://dx.doi.org/10.1016/0006-3207(95)00108-5

Shine, R., & Madsen, T. (1996). Is thermoregulation unimportant for most reptiles? An example using water pythons (*Liasis fuscus*) in Tropical Australia. *Physiological Zoology, 69*, 252–269. http://dx.doi.org/10.1086/physzool.69.2.30164182

Shine, R., Madsen, T. R., Elphick, M. J., & Harlow, P. S. (1997). The influence of nest temperatures and maternal brooding on hatchling phenotypes in water pythons. *Ecology, 78*, 1713–1721. http://dx.doi.org/10.1890/0012-9658(1997)078[1713:TIONTA]2.0.CO;2

Slip, D. J., & Shine, R. (1988). Thermoregulation of free-ranging diamond pythons, Morelia spilota (*Serpentes, Boidae*). *Copeia, 1988*, 984–995. http://dx.doi.org/10.2307/1445722

Snow, R. W., Krysko, K. L., Enge, E. M., & Oberhoffer, A. (2007). Introduced populations of Boa constrictor (Boidae) and Python molurus bivittatus (Pythonidae) in southern Florida. In R. W. Henderson & R. Powell (Eds.), *The Biology of Boas and Pythons*. (pp. 416–438). UT: Eagle Mountain.

Snyder, R. L., & De Melo-Abreu, J. P. (2005). *Frost protection: Fundamentals, practice and economics* (Vol. 1). Rome: Food and Agriculture Organization of the United Nations.

Storey, K. B., & Storey, J. M. (1992). Natural freeze tolerance in ectothermic vertebrates. *Annual Review of Physiology, 54*, 619–637. http://dx.doi.org/10.1146/annurev.ph.54.030192.003155

Walters, T. M., Mazzotti, F. J., & Fitz, H. C. (2016). Habitat selection by the invasive species burmese python in Southern Florida. *Journal of Herpetology, 50*, 50–56. http://dx.doi.org/10.1670/14-098

Whitaker, R., & Captain, A. (2004). *Snakes of India: The field guide*. Chengapattu: Draco Books.

Willson, J. D., Dorcas, M. E., & Snow, R. W. (2011). Identifying plausible scenarios for the establishment of invasive Burmese pythons (Python molurus) in Southern Florida. *Biological Invasions, 13*, 1493–1504. http://dx.doi.org/10.1007/s10530-010-9908-3

Zhao, E., & Alder, K. (1993). *Herpetology of China*. St Louis, MO: Society for the Study of Amphibians and Reptiles.

3

The spontaneous juvenile alopecia (*jal*) mutation in mice is associated with the insertion of an IAP element in the *Gata3* gene

Malcolm E. Connor[1] and Thomas R. King[1*]

*Corresponding author: Thomas R. King, Biomolecular Sciences, Central Connecticut State University, 1615 Stanley Street, New Britain, CT 06053, USA
E-mail: kingt@ccsu.edu

Reviewing editor: Jurg Bahler, University College London, UK
Additional information is available at the end of the article

Abstract: *Background:* A combination of genetic fine-mapping and complementation testing was used previously to assign the juvenile alopecia mutation (abbreviated *jal*) to the GATA binding protein 3 (*Gata3*) gene on Chromosome 2 in mice. However, sequence analysis of *Gata3* exons (including coding and noncoding regions) revealed no differences between wild type C3H/HeJ and co-isogenic C3H/HeJ-*jal*/J mutant mice. *Results:* Using a PCR-based scanning method, here we have tested the hypothesis that *jal* might result from insertion of a transposable element in or near the *Gata3* gene. We show that the *jal* mutation is specifically associated with an intracisternal A particle (IAP) element of the IΔ1 subtype that has transposed to Intron 3–4 in the *Gata3* gene, and use the same panel of recombinants used previously to fine-map *jal* to show that this IAP element and *jal* are located within the same small genetic interval. *Conclusion:* Transposition of an IAP element of the IΔ1 subtype into Intron 3–4 of the mutant *Gata3*^{jal} allele is the likely cause of the juvenile alopecia phenotype in mutant mice.

Subjects: Developmental Biology; Genetics; Molecular Biology

Keywords: retrotransposition; germline mutation; insertional mutagenesis; focal alopecia; mouse model

1. Introduction

A significant amount of the mouse genome (8–10%) is composed of endogenous retroviruses (ERVs) and other DNA elements with long terminal repeats (LTRs) (Mouse Genome Sequencing Consortium (MGSC), 2002), and many of these remain "active," in that their RNA can be reverse-transcribed and integrated into new genomic sites by a retrovirus-like transposition mechanism. Indeed, it has been estimated that some 10–15% of new mutations in mice are due to transposition of an ERV element into or near the altered gene (Maksakova et al., 2006). One family of ERV

ABOUT THE AUTHORS

The research group led by Thomas R. King at Central Connecticut State University (New Britain, CT) aims to identify the genetic basis of spontaneous hair variants in mice and rats. Identification of such causative gene mutations can make probes available for the molecular analysis of both normal and disrupted development of the mammalian integument, and can help to identify human conditions these animal resources might model. Malcolm E. Connor is a graduate student in the Department of Biomolecular Sciences.

PUBLIC INTEREST STATEMENT

In previous work, a recessive form of progressive hair loss known as juvenile alopecia was assigned to the *Gata3* gene on mouse Chromosome 2, but no DNA defect was discovered. In this report, we identify a retrovirus-like element that is integrated within the *Gata3*^{jal} allele and is the likely cause of the mutant phenotype. This mutational mechanism would explain the variable expressivity of the hair loss phenotype in juvenile alopecia mice, and suggests that these animals may serve as a living resource for the further study of epigenetic gene regulation.

elements, the intracisternal A particle (IAP), is present in about 1,000 full-length or partially deleted copies in the haploid mouse genome (Kuff & Lueders, 1988), and one subclass of deleted IAP elements, the IΔ1 subclass (which has a 1.9 kb deletion in *gag-pol*), is responsible for nearly all of the known, ERV-induced, *de novo* germline mutations in mice (Maksakova et al., 2006). Most of these IΔ1 insertional mutations have occurred in the C3H/HeJ standard inbred mouse strain, suggesting that one or a small number of IΔ1, IAP elements in this strain must persist in evading host suppression mechanisms (Maksakova et al., 2006).

Our group has previously assigned (by a positional-candidate approach) the juvenile alopecia mutation (abbreviated *jal*)—which arose spontaneously on the C3H/HeJ genetic background—to the *Gata3* gene in mice (Ramirez et al., 2013). While complementation testing between *jal* and the engineered *Gata3*^{tm1Gsv} null allele (van Doorninck et al., 1999) verified allelism, we could find no DNA defect in the exonic portions of *Gata3* in *jal* mutants (Ramirez et al., 2013). The *jal* mutation's C3H/HeJ strain-of-origin, the lack of an obvious coding defect in *Gata3*^{jal}, and the markedly variable expressivity of the mutant phenotype (vibrissae defects and focal alopecia that may range from global to undetectable), combine to suggest that an IAP integration near or within an intron of *Gata3* might be the basis of this natural variant. Here we test that hypothesis, and report the detection, orientation, and precise location of an IΔ1, IAP element in Exon 3-4 of *Gata3* that appears to be specific to the C3H/HeJ-*jal*/J strain and maps to the same small region as the *jal* mutation itself.

2. Methods

Animals were housed and fed according to Federal guidelines, and the Institutional Animal Care and Use Committee at CCSU approved of all procedures involving mice (Animal Protocol Application numbers 101, 119, and 122). Standard inbred mice (from strains C57BL/6 J, A/J, C3H/HeJ) and C3H/HeJ-*jal*/J mutant mice were obtained from The Jackson Laboratory (Bar Harbor, ME, USA). We maintained our coisogenic C3H/HeJ-*jal*/J line by crossing heterozygous females with homozygous males, to produce segregating litters. Mutant *jal/jal* mice were mostly detected by vibrissae defects that were observable shortly after birth, and those homozygotes displayed distinct patches of hair loss by two weeks of age that persisted throughout life (Figure 1). The mutant phenotype was not fully penetrant, and was highly expressive such that the amount of body surface affected by baldness varied from less than 5% to greater than 95% (see additional photographs shown in Figure 1 and Additional file 1 in Ramirez et al., 2013).

Genomic DNA was isolated from 3 mm tail tip biopsies taken from two-week-old mice using Nucleospin kits from BD Biosciences (Palo Alto, CA, USA). The polymerase chain reaction (PCR) was

Figure 1. A *jal/jal* mutant at 3 months of age.

performed using the Titanium PCR kit from Clontech (Mountain View, CA, USA), and oligonucleotide primers for PCR were synthesized by Integrated DNA Technologies, Inc. (Coralville, IA, USA). *Gata3*-specific primers were based on sequence information available online (Ensembl Mouse Genome Server (EMGS), 2016) and IAP-element-specific primers were based on the IAP LTR sequence (GenBank D63767) published by Ishihara, Tanaka, Wan, Nojima, and Yoshida (2004). Specifically, the primer designated ISP-R1 or ISP-F4 herein was 5' GAGCTGACGTTCACGGGAAAAAC 3'; the primer designated ISP-F2 or ISP-R3 was 5' ACGACCACTTGTACTCTGTTTTTC 3'. Primers designed to overlap or flank the *gag-pol* deletion that is characteristic of the IΔ1 subtype were based on the widely-studied, full-length IAP element MIA14 (GenBank M17551), published by Miets, Grossman, Lueders, and Kuff (1987). The specific primers designated ISP-Fa, ISP-Fb and ISP-Fc herein are shown in Supplementary Figure S1. PCR products were visualized by electrophoresis through 3.5% NuSieve 3:1 agarose gels (Lonza, Rockland, ME, USA) for products under 1,000 bp and through 1.5% SeaKem agarose gels (Lonza) for products over 1,000 bp in length. Gels were stained with ethidium bromide (0.5 µg/mL) and photographed under ultraviolet light. For sequence analysis, 1.5 µg of individual PCR amplimers were concentrated into 30 µl using QIAquick PCR Purification kits (Qiagen, Valencia, CA, USA). Purified amplimers were shipped to the Keck DNA Sequencing Lab at Yale University (New Haven, CT, USA) for primer-extension analysis. All datasets supporting the results of this article are included here in the five formal Figures and single Supplementary Figure. The nucleotide sequences described in this study are also accessioned in GenBank, NCBI (accession numbers KX555562 and KX555563).

The creation of a 374-member [(A/J x C3H/HeJ-*jal*/J)F1-*jal*/+ x C3H/HeJ-*jal*/J-*jal*/*jal*] backcross (N2) "mapping panel" that was used previously to locate *jal* on proximal Chr 2 was described in Ramirez et al. (2013). DNA samples from recombinant panel members, which had been previously characterized for numerous microsatellite and single-nucleotide polymorphisms (SNPs) on proximal Chr 2 (see Ramirez et al., 2013) and stored at −80° C for retrospective analysis, were used here to locate the IAP element integrated with *Gata3* with respect to *jal* and to these other marker sites.

3. Results

To detect an IAP element associated with the *Gata3*^{jal} allele, gene-specific primer (GSP) pairs were designed that lay about 1,000 bp apart in the noncoding regions of *Gata3*. A GSP pair might fail to generate amplimers of the expected (wild type) length in PCRs using *jal*/*jal* template DNAs if a large IAP element (which range in size from 5.4 to 7.3 kb) were integrated between their annealing sites (Figure 2(a)). One such GSP pair (which directed amplification of wild type but not *jal* templates) was found (see Figure 2(b)) and these primers were tested individually with primers designed to anneal within the LTR region of the IAP element (based on the sequence published by Ishihara et al., 2004; GenBank D63767), designated here as ISP-R1, -F2, -R3, and -R4 (see Figure 2(c)). For example, the "failed" GSP in the forward orientation (GSP-F) might generate an amplimer with ISP-R1 if an IAP in the sense orientation was integrated downstream of the GSP, or with ISP-R3 if an IAP element in the antisense orientation was integrated downstream of the GSP. The "failed" GSP in the reverse orientation (GSP-R) might generate an amplimer with ISP-F2 if an IAP in the sense orientation was integrated upstream of that GSP, or with ISP-F4 if an IAP element in the antisense orientation was integrated upstream of the GSP. Figure 2(d) shows the results of PCRs that reveal the integration of an IAP element in the antisense orientation at a position about 700 bp downstream of GSP-F and about 250 bp upstream of GSP-R. The amplimers produced in these positive reactions were sequenced by primer-extension analysis, and the precise integration site of an IAP element in Intron 3–4 of the *Gata3*^{jal} allele, designated *Gata3*^{IAP}, was determined, as shown in Figure 3.

The characterization of a panel of 374 [(A/J × C3H/HeJ-*jal*/J)F_1 × C3H/H3 J-*jal*/J] backcross offspring for the inheritance of *jal* and various molecular markers on proximal mouse Chromosome 2 established a location for the *jal* mutation between positions designated as *SNP1* and *SNP2* by Ramirez et al. (2013), SNPs officially known as *rs27112885* and *rs27131573*, respectively (Ensembl

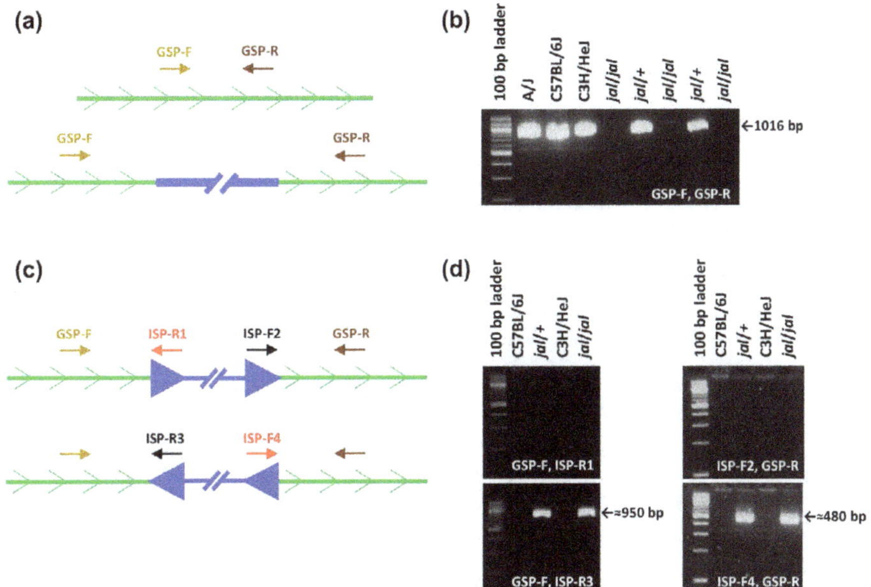

Figure 2. Scanning the *Gata3*^{jal} allele for an integrated intracisternal A particle (IAP) element. (a) Annealing sites for a gene-specific primer (GSP) pair, one in the forward and one in the reverse orientation, are shown within an intron of the *Gata3* gene (represented by a green line with the sense orientation indicated). In the top diagram, the GSP are close enough together to yield a PCR product of the wild-type length. In the bottom diagram, an IAP (represented by a thick blue line) has integrated between the primer-annealing sites, moving the GSP pair so far apart that no product is generated in the PCR. (b) One GSP pair, designated GSP-F and GSP-R (sequences given in Figure 3), produced a 1,016 bp amplimer from wild-type DNA templates, but no products from *jal/jal* mutant DNA templates, as expected if an IAP integration occurred between the primer-annealing sites. (c) Here an intron with an IAP insertion is depicted in each of the two possible orientations, sense (top diagram) or antisense (bottom diagram), with respect to the *Gata3* gene. If the IAP is integrated in the sense orientation, GSP-F should yield a PCR product with IAP-specific primer (ISP) -R1, and GSP-R should yield a product with ISP-F2. If the IAP is integrated in the antisense orientation, GSP-F should yield a PCR product with ISP-R3, and GSP-R should yield a product with ISP-F4. (d) GSP-F yielded a 959 bp product from *jal* templates with ISP-R3 but none with ISP-R1. GSP-R yielded a 486 bp product from *jal* templates with ISP-F4 but none with ISP-F2. Therefore, the IAP integrated in Intron 3–4 in the *Gata3*^{jal} allele is in the antisense orientation with respect to the gene.

Mouse Genome Server, 2016). Specifically, 6 meiotic crossovers were identified in this 374-member panel that fell between *SNP1* and *jal*, and 1 crossover was identified that fell between *jal* and *SNP2* (summarized in Figure 4(a)). If the IAP element inserted in Intron 3-4 of *Gata3*^{jal} is the mutational basis of the *jal* mutant phenotype, then this IAP element should map to the same site as *jal*, that is, *Gata3*^{IAP} should co-segregate with *jal* in the 374-member backcross panel. As shown in Figure 4(b), the *Gata3*^{IAP} genotype of each rare, recombinant panel member matches its *jal* genotype, suggesting that the *Gata3*^{IAP} and the *jal* mutation are at least very closely linked (mapping less than 0.3 cM apart), if not identical.

The IΔ1 subtype of IAP elements is known to have a 1.9 kb deletion that extends from position 1708 in *gag* to position 3608 in *pol* (Kuff & Lueders, 1988). To determine if the IAP element integrated in the *Gata3* gene in C3H/HeJ-*jal*/J mice might be of the IΔ1 subtype (which is known to be most frequently associated with *de novo* insertional mutations in mice, Ishihara et al., 2004; Maksakova et al., 2006), a set of forward primers for PCR were designed that would anneal just before (ISP-Fa), within (ISP-Fb), and just after (ISP-Fc) this characteristic deletion in the IAP sequence. These primers were tested individually with a gene-specific reverse primer (GSP-Rd) that would anneal just downstream of the integration site of the IAP element in Intron 3–4 of *Gata3*^{jal} (see Figure 5(a)). The presence and length of the PCR products obtained (see Figure 5(b)) were consistent with the *Gata3*^{IAP} being of the IΔ1 subtype, and primer-extension sequence analysis of the entire interval defined by the annealing sites for ISP-Fa and GSP-Rd in the *Gata3*^{jal} allele (see Supplementary Figure S1) confirmed this classification.

```
               GSP-F→
2:9,869,836 tctgttctgccctcttctctacttgcttttcccctcttgcagatctcctgcct
aaaggctgatctgtgctttcttccatcagccgttggctggccaggctggggctaaagagctgact
tcaattcctctaacagcagctgcaaacctatgggagatgttggcagtttttgaaagggagactgt
gaggatagtggatgagtcgatcctgctatgtcttaattgatgactttgtaactaggggcatttat
agttttattggaagttaaaaaatgtgacgttttagctactcaatcagccaattgatcaatgtata
cgtccatctattatctgtttatctctctagccattggttaatctttcttttcttcccatcctgat
ctatgttacctagtcggtctatatagaagggcccggtgacgaactggatcaccgtcccttgcttc
ttttacttttcctttggaggaaaatgaaagcagtctttggctttggccacagactgatctgtgca
cagcatggtcttagggaaatggttaagattcagtctccctccttcttccccccttcccggctct
atcttcctcagcaagtgggacatgtcaaattcactcacctgaggtctgtttgtatcctggaagag
ttatgtcagggcactaagggttgttaactttggatcatgttattcgacgcgttctcacgaccggc
caggaagaacaccacagaccagaatcttctgcgacaaagctttattcttacatcttcaggaaaag
agagcaagaagcaagagagagcaagaagcaagagagggaagcaagagagagcaagaagcaagaga
gagagaaaaacgaaaccccttctatttaaagagaacaaccattgcctagggcgcatcactccct
gattggctgcagcccatggccgagctgacgttcacgggaaaaacagagtacaagtggtcgtaaat
                                                      ←ISP-R3
acccttggctcatgcgcagattatttgtttaccaacttagaacacaggatgtcagcgccatcttg
tgacggcgaatgtgggggcggcttcccaca********tgttattcgacgcgttctcacgaccgg
                          3' LTR
ccaggaagaacaccacagaccagaatcttctgcgacaaagctttattcttacatcttcaggaaaa
gagagcaagaagcaagagagagcaagaagcaagagagggaagcaagagagagcaagaagcaagag
agagagaaaaacgaaaccccttctatttaaagagaacaaccattgcctagggcgcatcactccc
                         ISP-F4→
tgattggctgcagcccatggccgagctgacgttcacgggaaaaacagagtacaagtggtcgtaaa
tacccttggctcatgcgcagattatttgtttaccaacttagaacacaggatgtcagcgccatctt
gtgacggcgaatgtgggggcggcttcccacaggatcactttgtaactattttcctcagtcgtaag
                           5' LTR
agtttggaagtgcttttggttttattttcccttctgatctcaatgcatgtctaggggtttgtttt
tctgtttttgtgtgtctcaggatggaattcagattttcaggctaggggcaagtgctaggttgc
tgagggatctcacagaccctccctcatcattctaagtgttacctcgttttcacacgctttccttc
                                    Exon 4
cctaagtgacttatctgtgaccttgtttccagAAGGCAGGGAGTGTGTGAACTGCGGGGCAACCT
CTACCCCACTGTGGCGGCGAGATGGTACCGGGCACTACCTTTGCAATGCCTGC 2:9,868,821
                                                   ←GSP-R
```

Figure 3. Nucleotide sequence at the site of IAP integration in Intron 3–4 of the mutant $Gata3^{jal}$ allele. Sequences in lower case are located within *Gata3*, Intron 3–4 (beginning with nucleotide number 2:9869836); nucleotides from Exon 4 are shown in blue, upper case letters (ending with nucleotide number 2:9868821). Sequences that correspond to the primers described in the text and in Figure 2 as GSP-F and GSP-R are highlighted in yellow. The six-nucleotide direct repeat created in mouse genomic DNA by IAP element insertion (the target site duplication, TSD) is shown in red (and corresponds to nucleotides 2:9869167–9869162). The IAP element's 5' and 3' LTRs, in antisense orientation, are shown in green; IAP sequences located between the LTRs are abbreviated by blue stars. Sequences that correspond to IAP-specific primers described in the text and in Figure 2 as ISP-R3 and ISP-F4 are highlighted in blue and pink, respectively. Base-pair positions on mouse Chromosome 2 are from NCBI Build 37.2 (Ensembl Mouse Genome Server, 2016; Mouse Genome Database, 2016).

4. Discussion

Although mice homozygous for germline null alleles of *Gata3* die around embryonic day 11 (Lim et al., 2000; Pandolfi et al., 1995; van Doorninck et al., 1999), conditional ablation in the epidermis has revealed a role for *Gata3* in hair follicle development and skin cell lineage determination (Kaufman et al., 2003; Kurek, Garinis, van Doorninck, van der Wees, & Grosveld, 2007). These null mutants, however, show a complete absence of hair, while mice homozygous for the spontaneous juvenile alopecia allele of *Gata3*, which show patchy hair loss, typically display at least some normally furred sectors. On this basis, we have previously suggested that $Gata3^{jal}$ likely encodes a normal primary protein sequence that is, by some stochastic mechanism, improperly regulated (Bisaillon et al., 2014; Ramirez et al., 2013).

While we have not attempted here to directly elucidate the mutagenic mechanism of the $Gata3^{IAP}$ insertion that we now propose to be the basis of the juvenile alopecia phenotype, several well-characterized examples of antisense integration of IΔ1, IAP elements into the introns of other mouse genes (reviewed by Maksakova et al., 2006), suggest a few likely possibilities. For example, IAP insertions within introns can cause aberrant splicing or premature termination of the disrupted gene's mRNA (Druker, Bruxner, Lehrbach, & Whitelaw, 2004; Gunn et al., 2001; Vasicek et al., 1997; Ware et al., 1997), but our previous investigation of *Gata3* cDNA did not reveal any anomalous, *jal*-specific transcripts (Ramirez et al., 2013). Most commonly, antisense integrations of IΔ1, IAP

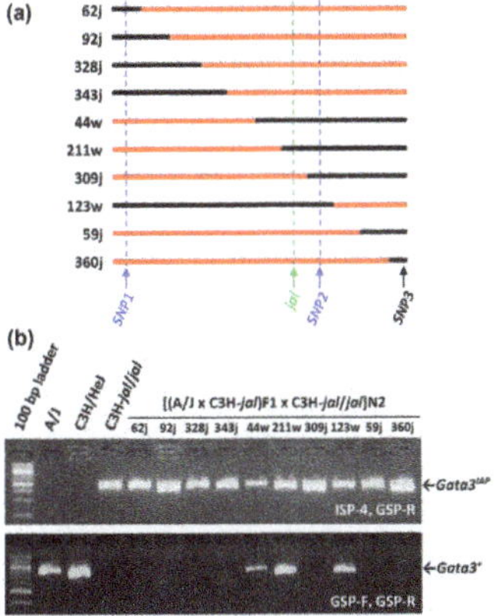

Figure 4. *Gata3^{IAP}* and *jal* are meiotically inseparable among a panel of 374 backcross mice. (a) Ramirez et al. (2013) described the production of a panel of 374 backcross mice from a cross of (A/J x C3H/HeJ-*jal*/J)F1 females with C3H/HeJ-*jal*/J mutant males. These panel members segregated for *jal* and for various other markers on proximal Chromosome 2. Here, the most salient 10 members of that panel (i.e., those with crossovers nearest the *jal* mutation) are each represented by a line that depicts the region of Chr 2 between markers *SNP1* (officially designated *rs27112885*) and *SNP3* (a.k.a. *rs27100936*) that was inherited from their F1 parent. A/J-derived sequences are drawn in black, and C3H-derived sequences are drawn in red. Each recombinant is identified by its pedigree number, followed by a letter to indicate its phenotype as wild type (w) or mutant (j). Six crossovers (found in recombinants 62, 92, 328, 343, 44, and 211) lie telomeric to *SNP1* but centromeric to *jal*; one crossover (found in recombinant 309) lies telomeric to *jal* but centromeric to *SNP2* (a.k.a. *rs27131571*). Thus, *jal* must lie between *SNP1* and *SNP2*, an interval that includes only one gene known to be expressed in skin, *Gata3*. Official SNP designations are from dbSNP Build 142 (Ensembl Mouse Genome Server, 2016; Mouse Genome Database, XXXX). (b) Recombinant DNA samples from the panel described by Ramirez et al. (2013) and controls were characterized for the presence of none, one, or two copies of *Gata3^{IAP}*. All recombinants that were phenotypically mutant were homozygous for *Gata3^{IAP}*, and the wild-type recombinants, heterozygous *jal*/+, were heterozygous for *Gata3^{IAP}*/*Gata3^{+}*. Co-segregation of *Gata3^{IAP}* and *jal* in this backcross suggests that they are very close or identical.

elements have been seen to cause ectopic gene expression from a promoter located in the 5' LTR (Duhl, Vrieling, Miller, Wolff, & Barsh, 1994; Michaud et al., 1994; Vasicek et al., 1997). Many of these mutant alleles show variable expressivity among genetically identical mice (Druker et al., 2004; Michaud et al., 1994; Rakyan et al., 2003), as we have documented for *Gata3^{jal}* (see Additional file 1 in Ramirez et al., 2013). Such variable expressivity has been shown to correlate with the methylation state of the 5' LTR, which seems to be established somewhat stochastically (Druker et al., 2004; Michaud et al., 1994; Rakyan et al., 2003). Perhaps if the 5'LTR is mostly methylated, its antisense promoter will be inactive, and little effect on Gata3 expression in that skin patch leads to normal hair development and maintenance in that sector. However, if the LTR is unmethylated, its promoter may drive aberrant gene expression that results in follicular dysgenesis. For example, an active LTR promoter may cause ectopic expression of the functional domains of Gata3 protein that are encoded by Exons 4–6 in hair-follicle stem cells, leading to their terminal differentiation and loss in that skin patch. Alternatively, host silencing mechanisms (such as hypermethylation of DNA; Walsh, Chaillet, & Bestor, 1998) directed at the intron-integrated IAP element may interfere with the production of a normal *Gata3* transcript, leading to alopecia in those skin sectors. In any case, the *Gata3^{jal}* mutation appears to offer a valuable and convenient model for the study of a "metastable epiallele" (Rakyan, Blewitt, Druker, Preis, & Whitelaw, 2002), the expression of which appears to be regulated by a stochastic, epigenetic mechanism that can vary between genetically matched individuals or even between different skin patches on the same individual.

As in mice, a large (8%) portion of the human genome is of retroviral origin (International Human Genome Mapping Consortium (IHGMC), 2001). While the majority of human ERV's represent ancient integrations and lack function due to accumulated mutations and deletions, recent work has identified a few nearly intact ERV integrations in humans (including at least one HERV-K provirus that may retain the potential for infectivity), and expression of such proviruses in tissues associated with cancer and autoimmune disease suggests at least the potential for pathogenic effects

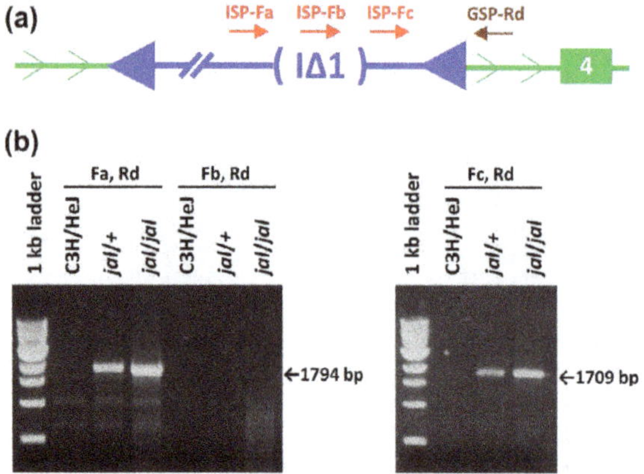

Figure 5. Characterization of the IAP subtype associated with the *Gata3^jal^* insertion. (a) The relative location of the ISP-Fa, -Fb, -Fc and GSP-Rd annealing sites are shown. Note that each of these primer sequences is displayed in Supplementary Figure S1. (b) The ISP-Fa, GSP-Rd primer pair should produce a *jal*-specific amplimer only if the 1.9 kb *gag-pol* deletion characteristic of IAP elements of the IΔ1 subtype should position primer-annealing sites for -Fa and -Rd near one another, similar to the ISP-Fc to GSP-Rd distance. The IΔ1 deletion should remove the annealing site for ISP-Fb, so a deleted IAP template is not expected to produce an amplimer with the -Fb, -Rd pair. The -Fa and -Rd pair directed the amplification of a 1,794 bp product from *jal* templates, only 85 bp longer than the 1,709 bp *jal*-specific product yielded by the -Fc, -Rd primer pair, suggesting that the *Gata3^jal^*-associated IAP is of the IΔ1 subtype. The full DNA sequence of the *jal*-specific, -Fa to -Rd amplimer (which includes the IΔ1 deletion breakpoints in the *Gata3^jal^*-associated IAP insertion) is shown in Supplementary Figure S1.

(Wildschutte et al., 2016). Although ERV insertions in humans are not an important source of new germline mutations (as they are known to be in mice), study of ERV-mediated biological effects in animal models like the juvenile alopecia mouse are still needed to better understand the regulatory effects of existing ERVs and LTRs on the many human genes they are integrated into or near (Jern & Coffin, 2008; van de Lagemaat, Landry, Mager, & Medstrand, 2003; Medstrand, van de Lagemaat, & Mager, 2002), to anticipate and control the potential side effects of retroviral vectors that might be used therapeutically (Dahl et al., 2015; Kagiava et al., 2016; Yi, Jong Noh, & Hee Lee, 2011), and to further explore epigenetic silencing of proviruses (mediated by RNA or drug action, for example) (Herrera-Carrillo & Berkhout, 2015; Lee et al., 2002; Tyagi & Karn, 2007) as an approach to finding a functional cure for retroviral infection.

5. Conclusions

We describe an intracisternal A-type particle element that has integrated into Intron 3–4 of the mouse *Gata3* gene and is specifically associated with the *Gata3^jal^* mutation, which arose spontaneously on the standard inbred C3H/HeJ genetic background and was first described in 1999. This IAP element is in the antisense orientation with respect to the gene, is flanked by a six-base direct repeat, and is shown by sequence analysis to be a deleted element of the IΔ1 subtype. The *Gata3^IAP^* element was mapped to the same small genetic interval as the coisogenic *jal* mutation itself, suggesting that this IAP insertion is the molecular basis of the juvenile alopecia mutant phenotype.

Abbreviations

ERV endogenous retrovirus

IAP intracisternal A particle

PCR polymerase chain reaction

LTR long terminal repeat

SNP single nucleotide polymorphism

GSP gene-specific primer

ISP IAP-specific primer

TSD target site duplication

Acknowledgments
The authors thank Mary Mantzaris for excellent animal care.

Funding
This study was supported by a small research grant from the Connecticut State University System [grant number ARKINJ].

Competing interests
The authors declare no competing interests.

Author details
Malcolm E. Connor[1]
E-mail: malcolm.connor@myccsu.edu
Thomas R. King[1]
E-mail: kingt@ccsu.edu
ORCID ID: http://orcid.org/0000-0001-6332-5484
[1] Department of Biomolecular Sciences, Central Connecticut State University, 1615 Stanley Street, New Britain, CT 06053, USA.

Author's contributions
MEC led all aspects of this study, including experimental design, data acquisition and interpretation. TRK conceived of the study, carried out all procedures involving mice, and drafted the manuscript. Both authors read, edited, and approved the final manuscript.

References
Bisaillon, J. J., Radden II, L. A., Szabo, E. T., Hughes, S. R., Feliciano, A. M., Nesta, A. V., & ... King, T. R. (2014). The retarded hair growth (*rhg*) mutation in mice is an allele of ornithine aminotransferase (*Oat*). *Molecular Genetics and Metabolism Reports*, 1, 378–390. http://dx.doi.org/10.1016/j.ymgmr.2014.08.002

Dahl, M., Doyle, A., Olsson, K., Månsson, J. E., Marques, A. R., Mirzaian, M., & ... Karlsson, S. (2015). Lentiviral gene therapy using cellular promoters cures Type 1 gaucher disease in mice. *Molecular Therapy, 23*, 835–844. http://dx.doi.org/10.1038/mt.2015.16

Druker, R., Bruxner, T. J., Lehrbach, N. J., & Whitelaw, E. (2004). Complex patterns of transcription at the insertion site of a retrotransposon in the mouse. *Nucleic Acids Research, 32*, 5800–5808. http://dx.doi.org/10.1093/nar/gkh914

Duhl, D. M., Vrieling, H., Miller, K. A., Wolff, G. L., & Barsh, G. S. (1994). Neomorphic agouti mutations in obese yellow mice. *Nature Genetics, 8*, 59–65. http://dx.doi.org/10.1038/ng0994-59

Ensembl Mouse Genome Server. (2016). *Mouse genome sequencing consortium: The European Bioinformatics Institute (EBI) and Welcome Trust Sanger Institute (WTSI)*. Release 84. Retrieved 2016, from http://www.ensembl.org

Gunn, T. M., Inui, T., Kitada, K., Ito, S., Wakamatsu, K., He, L., ... Barsh, G. S. (2001). Molecular and phenotypic analysis of *Attractin* mutant mice. *Genetics, 158*, 1683–1695.

Herrera-Carrillo, E., & Berkhout, B. (2015). The impact of HIV-1 genetic diversity on the efficacy of a combinatorial RNAi-based gene therapy. *Gene Therapy, 22*, 485–495. http://dx.doi.org/10.1038/gt.2015.11

International Human Genome Mapping Consortium. (2001). A physical map of the human genome. *Nature, 409*, 934–941.

Ishihara, H., Tanaka, I., Wan, H., Nojima, K., & Yoshida, K. (2004). Retrotransposition of limited deletion type intracisternal A-particle elements in the myeloid leukemia cells of C3H/He mice. *Journal of Radiation Research, 45*, 25–32. http://dx.doi.org/10.1269/jrr.45.25

Jern, P., & Coffin, J. M. (2008). Effects of retroviruses on host genome function. *Annual Review of Genetics, 42*, 709–732. http://dx.doi.org/10.1146/annurev.genet.42.110807.091501

Kagiava, A., Sargiannidou, I., Theophilidis, G., Karaiskos, C., Richter, J., Bashiardes, S., ... Kleopa, K. A. (2016). Intrathecal gene therapy rescues a model of demyelinating peripheral neuropathy. *Proceedings of the National Academy of Sciences, 113*, E2421–E2429. http://dx.doi.org/10.1073/pnas.1522202113

Kaufman, C. K., Zhou, P., Pasolli, H. A., Rendl, M., Bolotin, D., Lim, K.-C., ... Fuchs, E. (2003). GATA-3: An unexpected regulator of cell lineage determination in skin. *Genes & Development, 17*, 2108–2122. http://dx.doi.org/10.1101/gad.1115203

Kuff, E. L., & Lueders, K. K. (1988). The intracisternal A-particle gene family: Structure and functional aspects. *Advances in Cancer Research, 51*, 183–276. http://dx.doi.org/10.1016/S0065-230X(08)60223-7

Kurek, D., Garinis, G. A., van Doorninck, J. H., van der Wees, J., & Grosveld, F. G. (2007). Transcriptome and phenotypic analysis reveals Gata3-dependent signalling pathways in murine hair follicles. *Development, 134*, 261–272. http://dx.doi.org/10.1242/dev.02721

Lee, N. S., Dohjima, T., Bauer, G., Li, H., Li, M.-J., Ehsani, A., ... Rossi, J. (2002). Expression of small interfering RNAs targeted against HIV-1 *rev* transcripts in human cells. *Nature Biochemistry, 19*, 500–505.

Lim, K. C., Lakshmanan, G., Crawford, S. E., Gu, Y., Grosveld, F., & Engel, J. D. (2000). Gata3 loss leads to embryonic lethality due to noradrenaline deficiency of the sympathetic nervous system. *Nature Genetics, 25*, 209–212.

Maksakova, I. A., Romanish, M. T., Gagnier, L., Dunn, C. A., van de Lagemaat, L. N., & Mager, D. L. (2006). Retroviral elements and their hosts: Insertional mutagenesis in the mouse germ line. *PLoS Genetics, 2*, e2. http://dx.doi.org/10.1371/journal.pgen.0020002

Medstrand, P., van de Lagemaat, L. N., & Mager, D. L. (2002). Retroelement distributions in the human genome: Variations associated with age and proximity to genes. *Genome Research, 12*, 1483–1495. http://dx.doi.org/10.1101/gr.388902

Michaud, E. J., van Vugt, M. J., Bultman, S. J., Sweet, H. O., Davisson, M. T., & Woychik, R. P. (1994). Differential expression of a new dominant agouti allele (*Aiapy*) is correlated with methylation state and is influenced by parental lineage. *Genes & Development, 8*, 1463–1472. http://dx.doi.org/10.1101/gad.8.12.1463

Miets, J. A., Grossman, Z., Lueders, K. K., & Kuff, E. L. (1987). Nucleotide sequence of a complete mouse intracisternal A-particle genome: Relationship to known aspects of particle assembly and function. *Journal of Virology, 61*, 3020–3029.

Mouse Genome Database. (2016). *Mouse genome database group: The mouse genome Informatics website*. The Jackson Laboratory, Bar Harbor, ME. Retrieved July 2016, from http://www.informatics.jax.org

Mouse Genome Sequencing Consortium. (2002). Initial sequencing and comparative analysis of the mouse genome. *Nature, 420*, 520–562.

Pandolfi, P. P., Roth, M. E., Karis, A., Leonard, M. W., Dzierzak, E., Grosveld, F. G., ... Lindenbaum, M. H. (1995). Targeted disruption of the GATA3 gene causes severe abnormalities in the nervous system and in fetal liver haematopoiesis. *Nature Genetics, 11*, 40–44. http://dx.doi.org/10.1038/ng0995-40

Rakyan, V. K., Blewitt, M. E., Druker, R., Preis, J. I., & Whitelaw, E. (2002). Metastable epialleles in mammals. *Trends in Genetics, 18*, 348–351. http://dx.doi.org/10.1016/S0168-9525(02)02709-9

Rakyan, V. K., Chong, S., Champ, M. E., Cuthbert, P. C., Morgan, H. D., Luu, K. V., & Whitelaw, E. (2003). Transgenerational inheritance of epigenetic states at the murine *AxinFu* allele occurs after maternal and paternal transmission. *Proceedings of the National Academy of Sciences, 100*, 2538–2543. http://dx.doi.org/10.1073/pnas.0436776100

Ramirez, F., Feliciano, A. M., Adkins, E. B., Child, K. M., Radden II, L. A., Salas, A., ... King, T. R. (2013). The juvenile alopecia mutation (*jal*) maps to mouse Chromosome 2, and is an allele of GATA binding protein 3 (*Gata3*). *BMC Genetics, 14*, 40. http://dx.doi.org/10.1186/1471-2156-14-40

Tyagi, M., & Karn, J. (2007). CBF-1 promotes transcriptional silencing during the establishment of HIV-1 latency. *The EMBO Journal, 26*, 4985–4995. http://dx.doi.org/10.1038/sj.emboj.7601928

van de Lagemaat, L. N., Landry, J. R., Mager, D. L., & Medstrand, P. (2003). Transposable elements in mammals promote regulatory variation and diversification of genes with specialized functions. *Trends in Genetics, 19*, 530–536. http://dx.doi.org/10.1016/j.tig.2003.08.004

van Doorninck, J. H., van der Wees, J., Karis, A., Goedknegt, E., Engel, J. D., Coesmans, M., ... De Zeeuw, C. I. (1999). GATA-3 is involved in the development of serotonergic neurons in the caudal raphe nuclei. *Journal of Neuroscience, 19*, 1–8.

Vasicek, T. J., Zeng, L., Guan, X.-J., Zhang, T., Costantini, F., & Tilghman, S. M. (1997). Two dominant mutations in the mouse Fused gene are the result of transposon insertions. *Genetics, 147*, 777–786.

Walsh, C. P., Chaillet, J. R., & Bestor, T. H. (1998). Transcription of IAP endogenous retroviruses is constrained by cytosine methylation. *Nature Genetics, 20*, 116–117. http://dx.doi.org/10.1038/2413

Ware, M. L., Fox, J. W., González, J. L., Davis, N. M., Lambert de Rouvroit, C. L., Russo, C. J., ... Walsh, C. A. (1997). Aberrant splicing of a mouse disabled homolog, *mdab1*, in the *scrambler* mouse. *Neuron, 19*, 239–249. http://dx.doi.org/10.1016/S0896-6273(00)80936-8

Wildschutte, J. H., Williams, Z. H., Montesion, M., Subramanian, R. P., Kidd, J. M., & Coffin, J. M. (2016). Discovery of unfixed endogenous retrovirus insertions in diverse human populations. *Proceedings of the National Academy of Sciences, 113*, E2326–E2334. http://dx.doi.org/10.1073/pnas.1602336113

Yi, Y., Jong Noh, M., & Hee Lee, K. (2011). Current advances in retroviral gene therapy. *Current Gene Therapy, 11*, 218–228. http://dx.doi.org/10.2174/156652311795684740

4

Analysis of influenza A viruses from gulls: An evaluation of inter-regional movements and interactions with other avian and mammalian influenza A viruses

Jessica Benkaroun[1§], Dany Shoham[2§], Ashley N.K. Kroyer[1], Hugh Whitney[3] and Andrew S. Lang[1*]

*Corresponding author: Andrew S. Lang, Department of Biology, Memorial University of Newfoundland, 232 Elizabeth Ave., St. John's, NL, Canada A1B 3X9

E-mail: aslang@mun.ca

Reviewing editor: Stephen Mark Tompkins, University of Georgia, USA

Additional information is available at the end of the article

Abstract: Birds, including members of the families Anatidae (waterfowl) and Laridae (gulls and terns), serve as the major reservoir of influenza A viruses (IAVs). The ecogeographic contributions of gulls to global IAV dynamics, in terms of geographic scale and virus movements, are important and are distinct from those of waterfowl. Gulls primarily carry the H13 and H16 subtypes, yet can be infected by additional subtypes. Also, gulls are frequently infected by IAVs that contain mixtures of genes from different geographic phylogenetic lineages (e.g. North American and Eurasian). The present analysis examines a variety of viruses isolated from gulls and terns across the world that exhibit particularly high phylogenetic affinities to viruses found in other hosts. This illustrates the potential for gulls to act as highly pathogenic virus carriers, disseminators of viruses over long distances, and contributors in the genesis of pandemic strains. The historical evolution of an entirely Eurasian gull virus isolated in North America was also traced and indicates the Caspian Sea, in southwestern Asia, was an important area for the generation of this virus, and analysis of IAVs from terns also points to this region as relevant for the generation of novel strains.

Subjects: Genetics; Natural History; Virology

Keywords: IAV; gene migration; virus evolution; inter-species transmission

1. Introduction

Wild birds, including waterfowl, shorebirds, and gulls, harbor the vast majority of known influenza A virus (IAV) genetic diversity. Within these influenza IAVs, there is clear distinction between Eurasian and North American gene pools (Olsen et al., 2006) and there is increasing evidence of additional

ABOUT THE AUTHORS

Research on influenza A viruses in the laboratory of Andrew Lang is focused on identification and characterization of viruses from wild birds. There is a particular emphasis on the study of gulls and seabirds as hosts of influenza A viruses, and this research is focused in the North Atlantic region of North America. These viruses are analyzed for their evolutionary relationships with those from other species and locations to follow virus transmission among different hosts and regions.

PUBLIC INTEREST STATEMENT

Influenza A viruses are most well known for their effects on human health and the agriculture industry. However, the movement of these viruses among their natural reservoir species, which are wild birds, plays an important role in these human society-affecting outcomes. Within wild birds, gulls are one important host group and the dynamics of influenza A virus transmission and evolution within this group need to be considered alongside the dynamics in other host groups. This paper presents an analysis of gull viruses and gull biology relevant to this topic.

phylogeographic lineages (Gonzalez-Reiche & Perez, 2012; Hurt et al., 2014; Pereda et al., 2008). However, bird movements lead to inter-regional transmission of IAV genes and their establishment in new regions. Intercontinental transmission of IAV genes from Eurasia to America has been frequently observed, and isolation of reassortant viruses that contain gene segments of different continental origins is particularly common from gulls in these regions, which is presumably facilitated by gull movements across the oceans. The North Atlantic Ocean is one location where gull and IAV movements between North America and Europe have been documented (Wille, Robertson, Whitney, Ojkic, & Lang, 2011), and intercontinental IAV reassortants have also been identified in high proportions (Huang et al., 2014). Additionally, both entirely Eurasian and North American gull IAVs have been found in Iceland (Dusek et al., 2014), suggesting this location as important for the mixing of viruses between regions. An equivalent position has also been proposed for Greenland (Shoham & Rogers, 2006). In North America, only two wholly Eurasian avian IAVs have been found to date. One was an H16N3 virus from an American herring gull (*Larus smithsonianus*) in eastern Canada (Huang et al., 2014). The other was the highly pathogenic avian influenza (HPAI) H5N8 virus that recently emerged in China, spread to South Korea, Japan, and Siberia, and is proposed to have been transmitted through the Beringia region by migratory waterfowl into Pacific North America (Lee et al., 2015). This H5N8 arrival ended the long speculation about whether or not virulent Asian avian strains, such as H5N1, could be transmitted into the Americas by migrating birds. There is also a third case, involving H9N2 viruses from Alaska that were nearly identical (>99%) to viruses in China and South Korea, which were proposed to be carried to Alaska by migratory waterfowl (Ramey et al., 2015). In this case, however, the gene segments of the viruses showed mixed geographic origins.

In addition to geographic divisions in IAV genes, distinct host group lineages are found. These include mainly the avian, gull, human, swine, and equine lineages (Olsen et al., 2006). These distinctions make it possible to recognize that inter-species transmissions of IAVs between gulls and other hosts have occurred. Although most IAVs from gulls contain gull-lineage gene segments, and most are of the hemagglutinin (HA) subtypes H13 and H16, various viruses found in gulls contain HA genes and other gene segments that do not appear to be gull-specific, but rather fall within avian clades and are closely related to waterfowl viruses in phylogenetic analyses (Hall et al., 2013; Van Borm et al., 2012; Wille et al., 2011). The opportunities for these transmissions certainly exist, as gulls share terrestrial, freshwater, and marine habitats with other birds and mammals, and therefore gulls could be important for moving IAVs among ecosystems and for contributing to the transmission of highly pathogenic IAVs to other hosts.

The present analysis is an investigation of the potential for gulls to act as carriers of highly pathogenic viruses, disseminators of viruses over long distances, and contributors in the genesis of pandemic strains. This is done through examination of a variety of gull and tern viruses isolated from different locations that exhibit particularly high phylogenetic affinities to viruses found in other hosts.

2. Results and discussion

2.1. Ecogenetic aspects of gull IAVs

Avian IAVs constitute the major, and likely the primal, component within the total IAV gene pool worldwide (Olsen et al., 2006; Webster, Bean, Gorman, Chambers, & Kawaoka, 1992). Distinct phylogeographic lineages, such as those representing Eurasia and North America, can be identified within avian IAVs, mainly as a result of partial segregation of migratory birds (Olsen et al., 2006). The first indication that gulls might represent a distinct reservoir for IAVs came from surveillance work on the Atlantic coast of North America, with the identification of a new HA antigen, H13, distinct from the previous 12 HA antigens identified in other avian species (Hinshaw et al., 1982). Two decades later, an additional novel HA subtype, H16, was identified and characterized from gulls in Sweden (Fouchier et al., 2005). Continued surveillance and research have shown that gulls are predominantly infected by these H13 and H16 subtypes (Figure 1), and these subtypes are not often found outside of gulls (other than in shorebirds) (Arnal et al., 2015). We performed phylogenetic and time of most recent common ancestor (TMRCA) analyses of H13 and H16 nucleotide sequences. These showed three

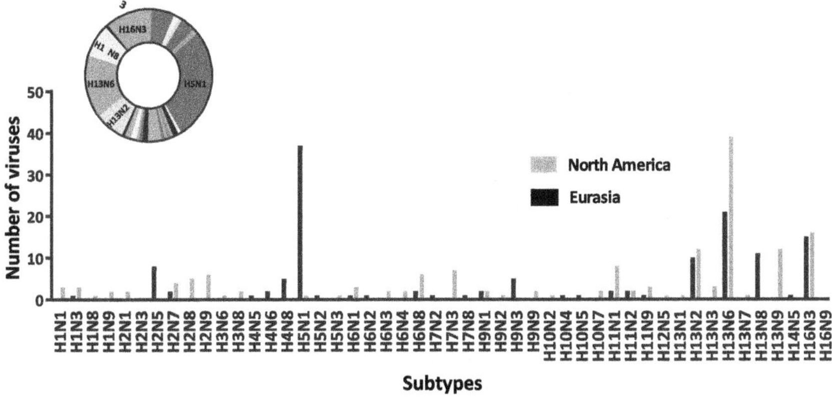

Figure 1. Distribution of IAV subtypes isolated from gulls.

Notes: The numbers of viruses for which sequence data are available with the indicated subtypes isolated from gulls in North America and Eurasia are shown. The circular representation shows the total proportions when both regions are combined. Gull viruses were located by presence of M gene segment sequences in the NCBI Influenza Virus Resource (http://www.ncbi. nlm.nih.gov/genomes/FLU/FLU. html), as of September 2015.

distinct clades for each subtype that were estimated to originate from a common ancestor in the 1930s for the H13 lineage (Figure 2) and in the 1920s for the H16 lineage (Figure 3), similar to previous estimates (Worobey, Han, & Rambaut, 2014). Both trees show some segregation of sequences according to geographic origin as well as clades that contain mixtures of viruses from both Eurasia and North America. These gull lineage HA genes have been transferred to waterfowl, poultry, and marine mammals but there is no evidence of persistence of these sequences in these other host groups.

Inter-hemispheric movements of avian IAV genes from Eurasia to America have repeatedly been documented (Dusek et al., 2014; Fries et al., 2013; Ip et al., 2008). However, observations of IAV movements from America to Eurasia have been less common. Possible explanations for this include a higher rate of infection in birds during movements from Eurasia to America, differences in net avian biomass traveling in the two directions, incompatibility of the Eurasian IAV gene pool with North American genes introduced into Eurasia, or uneven virological monitoring. Regardless, the North American avian IAV gene pool appears appreciably permissive for Eurasian genes, which are frequently assimilated (Bahl, Vijaykrishna, Holmes, Smith, & Guan, 2009). This is exemplified by the case of the H6 subtype, where it was shown that IAV movement from Eurasia to North America led to the replacement of the North American H6 subtype, continent-wide, within a decade of its introduction (Zu Dohna, Li, Cardona, Miller, & Carpenter, 2009). However, this is an extreme case, and Eurasian gene introductions do not usually appear to be so dramatic and successful. As mentioned above, there have been only two discoveries of entirely Eurasian IAVs in North America, the H16N3 gull virus and the highly pathogenic H5N8 virus. Examining the evolutionary histories of the entirely Eurasian H16N3 gull virus' genes reveals that the Caspian Sea appears to be a source for the generation of this virus and others. Indeed, genetic analyses illustrate the spatiotemporal distribution of viruses that traces its gene segment origins from Astrakhan, in Russia, in 1983 to Ukraine in 2005, Scandinavia between 2006–2009, then to Iceland and Eastern Canada in 2010 (Figure 4 and Table S1). Therefore, it appears that the Eurasian isolate from Eastern Canada has been generated through the gradual accumulation of different genes (through reassortment events) at different locations from 1983 to 2010, although we recognize that there are limitations in the consistency of availability of data through time and from all regions. This was presumably facilitated through the migration of gulls across these regions and the spatiotemporal pattern suggests a predominantly east to west movement and evolutionary trend. While only single gene segments were found closely related to the Eurasian gull virus preceding Iceland, this recent virus from Iceland showed a close relationship to the virus from Eastern Canada for four gene segments (HA, PB2, PA, and NP). It is also notable that an H16 segment similar to the North American virus was detected three years later in a duck in Japan, which points towards an active inter-hemispheric gull-duck interface (Table S1). Globally, the H16N3 North American gull virus is presently the first and only whole IAV of which a trans-Atlantic

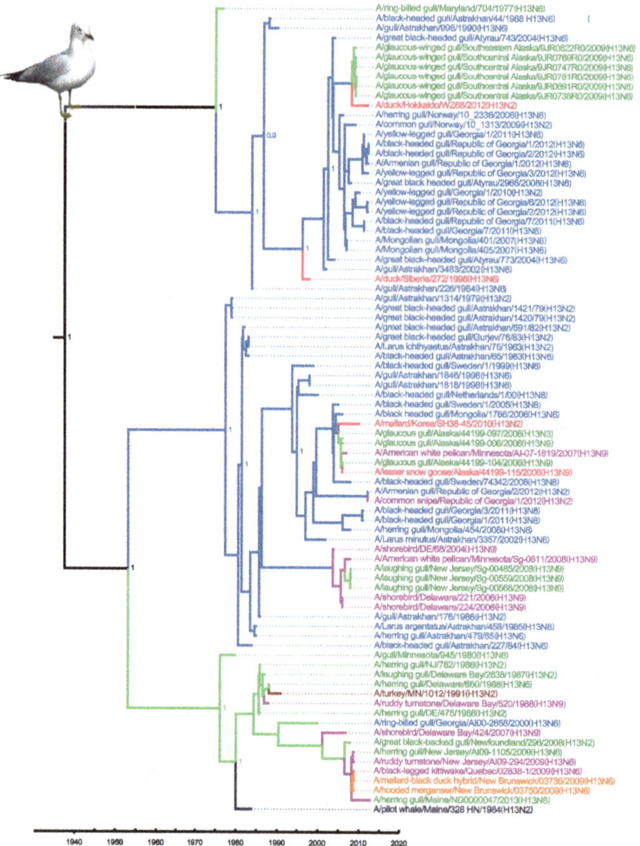

Figure 2. Time-clock Bayesian inference analysis of H13 nucleotide sequences from North America and Eurasia.

Notes: All H13 nucleotide sequences were downloaded from the NCBI Influenza Virus Resource (Bao et al., 2008) and aligned using MUSCLE integrated in MEGA6 (Tamura, Stecher, Peterson, Filipski, & Kumar, 2013). The aligned sequences were quality-trimmed, resulting in an alignment of 1133 nts (corresponding to nt positions 465–1597). This region was chosen because it allowed the inclusion of the largest number of virus sequences. We compared this tree to one constructed with a subset of viruses for which the complete segment sequences were available and the overall topology is not affected by use of this portion (not shown). The maximum credibility tree was generated from the trimmed alignment by the Bayesian inference method implemented in BEAST v1.8.0 (Drummond, Suchard, Xie, & Rambaut, 2012) using the SRD06 substitution model with a strict molecular clock and with the use of the GMRF Bayesian sky ride coalescent prior distribution. These parameters were chosen because they gave the best distribution of posterior probabilities. A strict molecular clock was chosen as appropriate to estimate the time of the most recent ancestors at each node because the sequences are almost all from the same host species. The branches are colored according to the host group: green, North American gulls; blue, Eurasian gulls; brown, poultry; dark blue, marine mammals; dark pink, Eurasian waterfowl; orange, North American waterfowl; purple, shorebirds and other wild birds excluding gulls, waterfowl, and poultry. The posterior probabilities for the support of the branches are given at major nodes. Photo credit: http://www.freeimages.com.

ecophylogenetic course has been concretely traced. Its origin and spatiotemporal phylogenetic history are analogous to the apparent natural history of the H14 viruses that were originally isolated from a gull and mallard (*Anas platyrhynchos*) in Astrakhan, and subsequently recovered 28 years later in ducks in North America (Fries et al., 2013). The H14 subtype was originally isolated in 1982 from both European herring gulls (*Larus argentatus*) and mallards sampled in Central Asia along the northern shore of the Caspian Sea. Subsequent identifications of H14 viruses have occurred solely in North and Central America (USA and Guatemala) during 2010 to 2012 (Boyce et al., 2013; Ramey et al., 2014). These viruses originated from sea ducks and dabbling ducks, which are most frequently associated with marine and freshwater habitats, respectively. This unusual spatiotemporal profile might indicate a role played by gulls in inter-hemispheric and marine-freshwater avifauna IAV transmission. The gap in H14 virus isolation from 1982 to 2010 could be due to a lack of fortuitous

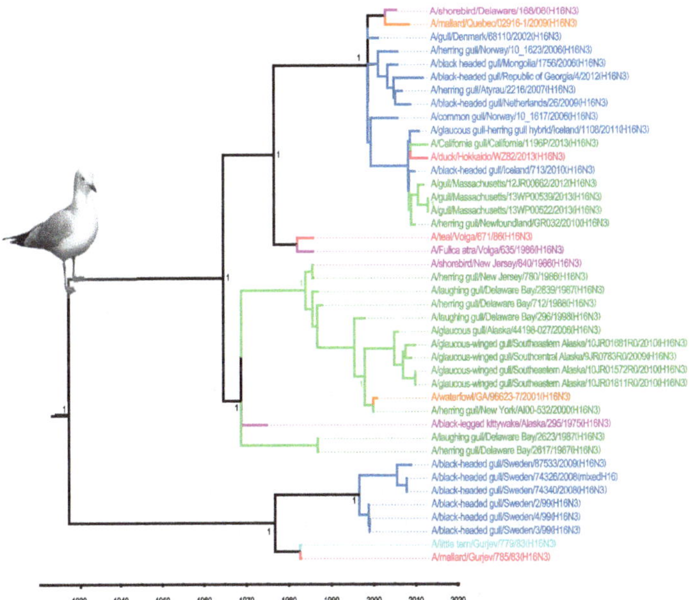

Figure 3. Time-clock Bayesian inference analysis of H16 nucleotide sequences from North America and Eurasia.

Notes: All complete H16 nucleotide sequences were downloaded from the NCBI Influenza Virus Resource (Bao et al., 2008) and aligned using MUSCLE integrated in MEGA6 (Tamura et al., 2013). The aligned sequences were quality-trimmed, resulting in an alignment of 1562 nts (corresponding to nt positions 94–1655). The maximum credibility tree was generated from the trimmed alignment by the Bayesian inference method implemented in BEAST v1.8.0 (Drummond et al., 2012) using the SRD06 substitution model with a strict molecular clock and with the use of the GMRF Bayesian sky ride coalescent prior distribution. The branches are colored according to the host group: green, North American gulls; blue, Eurasian gulls; brown, poultry; dark pink, Eurasian waterfowl; orange, North American waterfowl; purple, shorebirds and other wild birds excluding gulls and waterfowl. The posterior probabilities for the support of the branches are given at major nodes. Photo credit: http://www.freeimages.com.

sampling during those periods and the same applies for the large gap between the identification of gene segments similar to the North American H16N3 gull virus. Regardless, the overall movements of the involved host species are compatible with the viral migration routes. Broadly, such analyses might be useful for detailed reconstruction of evolutionary processes in other IAVs.

2.2. Ecogeographic aspects of gull IAVs

The global distribution of gulls is vast and nearly unlimited worldwide (Olsen & Larsson, 2004). They breed on every continent, from the margins of Antarctica to the high Arctic. Many species breed in coastal colonies and although they are less common on tropical islands, where terns are often very common, a few species do inhabit islands such as the Galapagos and New Caledonia. There is considerable variety within the group, and species may breed and feed in marine, freshwater and/or terrestrial habitats. Many gull species migrate, but the extent to which they migrate varies among species. Some species migrate long distances, such as Franklin's gull (*Leucophaeus pipixcan*) that migrates from Canada to wintering grounds in southern South America. Two examples illustrating inter-hemispheric distribution are black-headed gull (*Chroicocephalus ridibundus*) and laughing gull (*Leucophaeus atricilla*). The black-headed gull mostly breeds in Europe and Asia, and most of the population is migratory, wintering further south, but some birds in the milder westernmost areas of Europe are non-migratory. Some birds also spend the winter in northeastern North America. Additionally, black-headed gulls have an impressive ability to adapt to man-made environments. The laughing gull is largely a species of North and South America although it also occurs as a rare vagrant to western Europe. Herring gulls exhibit intercontinental distribution and connectivity and there is overlap in the ranges of closely related species/subspecies within a larger species complex that includes American herring gull, European herring gull, and East Siberian gull (*Larus vegae*) (Sangster, Collinson, Knox, Parkin, & Svensson, 2007). However, it is likely that migratory connectivity

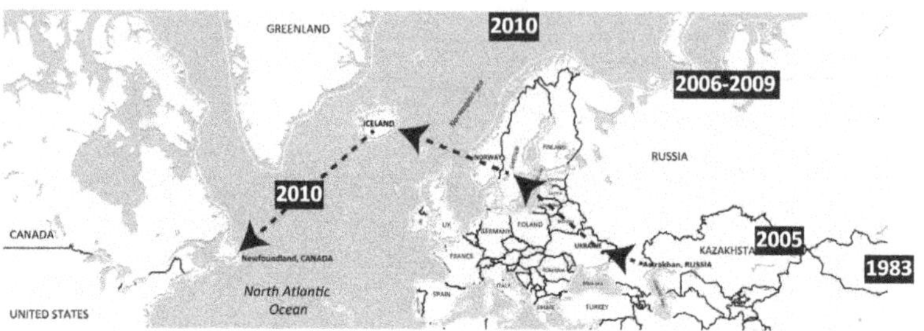

Figure 4. Putative geographic progression of gull virus genes that contributed to the genesis of the Eurasian H16N3 gull virus identified in Newfoundland, Canada.

Notes: The background map was downloaded from Mapbox Editor online tool (https://www.mapbox.com/editor).

exists among these related American, East Eurasian, and European populations, and this may serve as a platform for virus movements across both the Pacific and Atlantic Oceans.

2.3. Clinical and subclinical IAV infections of gulls

Historically, IAVs have been classified as virulent in gulls and terns in only two natural episodes. These are the first wild bird IAV isolate, A/Tern/South Africa/61(H5N3) (Becker, 1966), and the Asian highly pathogenic H5N1 lineage that infects many species of mammals and birds (Amonsin et al., 2007; Chen et al., 2005; Claas et al., 1998; Mushtaq et al., 2008; Subbarao et al., 1998). However, an H13 virus circulating in a Canadian colony of dense-breeding ring-billed gulls (*Larus delawarensis*) caused inflammation of the heart, kidney, pancreas, and liver in chicks but no clinical symptoms (Velarde, Calvin, Ojkic, Barker, & Nagy, 2010). Details from some of the experimental infection studies involving gulls or other hosts infected with avian strains or with gull viruses are summarized in Tables 1–3, and some details are expanded below.

Experimental infection with the A/chicken/Hong Kong/220/97(H5N1) strain produced mild disease in laughing gulls (Perkins & Swayne, 2002), whereas severe disease and mortality were reported in laughing gulls after infection with highly pathogenic H5N1 strains from 2001 and 2006 (Brown, Stallknecht, Beck, Suarez, & Swayne, 2006). Another highly pathogenic H5 strain, A/Chicken/Penn/1370/83(H5N2), which caused 80% mortality in chickens in a Pennsylvania outbreak, produced little to no clinical signs in experimentally infected ring-billed gulls (Wood, Webster, & Nettles, 1985). Franklin's gulls experimentally infected with a turkey virus that was pathogenic for turkeys did not develop signs of disease, yet were capable of shedding the virus (Bahl & Pomeroy, 1977). These studies suggest that gulls could be infected with virulent strains from poultry and serve as asymptomatic carriers for subsequent virus transmission. In another study, the virus A/laughing gull/NJ/AI08-1460/2008(H13N9) was used to experimentally infect ring-billed gulls by the respiratory route, and high susceptibility to the virus was observed without developing any clinical symptoms (Brown et al., 2012). Most importantly, the virus was excreted via the oropharynx and cloaca for several days. Collectively, *in vivo* experimental infection studies show fairly consistent clinical sensitivity of gulls to different highly pathogenic viruses, and subclinical sensitivity to low pathogenicity viruses, with consistent evidence of virus shedding.

Studies involving experimental infections of mammals with gull viruses have been limited, but several are of particular interest. In one study (Hinshaw et al., 1982), the H13 gull virus A/gull/Massachusetts/26/1980(H13N6) was used in experimental infection of ferrets, which are generally considered as the best animal model for human influenza infection. When challenged intranasally, the ferrets contracted the virus and shed it in nasal secretions. Additionally, attachment of the strain A/black-headed gull/Sweden/2/99(H16N3) to human, gull, and mallard tissues was investigated using tissue microarrays and virus histochemistry, using mallard and human viruses as references (Lindskog et al., 2013). The H16N3 virus attached more readily to the human respiratory tract than the other avian IAVs, and it could also bind to the human cornea and conjunctiva. The human virus, A/Netherlands/213/03(H3N2), also attached to the trachea of both gull species investigated, which demonstrates the presence of the first necessary characteristic for respiratory transmission. It was

Table 1. Experimental infections of gulls with avian strains

Challenged species	Virus strain	Pathogenicity	Route of inoculation	Inoculum titer[a]	Clinical and pathological findings	Virus shedding or presence	Reference
Laughing gull (*Larus atricilla*)	A/tern/South Africa/61/H5N3	HPAI	Intranasal	10^6 ELD_{50}	Mild lesions	Oropharynx and cloaca	(Perkins & Swayne, 2002)
	A/chicken/Hong Kong/220/97/H5N1	HPAI	Intranasal	10^6 ELD_{50}	Mild lesions	Oropharynx and cloaca	
Laughing gull	A/Duck Meat/Anyang/01/H5N1	HPAI	Intranasal	10^6 EID_{50}	Severe disease, death (2/3), histological lesions	Oropharynx > cloaca	(Brown et al., 2006)
Herring gull (*Larus argentatus*)	A/whooper swan/Mongolia/244/05/H5N1	HPAI	Intranasal	10^6 EID_{50}	Sign of disease, death (2/3)	Oropharynx > cloaca	(Brown, Stallknecht, & Swayne, 2008)
	A/duck meat/Anyang/AVL-1/01/H5N1	HPAI	Intranasal	10^6 EID_{50}	Sign of disease, no death		
	Infected chicken meat (A/whooper swan/Mongolia/244/05(H5N1))	HPAI	Ingestion	10^6 EID_{50}	Sign of disease, death (1/3)		
Black-headed gull (*Chroicocephalus ridibundus*)	A/turkey/Turkey 1/2005/H5N1	HPAI	Intratracheal and intraoesophageal	10^4 $TCID_{50}$	Neurological disorder, spontaneous mortality, loss of body weight, no histological lesions	Pharynx and cloaca	(Ramis et al., 2014)
Ring-billed gull (*Larus delawarensis*)	A/Chicken/Pennsylvania/1370/83/H5N2	HPAI	Nasal and ocular	10^8 EID_{50}	Few or no clinical sign of disease	Upper respiratory tract	(Wood et al., 1985)
Franklin's gull (*Larus pipixcan*)	A/turkey/Minn/BF/72/Hav6Neq2/H6N8	LPAI	Intratracheal	3.1×10^3 EID_{50}	None, 1/5 died with no histological lesions observed	Trachea and cloaca	(Bahl & Pomeroy, 1977)
Ring-billed gull	A/laughing gull/NJ/AI08-1460/2008/H13N9	LPAI	Intranasal and intratracheal	10^4 EID_{50}	No sign of disease	Oropharynx and cloaca	(Brown et al., 2012)
Silver gull (*Chroicocephalus novaehollandiae*)	A/Eurasian coot/WA/2727/79/H6N2	LPAI	Oropharyngeal	$10^{7.95}$ EID_{50}	None	Trachea and cloaca	(Curran, Robertson, Ellis, Selleck, & O'Dea, 2013)

[a]ELD50 and EID50: amount of infectious virus that will cause the death of 50% of inoculated embryonated eggs; TCID50: amount of infectious virus that will cause the infection of 50% of tissue culture.

Table 2. Experimental infections of avian and mammalian hosts with gull strains

Challenged species	Virus strain	Pathogenicity	Route of inoculation	Inoculum titer[a]	Clinical and pathological findings	Virus shedding or presence	Reference
White Leghorn chicken	A/black-tailed gull/Tottori/61/80/Hav1Neq1/H7N7	LPAI	Intratracheal	$10^{7.8-108.3}$ EID_{50}	No sign of disease	Brain, lungs, liver, spleen,	(Otsuki et al., 1982)
			Intraperitoneal		1/9 died	Kidneys, jejunum, and rectum	
Duck	A/ring-billed gull/Maryland/704/77/H13N6	LPAI	Oral and intratracheal	10^8 EID_{50}	No sign of disease	No tracheal and cloacal shedding	(Hinshaw et al., 1982)
Chicken	A/ring-billed gull/Maryland/704/77/H13N6	LPAI	Oral and intratracheal	10^8 EID_{50}	No sign of disease	No tracheal and cloacal shedding	
Ferret	A/gull/Massachusetts/26/1980/H13N6	LPAI	Intranasal	10^6 EID_{50}	No sign of disease	Nasal shedding	
Domestic Lohmann white chicken	A/herring gull/Norway/10_1623/2006/H16N3	LPAI	Intranasal	10^6 EID_{50}	No sign of disease	Oropharynx (2/19) and nasal (1/19)	(Tønnessen, Valheim, Rimstad, Jonassen, & Germundsson, 2011)
Turkey	A/laughing gull /NJ/AI08–1460/2008/H13N9	LPAI	Intranasal and intratracheal	10^7 EID_{50}	No sign of disease	Oropharynx (1/8) and cloaca	(Brown et al., 2012)
	A/ring-billed gull/MN/AI10–1708/2010/H13N6	LPAI	Intranasal and intratracheal	10^7 EID_{50}	No sign of disease	None	
Mallard (*Anas platyrhynchos*)	A/Gull/Ontario/680–6/2001/H13N6	LPAI	Intratracheally and intraesophageally	10^8 EID_{50}	No sign of disease, some lung tissue lesions observed	Limited pharyngeal, cloacal, and respiratory tissue	(Daoust et al., 2013)
Mallard	A/herring gull/Germany/R3309/07/H16N3	LPAI	Oculo-nasal-oral	10^5 $TCID_{50}$	No sign of disease	None	(Fereidouni, Harder, Globig, & Starick, 2014)
Ferret	A/gull/Delaware/428/2009(H1N1)	LPAI	Airborne	10^6 EID_{50}	Lethargy, fever, sneezing, coughing	Nasal	(Koçer et al., 2015)

[a]ELD50 and EID50: amount of infectious virus that will cause the death of 50% of inoculated embryonated eggs; TCID50: amount of infectious virus that will cause the infection of 50% of tissue culture.

Table 3. Gull virus binding assays on avian and human tissues (Lindskog et al., 2013)

Species' tissue tested	Virus strain	Tissues bound
Franklin's gull (*Leucophaeus pipixcan*)	A/black-headed gull/Sweden/2/99/ H16N3	Trachea, duodenum, ileum, ileocecal junction, colon
Herring gull (*Larus argentatus*)	A/black-headed gull/Sweden/2/99/ H16N3	Duodenum, ileum, ileocecal junction, colon
Mallard (*Anas platyrhynchos*)	A/black-headed gull/Sweden/2/99/ H16N3	Trachea
Human	A/black-headed gull/Sweden/2/99/ H16N3	Cornea, conjunctiva, nasopharynx, bronchus, pulmonary alveolus, salivary gland

consequently suggested that H16 viruses might present a different cell tropism in humans and mallards than other bird-origin IAVs (Lindskog et al., 2013). Elsewhere, it was shown that several gull species display α2,6-linked sialic acid receptors, to which human IAVs usually bind, on the surface of their tracheal epithelium (Jourdain et al., 2011). Those findings indicate the potential for either direct or indirect gull-human IAV interface. Viruses from gulls were also among those identified in wild birds as encoding proteins that differed at relatively few amino acid positions from the 1918 pandemic strain (Watanabe et al., 2014), with gull viruses found that had <15 amino acid differences in their PB2, PB1, PA, NP, M1 or NS1 proteins. A virus identified in wild birds that contained such segments showed higher pathogenicity than typical avian viruses and was able to gain mammalian respiratory drop transmission after acquiring as few as three substitutions in the HA protein (Watanabe et al., 2014). This supports the notion that the potential to generate a pandemic virus exists within wild birds and, although the sequences used in the experimental study were not from gull viruses, gulls are a relevant part of this scenario.

2.4. *Inter-species transmission of gull IAVs*

Although the level of active infection found is much lower, seroprevalence for IAV in gulls is high, reaching 45–92% (Brown et al., 2010; Huang et al., 2014; Toennessen et al., 2011; Velarde et al., 2010). As mentioned above, gulls also exhibit notable capacity in terms of inter-hemispheric virus transmission (Hall et al., 2013; Huang et al., 2014; Perkins & Swayne, 2002; Ratanakorn et al., 2012; Van Borm et al., 2012; Wille, Robertson, Whitney, Bishop, et al., 2011, Wille et al., 2011). If gull-imported Eurasian gene segments persist within the American gene pool, they could subsequently be transmitted to domestic hosts such as chickens, turkeys and pigs. In consideration of experimental studies that frequently show subclinical infections of gulls by highly pathogenic avian viruses, these hosts should be regarded as candidates for inter-hemispheric transmission of highly pathogenic viruses. The following sections will evaluate whether gulls could play a major role in inter-species and long-range transmission of IAVs and specific IAV genes and whether they could be considered as contributors to the transmission of HPAI viruses.

2.4.1. The gull-poultry IAV interface

The first evidence of poultry infection by a gull IAV came from a study in which H13N2 viruses were isolated in North America from subclinically infected domestic turkey flocks living near accessible pond water and gulls (Sivanandan, Halvorson, Laudert, Senne, & Kumar, 1991). The H13 sequence of one of the resultant viruses, A/turkey/MN/1012/1991(H13N2), is included within the North American gull AIV lineage (Figure 2). These turkey H13N2 viruses did exhibit marked pathogenicity in experimentally infected turkeys (Laudert, Sivanandan, Halvorson, Shaw, & Webster, 1993). However, various H13 gull viruses used for experimental challenges of various poultry species infrequently resulted in infection and appeared only subclinical (Table 2). There have been no reports of H16 viruses infecting poultry. A low pathogenicity H7N7 gull strain was widely invasive and caused systemic, though subclinical, infection in experimental studies with chickens (Otsuki, Kawaoka, Nakamura, & Tsubokura, 1982). Another low pathogenicity gull virus of the H7 subtype, A/Laughing gull/DE/42/06(H7N3), is believed to have been naturally transmitted to chickens, with subsequent evolution to a highly pathogenic form therein (Krauss et al., 2007). This virus was also fully infectious, although mostly subclinically, in experimentally challenged chickens (Krauss et al., 2007). Inter-species transmission between gulls and chickens was also observed with A/brown-head gull/Thailand/vsmu-4/2008(H5N1), being highly similar to two highly pathogenic chicken isolates identified from the same region (Figure 5). Further sampling in turkey and chicken flocks, especially where there are potential interactions with gulls, would help confirm whether gulls could play a significant role in IAV transmission to poultry.

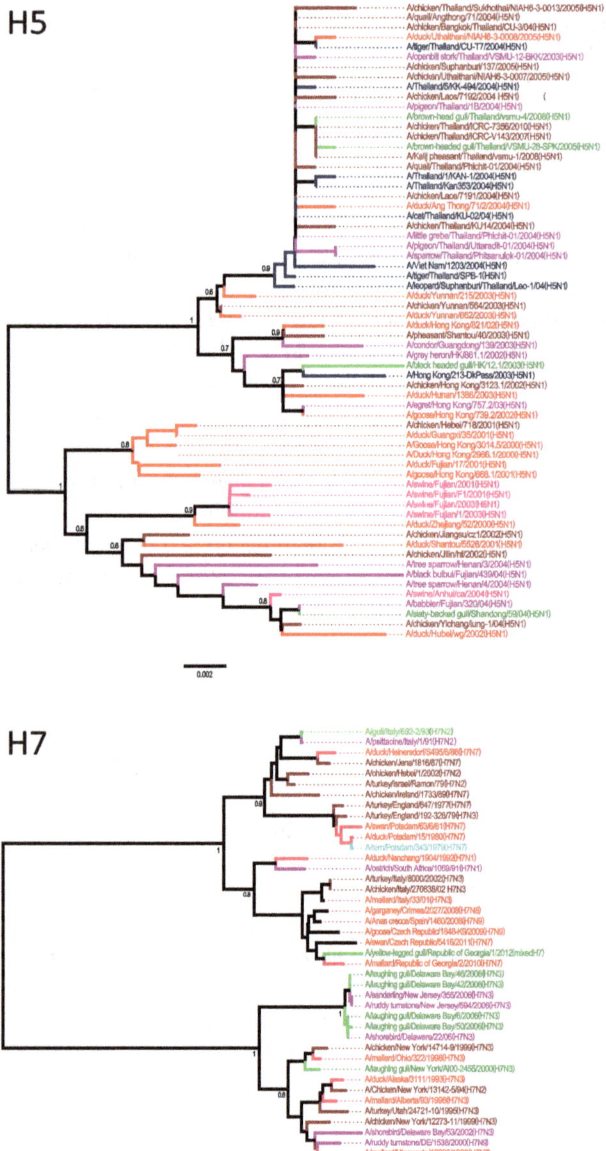

Figure 5. Phylogenetic analysis of gull H5 and H7 nucleotide sequences.

Notes: Sequences of H5 and H7 genes that showed the highest similarity in BLAST (Altschul, Gish, Miller, Myers, & Lipman, 1990) searches to gull gene sequences were downloaded from NCBI and aligned by MUSCLE with MEGA6 (Tamura et al., 2013). The aligned sequences were quality-trimmed, resulting in alignments of 1,073 nts (corresponding to nt positions 172–1,242) and 1,054 nts (corresponding to nt positions 32–1,085) for the H5 and H7 gene segments, respectively. The phylogenetic tree was inferred in MEGA6 with the neighbor-joining method with 1,000 bootstrap replications. Evolutionary distances were computed using the Maximum Composite Likelihood method and are represented by the number of base substitutions per site (scale bar). The branches are colored according to the host group: green, gulls; red, waterfowl from Eurasia; orange, waterfowl from North America; pink, swine; brown, poultry; purple, shorebirds and other wild birds excluding gulls and waterfowl; blue, humans and other mammals excluding swine.

2.4.2. The gull-waterfowl IAV interface

Sympatry exists among many gull species and waterfowl across freshwater and marine habitats. In addition to the H14 viruses discussed above, particularly high gene sequence identity can be found between various gull and duck viruses (Figures 4 and 5, Table S2). These include both wild (mostly mallard) and domestic duck viruses in America and Eurasia. These diverse affinities substantiate an ongoing, reciprocal gull-waterfowl interface, worldwide, which spans both natural ecosystems and farms containing domestic ducks. Additionally, one study showed that H16N3 gull viruses isolated in Russia readily replicated in respiratory and intestinal tissues in challenged ducks (Iamnikova et al., 2009).

2.4.3. The gull-marine mammal IAV interface

The movement of IAV between gulls and mammals was identified for the first time in 1981, when it was shown that all genomic segments from a seal H7N7 virus were highly similar to viral gene segments from avian species, including gulls (Webster et al., 1981). Interestingly, the seal isolate replicated poorly in avian species but well in mammals such as ferrets, pigs, and cats, which indicated this isolate had probably accumulated mammalian-adaptive mutations. Phylogenetic analysis of some seal and whale viruses has shown that these are frequently closely related to gull and shorebird viruses (Groth et al., 2014). More specifically, relatedness between NP gene segments of the strains A/gull/Maryland/704/77(H13N6) and A/whale/Maine/328/84(H13N2) from the Atlantic Coast of the USA was found (Mandler et al., 1990). Similarly, the HA gene segment from the same whale isolate falls within a North American gull H13 lineage, and presumably originated from gulls (Figure 2, Table S2). High sequence identity was also found between the H5 gene segments from a gull isolate, A/black_headed_gull/HK/12.1/2003(H5N1), and a human isolate, A/Hong_Kong/213/2003(H5N1), both obtained in Asia (Figure 5, Table S2).

2.4.4. The gull-swine IAV interface

Swine are considered one of the most important mixing vessels of IAV, as genetic reassortment between avian and mammalian viruses occurs in this host. Many of the viruses found in swine also contain genes that originated from human and bird viruses. H1N1, H1N2, and H3N2 viruses are endemic in the swine population worldwide (Brown, 2000), which also represent the subtypes circulating in the human population and the presence of those viruses or their genes within gulls might be important in terms of transmission of viruses across distances and host species. To date, several H1 viruses have been found in the gull population in North America and they all fall within a North American lineage that mainly comprises waterfowl viruses. Although no direct evidence of AIV transmission between gulls and swine has been found to date, we identified gull viruses that do exhibit phylogenetic affinities to swine viruses, in conjunction with related waterfowl and poultry viruses. One swine virus from 2002, A/swine/Saskatchewan/18789/02(H1N1), which was closely related to waterfowl viruses (Karasin, West, Carman, & Olsen, 2004), was also close to two gull viruses obtained in 2002 from the United States for the HA gene segment (Figure 6).

The H3N2 viruses currently found in swine are believed to have originated mostly from reassortment events involving avian and swine viruses or avian, swine and human viruses (Brown, 2000) and the triple reassortant virus has now become predominant in the swine population (Richt et al., 2003). Sequence analysis of H3 genes from gull viruses showed that two viruses from America and Asia, A/herring gull/New Jersey/159/1990(H3N6) and A/herring gull/Irkutsk/111/2005(H3), which belong to a Eurasian avian lineage, were related to swine and poultry viruses from England and China (Figure 6). Additionally, H3 gene segments from shorebirds related to A/herring gull/New Jersey/159/1990(H3N6) and found in North America are falling within a Eurasian lineage, which further illustrate the potential role of gulls in inter-hemispheric movement of IAVs (J. Bahl et al., 2013).

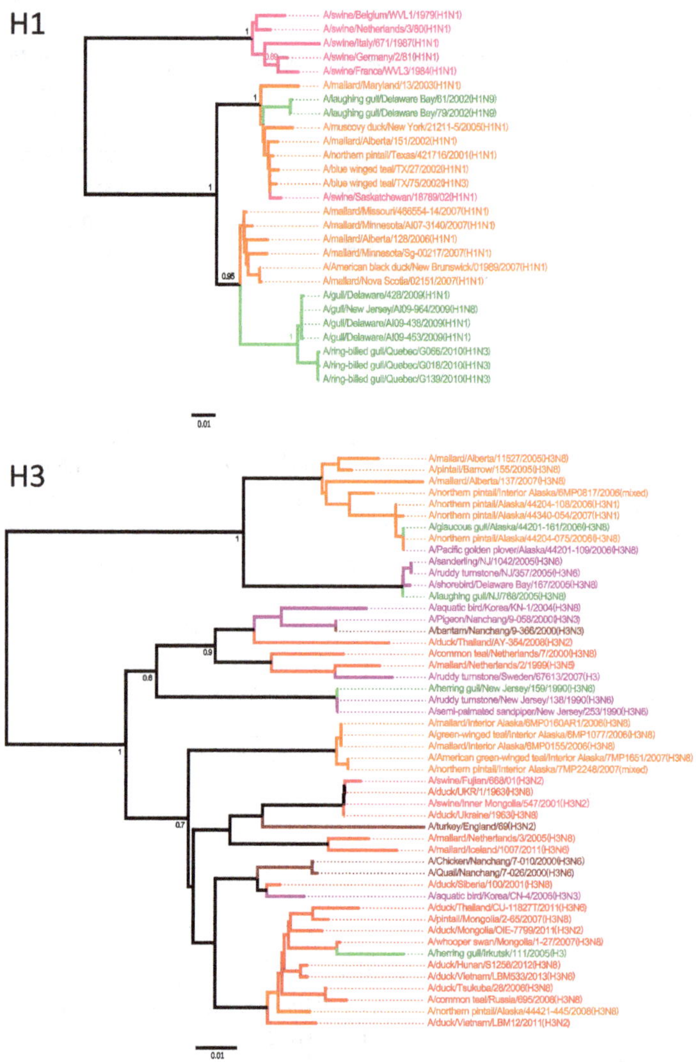

Figure 6. Phylogenetic analysis of gull H1 and H3 nucleotide sequences.

Notes: Sequences of H1 and H3 genes that showed the highest similarity in BLAST (Altschul et al., 1990) searches to gull gene segments were downloaded from NCBI and aligned by MUSCLE with MEGA6 (Tamura et al., 2013). The aligned sequences were quality-trimmed, resulting in alignments of 1,028 nts (corresponding to nt positions 10–1,037) and 641 nts (corresponding to nt positions 1,049–1,689) for the H1 and H3 gene segments, respectively. For the H3 sequences, this region was chosen because it allowed the inclusion of the largest number of virus sequences. We compared this tree to one constructed with a subset of viruses for which the complete segment sequences were available and the overall topology is not affected by use of this portion (not shown). The phylogenetic tree was inferred in MEGA6 with the neighbor-joining method with 1,000 bootstrap replications. Evolutionary distances were computed using the Maximum Composite Likelihood method and are represented by the number of base substitutions per site (scale bar). The branches are colored according to the host group: gulls, green; red, waterfowl from Eurasia; orange, waterfowl from North America; swine, pink; brown, poultry; purple, shorebirds and other wild birds excluding gulls and waterfowl.

2.5. The role of gulls as contributors to the transmission of HPAI viruses

Transmission of IAV from wild birds to poultry and swine is a major concern for the agriculture industry and must also be considered with respect to the possible generation of novel pandemic viruses. H5 and H7 subtypes have been commonly associated with high mortality in poultry (Stech & Mettenleiter, 2013). Analysis of H5 sequences from gull viruses showed that most of them are closely related to highly pathogenic H5N1 virus isolated from waterfowl, poultry, and swine (Figure 5). Specifically, the virus A/slaty-backed gull/Shandong/59/04(H5N1) is identical to the avian virus A/

babbler/Fujian/320/04(H5N1) and groups closely with a swine strain, A/swine/Anhui/ca/2004(H5N1), with all these isolates from 2004. This shows that there has been frequent inter-species transmission of this virus between avian species, including gulls, and swine. Similarly, phylogenetic analysis of H7 sequences from gulls showed that gull viruses from both North America and Eurasia are closely related to viruses associated with lethal H7 outbreaks in poultry (Figure 5).

The movements of viruses among gulls, waterfowl, poultry, and mammals indicate that gulls need to be considered as important contributors to IAV evolutionary dynamics and ecology. It is also important to recognize the potential contribution of gulls to the generation of pandemic viruses, which always carry gene segments of bird origin, and that could contribute to the emergence of pathogenic strains.

2.6. The role of terns in global IAV transmission, and relationships of tern IAVs to those from other hosts

Terns are the most closely related taxon to gulls; they exhibit shared ecological features and can also be sympatric with gulls. It is also noteworthy that the first isolate of IAV from wild birds was from terns, with A/tern/South Africa/1961(H5N3) representing a HPAI H5 virus that caused many deaths among terns in South Africa (Becker, 1966). Close relationships of viruses from terns to those from other hosts can also be found. A low pathogenicity tern isolate, A/tern/Potsdam/342–6/1979(H7N7), is similar to a virus that infected chickens in Germany and subsequently accrued several mutations to become a highly pathogenic virus that caused a severe outbreak in chickens (Röhm, Süss, Pohle, & Webster, 1996). A connection from terns to marine mammals has also been previously documented with the NP gene segment of the tern isolate A/tern/Turkmenistan/18/1972(H3N3) closely related to the whale isolate A/whale/Pacific Ocean/19/1976(H1N3) (Mandler et al., 1990).

Phylogenetic analyses with tern virus sequences showed high relatedness of these viruses to viruses identified in gulls, ducks, turkeys, and whales from around the world (Figures S1 and S2, Table S3). Broadly, these sequence comparisons reveal a series of IAVs isolated from two partially sympatric and highly migratory tern species, the common tern (*Sterna hirundo*) and the Arctic tern (*Sterna paradisaea*). In particular, the common tern isolate from South Africa showed inter-hemispheric reassortment, as well as affinities to isolates from turkey, seal, whale, and another common tern strain (Figure S2). This isolate from Turkmenistan is one of a group of common tern viruses from a region of Central Asia that show affinities to North American gull isolates, a European mallard isolate, a tufted duck (*Aythya fuligula*) isolate from southern Siberia, and some duck isolates from Taiwan. The phylogenetic analysis of M gene segments also showed significant homology between an Arctic tern virus (A/arctic tern/Alaska/300/1975(H5N3)) and gull and turkey viruses from North America (Figure S1). Finally, the three polymerase gene segments of the same A/arctic tern/Alaska/300/1975(H5N3) isolate also exhibit affinities to these same North American gull isolates, and the PB2 gene segment is also highly related to gull isolates from Astrakhan, Asia (Figure S1). Overall, these analyses show that terns also constitute an important IAV host. They share many characteristics and display migratory connectivity with gulls, and they should also be considered as contributors to IAV dynamics, diversity and transmission to different hosts.

2.7. Ecobiological aspects of gull IAVs

Two aspects of the ecobiology of gulls are particularly relevant for their role in IAV dynamics. These concern their utilization of both marine and freshwater habitats and their potential role in zoonotic transmission pathways. Gulls are generalist feeders and eat both live-captured prey and opportunistically scavenge, including on carcasses of birds and mammals. Presumably, one means by which gulls could potentially contract IAVs from other species in aquatic habitats is by eating IAV-infected dead birds or mammals. The utilization of both marine and freshwater habitats, and interactions with different species in both environments, raises the possibility for gulls to be exposed to viruses from both freshwater and marine sources. The ecology of IAVs is fairly well understood in freshwater habitats, but this has been explored less in marine environments. However, alongside gulls, additional marine birds such as murres and puffins (members of the family Alcidae) and sea ducks (family Anatidae, tribe Mergini) are known to host IAVs and be sympatric with gulls (Ip et al., 2008). Although some IAVs do not appear very stable in elevated salinity (Stallknecht, Kearney, Shane, & Zwank, 1990), marine birds drink seawater and it must be presumed that effective fecal-oral transmission can take place in seawater. Viral circulation within marine environments also involves seals and whales, which are known to host avian-derived IAVs and apparently mostly those originating from gulls and terns as our phylogenetic and phylogeographic findings concerning both gull and tern viruses indicate relationships with viruses from marine mammals globally.

There is also an important zoonotic dimension for IAV circulation in gulls. While the important role played by waterfowl as facilitators of the emergence and dispersal of epizootic and pandemic strains is soundly established, gulls are both cosmopolitan migrants and salient synanthropic birds adjacent to the duck-poultry-swine-human IAV interface. It has previously been articulated that there is a crucial epidemiological position occupied by gulls among wildlife, domestic birds as well as humans (Arnal et al., 2015), and the potential for gulls to act in transmission of micro-organisms from sewage plants or outflows to domestic animals has also been described (Teale, 2002). Natural interfaces between gull viruses and those from ducks, turkeys, chickens, and swine have been identified in the present analysis, either in terms of interchangeability of certain gene segments, or whole viruses, and IAV-related interactions of gulls with ducks, turkeys, chickens, and mammals, phylogenetically and ecologically, might contribute to emergence and spread of epizootic and pandemic strains. It is also of note that a connection was found between marine mammal and human infections with the 2009 H1N1 human pandemic strain, which naturally infected and spread, subclinically, among seals in California (Goldstein et al., 2013; Ohishi et al., 2004). However, it is not known if gulls or terns were directly involved in transmission of this virus.

3. Concluding remarks

Within the enormous spectrum of IAVs hosted by avian species, gull IAVs constitute an important component, somewhat parallel to waterfowl viruses, of the global ecophylogenetic system defining this protean virus. Gulls affect, both directly and indirectly, the evolutionary dynamics of low and high pathogenicity avian viruses and various mammalian IAVs worldwide. Analysis of the evolutionary history of an entirely Eurasian virus isolated in North America allowed the reconstruction of the gradual evolutionary process that generated this virus, starting in Central Asia, moving northwest, and ending in Iceland, with subsequent transmission to Canada. The origin of this virus appears to be associated with a region of Central Asia that also appears to have given rise to an important H14 strain that involved both gull and duck viruses. Such reverse ecophylogenetic analysis is useful for attaining a broadened perspective of the evolutionary history of a given virus and can allow identification of the presumed strains, genes, and hosts involved in the evolutionary and spatiotemporal pathways that gave rise to a strain in question. Additional gull viruses that were identified as phylogenetically interesting in this analysis were also from the same Central Asian geographic region. A similar scenario was found when analyzing a series of common tern viruses that were obtained in the same general region of Central Asia, where those viruses exhibited clear affinities to gull, duck, and whale viruses. Our analyses of the tern viruses reveal a broad phylogeographic contribution for these viruses beyond Central Asia, with gene transactions extending to Europe, South Africa, North America, and the Pacific Ocean.

The traits contributing to the importance of gulls as IAV hosts include their cosmopolitan distributions, migratory behaviors, synanthropic associations with humans, use of both marine and freshwater habitats, prevalence as a host of many different IAVs of a large variety of host group origins (including farm animals), and predominance of subclinical infections. As shown in the present study, the summary of these properties defines an important IAV host group that needs to be considered in the context of the perpetuation and emergence of enzootic, epizootic, and pandemic strains.

Acknowledgments
We thank M. Canuti for comments on the manuscript.

Funding
JEB and ANKK were partially supported by fellowships from the Memorial University School of Graduate Studies. Funding from the Newfoundland and Labrador Forestry and Agrifoods Agency and a Grant (341561) from the Natural Sciences and Engineering Research Council (NSERC) of Canada to ASL supported this research.

Competing Interests
The authors declare no competing interest.

Author details
Jessica Benkaroun[1]
E-mail: jb5163@mun.ca
Dany Shoham[2]
E-mail: shoham_d@013net.net
Ashley N.K. Kroyer[1]
E-mail: ankk24@mun.ca
Hugh Whitney[3]
E-mail: hughwhitneynl@gmail.com
Andrew S. Lang[1]
E-mail: aslang@mun.ca
ORCID ID: http://orcid.org/0000-0002-4510-7683
[1] Department of Biology, Memorial University of Newfoundland, 232 Elizabeth Ave., St. John's, NL, Canada A1B 3X9.
[2] Begin-Sadat Center for Strategic Studies, Bar Ilan University, Building 109, Ramat Gan 5290002, Israel.
[3] Newfoundland and Labrador Forestry and Agrifoods Agency, St. John's, NL A1E 3Y5, Canada.
[§]These authors contributed equally.

References

Altschul, S. F., Gish, W., Miller, W., Myers, E. W., & Lipman, D. J. (1990). Basic local alignment search tool. *Journal of Molecular Biology, 215*, 403–410. doi:10.1016/S0022-2836(05)80360-2

Amonsin, A., Songserm, T., Chutinimitkul, S., Jam-on, R., Sae-Heng, N., Pariyothorn, N., ... Poovorawan, Y. (2007). Genetic analysis of influenza A virus (H5N1) derived from domestic cat and dog in Thailand. *Archives of Virology, 152*, 1925–1933. doi:10.1007/s00705-007-1010-5

Arnal, A., Vittecoq, M., Pearce-Duvet, J., Gauthier-Clerc, M., Boulinier, T., & Jourdain, E. (2015). Laridae: A neglected reservoir that could play a major role in avian influenza virus epidemiological dynamics. *Critical Reviews in Microbiology, 41*, 508–519. doi:10.3109/1040841X.2013.870967

Bahl, A. K., & Pomeroy, B. S. (1977). Experimental exposure of Franklins' gulls (*Larvus pipixcan*) and mallards (*Anas platyrhynchos*) to a turkey influenza A virus A/Turkey/Minn/BF/72 (Hav6Neq2). *Journal of Wildlife Diseases, 13*, 420–426. http://dx.doi.org/10.7589/0090-3558-13.4.420

Bahl, J., Vijaykrishna, D., Holmes, E. C., Smith, G. J., & Guan, Y. (2009). Gene flow and competitive exclusion of avian influenza A virus in natural reservoir hosts. *Virology, 390*, 289–297. doi:10.1016/j.virol.2009.05.002

Bahl, J., Krauss, S., Kühnert, D., Fourment, M., Raven, G., Pryor, S. P., ... Webster, R. G. (2013). Influenza A virus migration and persistence in North American wild birds. *PLoS Pathogens, 9*, e1003570. doi:10.1371/journal.ppat.1003570

Bao, Y., Bolotov, P., Dernovoy, D., Kiryutin, B., Zaslavsky, L., Tatusova, T., ... Lipman, D. (2008). The Influenza Virus Resource at the National Center for Biotechnology Information. *Journal of Virology, 82*, 596–601. doi:10.1128/jvi.02005-07

Becker, W. B. (1966). The isolation and classification of Tern virus: Influenza virus A/Tern/South Africa/1961. *Journal of Hygiene, 64*, 309–320. http://dx.doi.org/10.1017/S0022172400040596

Boyce, W. M., Schobel, S., Dugan, V. G., Halpin, R. A., Lin, X., Wentworth, D. E., ... Plancarte, M. (2013). Complete genome sequence of a reassortant H14N2 avian influenza virus from California. *Genome Announcements, 1*, e00543–00513. doi:10.1128/genomeA.00543-13

Brown, I. H. (2000). The epidemiology and evolution of influenza viruses in pigs. *Veterinary Microbiology, 74*, 29–46. http://dx.doi.org/10.1016/S0378-1135(00)00164-4

Brown, J. D., Luttrell, M. P., Berghaus, R. D., Kistler, W., Keeler, S. P., Howey, A., ... Stallknecht, D. E. (2010). Prevalence of antibodies to type A influenza virus in wild avian species using two serologic assays. *Journal of Wildlife Diseases, 46*, 896–911. doi:10.7589/0090-3558-46.3.896

Brown, J. D., Stallknecht, D. E., Beck, J. R., Suarez, D. L., & Swayne, D. E. (2006). Susceptibility of North American ducks and gulls to H5N1 highly pathogenic avian influenza viruses. *Emerging Infectious Diseases, 12*, 1663–1670. http://dx.doi.org/10.3201/eid1211.060652

Brown, J. D., Stallknecht, D. E., & Swayne, D. E. (2008). Experimental infections of herring gulls (*Larus argentatus*) with H5N1 highly pathogenic avian influenza viruses by intranasal inoculation of virus and ingestion of virus-infected chicken meat. *Avian Pathology, 37*, 393–397. doi:10.1080/03079450802216595

Brown, J., Poulson, R., Carter, D., Lebarbenchon, C., Pantin-Jackwood, M., Spackman, E., ... Stallknecht, D. (2012). Susceptibility of avian species to North American H13 low pathogenic avian influenza viruses. *Avian Diseases, 56*, 969–975. http://dx.doi.org/10.1637/10158-040912-Reg.1

Chen, H., Smith, G. J., Zhang, S. Y., Qin, K., Wang, J., Li, K. S., ... Guan, Y. (2005). Avian flu: H5N1 virus outbreak in migratory waterfowl. *Nature, 436*, 191–192. doi:10.1038/nature03974

Claas, E. C., Osterhaus, A. D., van Beek, R., De Jong, J. C., Rimmelzwaan, G. F., Senne, D. A., ... Webster, R. G. (1998). Human influenza A H5N1 virus related to a highly pathogenic avian influenza virus. *The Lancet, 351*, 472–477. doi:10.1016/s0140-6736(97)11212-0

Curran, J. M., Robertson, I. D., Ellis, T. M., Selleck, P. W., & O'Dea, M. A. (2013). Variation in the responses of wild species of duck, gull, and wader to inoculation with a wild-bird-origin H6N2 low pathogenicity avian influenza virus. *Avian Diseases, 57*, 581–586. doi:10.1637/10458-112712-Reg.1

Daoust, P. Y., van de Bildt, M., van Riel, D., van Amerongen, G., Bestebroer, T., Vanderstichel, R., ... Kuiken, T. (2013). Replication of 2 subtypes of low-pathogenicity avian influenza virus of duck and gull origins in experimentally infected mallard ducks. *Veterinary Pathology, 50*, 548–559. doi:10.1177/0300985812469633

Drummond, A. J., Suchard, M. A., Xie, D., & Rambaut, A. (2012). Bayesian phylogenetics with BEAUti and the BEAST 1.7. *Molecular Biology and Evolution, 29*, 1969–1973. doi:10.1093/molbev/mss075

Dusek, R. J., Hallgrimsson, G. T., Ip, H. S., Jónsson, J. E., Sreevatsan, S., Nashold, S. W., ... Hall, J. S. (2014). North Atlantic migratory bird flyways provide routes for intercontinental movement of avian influenza viruses. *PLoS ONE, 9*, e92075. doi:10.1371/journal.pone.0092075

Fereidouni, S., Harder, T. C., Globig, A., & Starick, E. (2014). Failure of productive infection of mallards (*Anas platyrhynchos*) with H16 subtype of avian influenza viruses. *Influenza and Other Respiratory Viruses, 8*, 613–616. doi:10.1111/irv.12275

Fouchier, R. A. M., Munster, V., Wallensten, A., Bestebroer, T. M., Herfst, S., Smith, D., ... Osterhaus, A. D. M. E. (2005). Characterization of a novel influenza A virus Hemagglutinin subtype (H16) obtained from black-headed gulls. *Journal of Virology, 79*, 2814–2822. http://dx.doi.org/10.1128/JVI.79.5.2814-2822.2005

Fries, A. C., Nolting, J. M., Danner, A., Webster, R. G., Bowman, A. S., Krauss, S., ... Wang, Y. (2013). Evidence for the circulation and inter-hemispheric movement of the H14 subtype influenza A virus. *PLoS ONE, 8*, e59216. doi:10.1371/journal.pone.0059216

Goldstein, T., Mena, I., Anthony, S. J., Medina, R., Robinson, P. W., Greig, D. J., ... Boyce, W. M. (2013). Pandemic H1N1 influenza isolated from free-ranging northern elephant seals in 2010 off the Central California Coast. *PLoS ONE, 8*, e62259. doi:10.1371/journal.pone.0062259

Gonzalez-Reiche, A. S., & Perez, D. R. (2012). Where do avian influenza viruses meet in the Americas? *Avian Diseases, 56*, 1025–1033. doi:10.1637/10203-041412-Reg.1

Groth, M., Lange, J., Kanrai, P., Pleschka, S., Scholtissek, C., Krumbholz, A., ... Zell, R. (2014). The genome of an influenza virus from a pilot whale: Relation to influenza viruses of gulls and marine mammals. *Infection Genetics and Evolution, 24*, 183–186. doi:10.1016/j.meegid.2014.03.026

Hall, J. S., TeSlaa, J. L., Nashold, S. W., Halpin, R. A., Stockwell, T., Wentworth, D. E, ... Ip, H. S. (2013). Evolution of a reassortant North American gull influenza virus lineage: Drift, shift and stability. *Virology Journal, 10*, 179. doi:10.1186/1743-422X-10-179

Hinshaw, V. S., Air, G. M., Gibbs, A. J., Graves, L., Prescott, B., & Karunakaran, D. (1982). Antigenic and genetic characterization of a novel hemagglutinin subtype of influenza A viruses from gulls. *Journal of Virology, 42*, 865–872.

Huang, Y., Wille, M., Benkaroun, J., Munro, H., Bond, A. L., Fifield, D. A., ... Lang, A. S. (2014). Perpetuation and reassortment of gull influenza A viruses in Atlantic North America. *Virology, 456-457*, 353–363. doi:10.1016/j.virol.2014.04.009

Hurt, A. C., Vijaykrishna, D., Butler, J., Baas, C., Maurer-Stroh, S., Silva-de-la-Fuente, M. C., & Gonzalez-Acuna, D. (2014). Detection of evolutionarily distinct avian influenza A viruses in Antarctica. *mBio, 5*, e01098–01014. doi:10.1128/mBio.01098-14

Iamnikova, S. S., Gambarian, A. S., Aristova, V. A., L'Vov, D. K., Lomakina, N. F., Munster, V., & Foucher, R. A. (2009). A/H13 and A/H16 influenza viruses: Different lines of one precursors. *Voprosy virusologii, 54*, 10–18.

Ip, H. S., Flint, P. L., Franson, J. C., Dusek, R. J., Derksen, D. V., Gill, R. E., Jr, ... Rothe, T. C. (2008). Prevalence of influenza A viruses in wild migratory birds in Alaska: Patterns of variation in detection at a crossroads of intercontinental flyways. *Virology Journal, 5*, 71. doi:10.1186/1743-422x-5-71

Jourdain, E., van Riel, D., Munster, V. J., Kuiken, T., Waldenström, J., Olsen, B., & Ellström, P. (2011). The pattern of influenza virus attachment varies among wild bird species. *PLoS ONE, 6*, e24155. doi:10.1371/journal.pone.0024155

Karasin, A. I., West, K., Carman, S., & Olsen, C. W. (2004). Characterization of avian H3N3 and H1N1 influenza a viruses isolated from pigs in Canada. *Journal of Clinical Microbiology, 42*, 4349–4354. doi:10.1128/jcm.42.9.4349-4354.2004

Koçer, Z. A., Krauss, S., Zanin, M., Danner, A., Gulati, S., Jones, J. C., ... Webster, R. G. (2015). Possible basis for the emergence of H1N1 viruses with pandemic potential from avian hosts. *Emerging Microbes & Infections, 4*, e40. doi:10.1038/emi.2015.40

Krauss, S., Obert, C. A., Franks, J., Walker, D., Jones, K., Seiler, P., ... Webster, R. G. (2007). Influenza in migratory birds and evidence of limited intercontinental virus exchange. *PLoS Pathogens, 3*, e167. doi:10.1371/journal.ppat.0030167

Laudert, E., Sivanandan, V., Halvorson, D., Shaw, D., & Webster, R. G. (1993). Biological and molecular characterization of H13N2 influenza type A viruses isolated from turkeys and surface water. *Avian Diseases, 37*, 793–799. http://dx.doi.org/10.2307/1592031

Lee, D.-H., Swayne, D. E., Torchetti, M. K., Winker, K., Ip, H. S., & Song, C.-S. (2015). Intercontinental spread of asian-origin H5N8 to North America through Beringia by migratory birds. *Journal of Virology, 89*, 6521–6524. doi:10.1128/JVI.00728-15

Lindskog, C., Ellström, P., Olsen, B., Pontén, F., van Riel, D., Munster, V. J., ... Jourdain, E. (2013). European H16N3 gull influenza virus attaches to the human respiratory tract and eye. *PLoS ONE, 8*, e60757. doi:10.1371/journal.pone.0060757

Mandler, J., Gorman, O. T., Ludwig, S., Schroeder, E., Fitch, W. M., Webster, R. G., & Scholtissek, C. (1990). Derivation of the nucleoproteins (NP) of influenza A viruses isolated from marine mammals. *Virology, 176*, 255–261. doi:10.1016/0042-6822(90)90250-U

Mushtaq, M. H., Juan, H., Jiang, P., Li, Y., Li, T., Du, Y., & Mukhtar, M. M. (2008). Complete genome analysis of a highly pathogenic H5N1 influenza A virus isolated from a tiger in China. *Archives of Virology, 153*, 1569–1574. doi:10.1007/s00705-008-0145-3

Ohishi, K., Kishida, N., Ninomiya, A., Kida, H., Takada, Y., Miyazaki, N., ... Maruyama, T. (2004). Antibodies to human-related H3 influenza A virus in Baikal seals (*Phoca sibirica*) and ringed seals (*Phoca hispida*) in Russia. *Microbiology and Immunology, 48*, 905–909. http://dx.doi.org/10.1111/mim.2004.48.issue-11

Olsen, K. M., & Larsson, H. (2004). *Gulls of North America, Europe, and Asia*. Princeton: Princeton University Press.

Olsen, B., Munster, V. J., Wallensten, A., Waldenstrom, J., Osterhaus, A. D., & Fouchier, R. A. (2006). Global patterns of influenza A virus in wild birds. *Science, 312*, 384–388. doi:10.1126/science.1122438

Otsuki, K., Kawaoka, Y., Nakamura, T., & Tsubokura, M. (1982). Pathogenicity for chickens of avian influenza viruses isolated from whistling swans and a black-tailed gull in Japan. *Avian Diseases, 26*, 314–320. http://dx.doi.org/10.2307/1590100

Pereda, A. J., Uhart, M., Perez, A. A., Zaccagnini, M. E., La Sala, L., Decarre, J., ... Perez, D. R. (2008). Avian influenza virus isolated in wild waterfowl in Argentina: Evidence of a potentially unique phylogenetic lineage in South America. *Virology, 378*, 363–370. doi:10.1016/j.virol.2008.06.010

Perkins, L. E. L., & Swayne, D. E. (2002). Susceptibility of laughing gulls (*Larus atricilla*) to H5N1 and H5N3 highly pathogenic avian influenza viruses. *Avian Diseases, 46*, 877–885. doi:10.1637/0005-2086(2002)046[0877:SOLGLA]2.0.CO;2

Ramey, A. M., Poulson, R. L., González-Reiche, A. S., Perez, D. R., Stallknecht, D. E., & Brown, J. D. (2014). Genomic characterization of H14 subtype influenza A viruses in New World waterfowl and experimental infectivity in mallards (*Anas platyrhynchos*). *PLoS ONE, 9*, e95620. doi:10.1371/journal.pone.0095620

Ramey, A. M., Reeves, A. B., Sonsthagen, S. A., TeSlaa, J. L., Nashold, S., Donnelly, T., ... Hall, J. S. (2015). Dispersal of H9N2 influenza A viruses between East Asia and North America by wild birds. *Virology, 482*, 79–83. doi:10.1016/j.virol.2015.03.028

Ramis, A., van Amerongen, G., van de Bildt, M., Leijten, L., Vanderstichel, R., Osterhaus, A. D., & Kuiken, T. (2014). Experimental infection of highly pathogenic avian influenza virus H5N1 in black-headed gulls (*Chroicocephalus ridibundus*). *Veterinary Research, 45*, 84. doi:10.1186/s13567-014-0084-9

Ratanakorn, P., Wiratsudakul, A., Wiriyarat, W., Eiamampai, K., Farmer, A. H., Webster, R. G., ... Puthavathana, P. (2012). Satellite tracking on the flyways of brown-headed gulls and their potential role in the spread of highly pathogenic avian influenza H5N1 virus. *PLoS ONE, 7*, e49939. doi:10.1371/journal.pone.0049939

Richt, J. A., Lager, K. M., Janke, B. H., Woods, R. D., Webster, R. G., & Webby, R. J. (2003). Pathogenic and antigenic properties of phylogenetically distinct reassortant H3N2 swine influenza viruses cocirculating in the United States. *Journal of Clinical Microbiology, 41*, 3198–3205. http://dx.doi.org/10.1128/JCM.41.7.3198-3205.2003

Röhm, C., Süss, J., Pohle, V., & Webster, R. G. (1996). Different hemagglutinin cleavage site variants of H7N7 in an influenza outbreak in chickens in Leipzig, Germany. *Virology, 218*, 253–257. doi:10.1006/viro.1996.0187

Sangster, G., Collinson, J. M., Knox, A. G., Parkin, D. T., & Svensson, L. (2007). Taxonomic recommendations for British birds: Fourth report. *Ibis, 149*, 853–857. doi:10.1111/j.1474-919X.2007.00758.x

Shoham, D., & Rogers, S. (2006). Greenland as a plausible springboard for trans-Atlantic avian influenza spread. *Medical Hypotheses, 67*, 1460–1461. doi:10.1016/j.mehy.2006.05.039

Sivanandan, V., Halvorson, D. A., Laudert, E., Senne, D. A., & Kumar, M. C. (1991). Isolation of H13N2 influenza A virus from turkeys and surface water. *Avian Diseases, 35*, 974–977. doi:10.2307/1591638

Stallknecht, D. E., Kearney, M. T., Shane, S. M., & Zwank, P. J. (1990). Effects of pH, temperature, and salinity on persistence of avian influenza viruses in water. *Avian Diseases, 34*, 412–418. http://dx.doi.org/10.2307/1591429

Stech, J., & Mettenleiter, T. C. (2013). Virulence determinants of high-pathogenic avian influenza viruses in gallinaceous poultry. *Future Virology, 8*, 459–468. doi:10.2217/fvl.13.27

Subbarao, K., Klimov, A., Katz, J., Regnery, H., Lim, W., Hall, H., & Cox, N. (1998). Characterization of an avian influenza A (H5N1) virus isolated from a child with a fatal respiratory illness. *Science, 279*, 393–396. http://dx.doi.org/10.1126/science.279.5349.393

Tamura, K., Stecher, G., Peterson, D., Filipski, A., & Kumar, S. (2013). MEGA6: Molecular evolutionary genetics analysis version 6.0. *Molecular Biology and Evolution, 30*, 2725–2729. doi:10.1093/molbev/mst197

Teale, C. J. (2002). Antimicrobial resistance and the food chain. *Journal of Applied Microbiology, 92*, 85s–89s. http://dx.doi.org/10.1046/j.1365-2672.92.5s1.20.x

Toennessen, R., Germundsson, A., Jonassen, C. M., Haugen, I., Berg, K., Barrett, R. T., & Rimstad, E. (2011). Virological and serological surveillance for type A influenza in the black-legged kittiwake (*Rissa tridactyla*). *Virology Journal, 8*, 21. doi:10.1186/1743-422x-8-21

Tønnessen, R., Valheim, M., Rimstad, E., Jonassen, C. M., & Germundsson, A. (2011). Experimental inoculation of chickens with gull-derived low pathogenic avian influenza virus subtype H16N3 causes limited infection. *Avian Diseases, 55*, 680–685. http://dx.doi.org/10.1637/9701-030411-ResNote.1

Van Borm, S., Rosseel, T., Vangeluwe, D., Vandenbussche, F., van den Berg, T., & Lambrecht, B. (2012). Phylogeographic analysis of avian influenza viruses isolated from Charadriiformes in Belgium confirms intercontinental reassortment in gulls. *Archives of Virology, 157*, 1509–1522. doi:10.1007/s00705-012-1323-x

Velarde, R., Calvin, S. E., Ojkic, D., Barker, I. K., & Nagy, E. (2010). Avian influenza virus H13 circulating in ring-billed gulls (*Larus delawarensis*) in southern Ontario, Canada. *Avian Diseases, 54*, 411–419. doi:10.1637/8808-040109-Reg.1

Watanabe, T., Zhong, G., Russell, C. A., Nakajima, N., Hatta, M., Hanson, A., ... Kawaoka, Y. (2014). Circulating avian influenza viruses closely related to the 1918 virus have pandemic potential. *Cell Host & Microbe, 15*, 692–705. doi:10.1016/j.chom.2014.05.006

Webster, R. G., Hinshaw, V. S., Bean, W. J., Van Wyke, K. L., Geraci, J. R., St. Aubin, D. J., & Petursson, G. (1981). Characterization of an influenza A virus from seals. *Virology, 113*, 712–724. doi:10.1016/0042-6822(81)90200-2

Webster, R. G., Bean, W. J., Gorman, O. T., Chambers, T. M., & Kawaoka, Y. (1992). Evolution and ecology of influenza A viruses. *Microbiological Reviews, 56*, 152–179.

Wille, M., Robertson, G. J., Whitney, H., Bishop, M. A., Runstadler, J. A., & Lang, A. S. (2011). Extensive geographic mosaicism in avian influenza viruses from gulls in the Northern Hemisphere. *PLoS ONE, 6*, e20664. doi:10.1371/journal.pone.0020664

Wille, M., Robertson, G. J., Whitney, H., Ojkic, D., & Lang, A. S. (2011). Reassortment of American and Eurasian genes in an influenza A virus isolated from a great black-backed gull (*Larus marinus*), a species demonstrated to move between these regions. *Archives of Virology, 156*, 107–115. doi:10.1007/s00705-010-0839-1

Wood, J. M., Webster, R. G., & Nettles, V. F. (1985). Host range of A/Chicken/Pennsylvania/83 (H5N2) influenza virus. *Avian Diseases, 29*, 198–207. http://dx.doi.org/10.2307/1590708

Worobey, M., Han, G. Z., & Rambaut, A. (2014). A synchronized global sweep of the internal genes of modern avian influenza virus. *Nature, 508*, 254–257. doi:10.1038/nature13016

Zu Dohna, H., Li, J., Cardona, C. J., Miller, J., & Carpenter, T. E. (2009). Invasions by Eurasian avian influenza virus H6 genes and replacement of its North American clade. *Emerging Infectious Diseases, 15*, 1040–1045. doi:10.3201/eid1507.090245

5

First record of a *Leptus* Latreille mite (Trombidiformes, Erythraeidae) associated with a Neotropical trapdoor spider (Araneae: Mygalomorphae: Actinopodidae)

Lidianne Salvatierra[1]* and Marlus Q. Almeida[2]

*Corresponding author: Lidianne Salvatierra, Instituto Nacional de Pesquisas da Amazônia, Laboratório de Ecologia e Sistemática de Invertebrados do Solo, Av. André Araújo, 2.936, CEP 69067-375, Manaus, Amazonas, Brazil
E-mail: lidiannetrigueiro@gmail.com
Reviewing editor: Shelley Edwards, Rhodes University, South Africa
Additional information is available at the end of the article

Abstract: The first occurrence of a parasitic mite, *Leptus* Latreille (Trombidiformes, Erythraeidae) parasitizing an adult male of a trapdoor spider *Actinopus* Perty, 1833 (Araneae: Mygalomorphae: Actinopodidae) and the first occurrence of *Leptus* sp. larvae in the municipality of Manaus, Amazonas state, Brazil are reported.

Subjects: Environment & Agriculture; Zoology; Entomology & Acarology

Keywords: Acari; parasitism; mutualism; ectoparasite; Arachnida; Neotropical; Amazon

The cosmopolitan genus *Leptus* Latreille, 1796 (Prostigmata: Parasitengona: Erythraeidae) has a total of 10 species described for Brazil, 27 species for the Neotropical region, and more than 270 described species worldwide (Haitlinger, 2004; Mąkol & Wohltmann, 2012). *Leptus* mites have seven larval stages of which nymphs are ectoparasitic using a wide range of arthropods to feed and transport (Penney & Green, 2011).

After hatching from eggs, the mite larvae pierce the cuticle of the invertebrate host and gain access to the host's hemolymph and interstitial fluids via a straw-like stylostome. After engorging, larvae drop off the host and transform into eight-legged nymphs and then adults (Penney & Green, 2011). Both adults and deutonymphs are free-living predators of small invertebrates. Most common hosts are insects (Flechtmann, 1980; Kamran, Afzal, Bashir, & Raza, 2009; Teixeira, 2011; Wilson, Rubink, & Collins, 1990; Wilson, Wooley, Nunamaker, & Rubink, 1987) and arachnids such as opiliones, scorpions, and spiders (Fain, Gummer, & Whitaker, 1987; Mohamed & Mohamed, 2011; Townsend, Mulholland, Bradford, Proud, & Parent, 2006; Welbourn & Young, 1988).

This work reports the first occurrence of larva of the genus *Leptus* parasitizing an adult male of a trapdoor spider *Actinopus* (Araneae: Mygalomorphae: Actinopodidae) and the first occurrence of

ABOUT THE AUTHOR

Lidianne Salvatierra is a Brazilian invertebrate zoologist. Marlus Q. Almeida is a Brazilian invertebrate ecologist. The research in the Laboratory of Ecology and Systematic of Soil Invertabrates at The National Institute of Amazonian Research focuses on the systematics and evolutionary biology of Arachnida, with emphasis on the Amazonian fauna.

PUBLIC INTEREST STATEMENT

Mites are among the most diverse and successful of all the invertebrate groups. Many mites are parasitic on plants and animals. Most host-parasitic association is poorly described and understood. Investigations of the host-parasitic association can elucidate the biology and animal behavior, the link between parasites and pathogens, and define the role of the mite in transmission and pathogenesis. This article reports the first occurrence of larva of the genus *Leptus* parasitizing an adult male of a trapdoor spider *Actinopus*.

Leptus sp. larvae in the municipality of Manaus, Amazonas state. Specimens were collected near the "Universidade Federal do Amazonas," Manaus, Amazonas, Brazil, and are deposited into the "Coleção de Invertebrados" located at the Instituto Nacional de Pesquisas da Amazônia (INPA), Manaus, Amazonas, Brazil. Photos were taken using a stereomicroscope Leica MZ16, equipped with Leica camera M205c.

Actinopus Perty, 1833, also commonly known as trapdoor spiders, are distributed throughout Central and South America (Brescovit, Bonaldo, Bertani, & Rheims, 2002; Ríos, 2014) and can be easily identified by the thoracic groove procurved and rastellum on distinct process (Brescovit et al., 2002; Raven, 1985). The biology and ecology of the species are still little known, females and young males have cryptic habits, living most of the time in excavated burrows, closed by a trap door made of soil particles and silk, making difficult to detect these spiders in their natural habitat (Brescovit et al., 2002; Coyle, Goloboff, & Samson, 1990; Miglio, 2009). Few works reported associations between mites and mygalomorph spiders (Ebermann & Goloboff, 2002; Masan, Simpson, Perotti, & Braig, 2012; Welbourn & Young, 1988). The spider was identified as belonging to *Actinopus cucutaensis* Mello-Leitão, 1941 by the copulatory bulb with elongated and well-developed paraembolic apophysis.

The six-legged *Leptus* larva (body length 0.55 mm, width 0.18 mm) (Figure 1(A), (C–D)) was located near posterior margin on dorsal portion of the spider's carapace (body length 22.6 mm; carapace long 11.7 mm, wide 11.1 mm) (Figure 1(B)). The specimen was firmly attached to the carapace by its chelicerae, and when it was removed, no injury caused by its mouthparts was visibly detected. The larva is an undescribed species of *Leptus*.

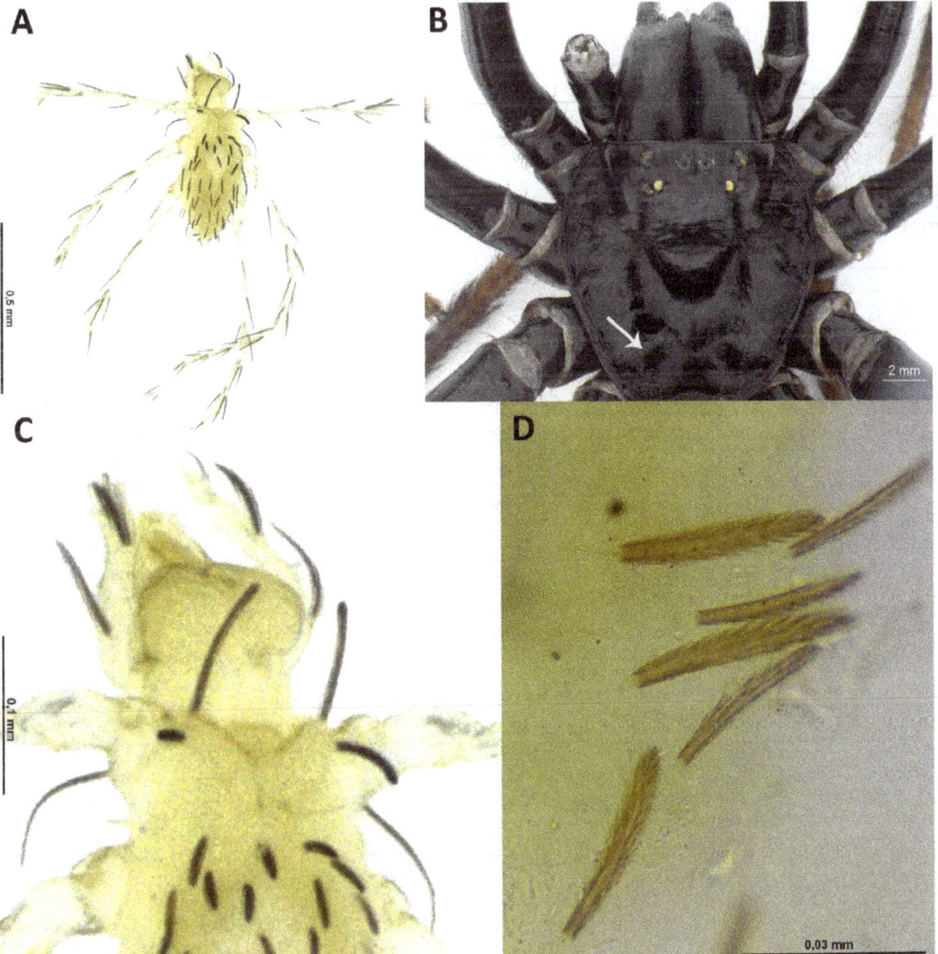

Figure 1. (A) Mite larva of *Leptus* sp., habitus, dorsal view; (B) spider of *Actinopus* sp., carapace, dorsal view (white arrow indicates where mite was attached); (C) details of larva's gnathosoma and dorsal scutum; (D) dorsal idiosomal setae.

Currently, 79 arachnid species among mites (9 spp.), spiders (17 spp.), scorpions (11 spp.), harvestmen (39 spp.), pseudoscorpions (2 spp.), and tailless whip scorpions (1 sp.) were reported as host of *Leptus* spp. (Table 1).

Table 1. Updated records of *Leptus* spp. parasitising arachnids

Host	Host species	*Leptus* species	Country	References
Subclass Acari				
Order Trombidiformes				
Family Erythraeidae	*Abrolophus* sp.	*L. trimaculatus*	Germany	Wendt, Olomski, Leimann, and Wohltmann (1992)
	Balaustium globigerum	*L. ignotus*	Netherlands	Oudemans (1912)
	Erythraeus sp.	*L. echinopus*	Denmark	Southcott (1992)
Family Anystidae	*Anystis baccarum*	*L. killingtoni*	UK	Turk (1945)
	A. baccarum	*L. trimaculatus*	Germany	Wendt et al. (1992)
	A. rosae			
Order Sarcoptiformes				
Family Damaeidae	*Damaeus grossmani*	*Leptus* spp.	USA	Norton, Welbourn, and Cave (1988)
	Spatiodamaeus verticillipes	*Leptus* spp.	USA	Norton et al. (1988) (as *Damaeus verticillipes*)
Family Oribatellidae	*Oribatella extensa*	*Leptus* spp.	USA	Norton et al. (1988)
Family Xenillidae	*Xenillus occultus*	*Leptus* spp.	USA	Norton et al. (1988)
Order Amblypygi				
Family Charinidae	*Phrynus kennidae*	*Leptus* sp.	Dominican Republic	Armas and Trueba (2003)
Order Araneae				
Family Actinopodidae	*Actinopus cucutaensis*	*Leptus* spp.	Brazil	This paper
Family Eutichuridae	*Cheiracanthium* sp.	*L. hidakai*	Japan	Kawashima (1958)
Family Sparassidae	*Delena cancerides*	*L. charon*	Australia	Southcott (1999)
	Isopeda frenchi	*L. minno*	Australia	Southcott (1999)
	Isopeda woodwardi	*L. charon*	Australia	Southcott (1999)
	Isopeda sp.	*L. charon*	Australia	Southcott (1999)
	Isopeda sp.	*L. faini*	Australia	Southcott (1999)
	Isopedella inola	*L. orthrius*	Australia	Southcott (1999)
	I.leai	*L. orthrius*	Australia	Southcott (1999)
Family Theridiidae	*Enoplognatha ovata*	*Leptus* sp.	USA	Reillo (1989)
	Holconia insignis	*L. faini*	Australia	Southcott (1999)
Family Lycosidae	*Lycosa* sp.	*L. gifuensis*	Japan	Kawashima (1958)
	Pardosa sp.	*Leptus* sp.	USA	Sorkin (1982)
Family Uloboridae	*Miagrammopes singaporensis*	*L. hidakai*	Singapore	Baker and Selden (1997)
Family Tetragnathidae	*Pachygnatha clerki*	*L. ignotus*	UK	Parker (1962)

(Continued)

Table 1. (Continued)

Host	Host species	*Leptus* species	Country	References
Family Philodromidae	*Philodromus imbecillus*	*Leptus* sp.	USA	Cokendolpher, Horner, and Jennings (1979)
Family Salticidae	*Saitis* sp.	*L. atticolus*	South Africa	Lawrence (1940)
Family Zodariidae	*Systenoplacis* sp.	*L. rwandae*	Rwanda	Fain and Jocqué (1996)
Undetermined spider	–	*L. ignotus*	France	Bruyant (1911)
Order Opiliones				
Family Manaosbiinae	*Cranellus montgomeryi*	*Leptus* sp.	Trinidad	Townsend et al. (2008)
Family Cosmetidae	*Cynorta* sp.	*L. gracilipes*	Surinam	Oudemans (1910a)
	Cynortula sp.	*Leptus* sp.	Trinidad	Townsend et al. (2008)
	Paecilaema inglei	*Leptus* sp.	Trinidad	Townsend et al. (2008)
Family Gonyleptidae	*Discocyrtus funestus*	*L. lomani*	Chile	Oudemans (1902)
Family Sclerosomatidae	*Gragellula niveata*	*L. phuketicus*	Thailand	Southcott (1994)
	Gragellula sp.	*L. phuketicus*	Thailand	Southcott (1994)
	Gagrella sp.	*L. gagrellae*	Indonesia	Oudemans (1910b)
	Leiobunum calcar	*L. indianensis*	USA	Fain et al. (1987)
	L. formosum	*L. indianensis*	USA	Cokendolpher (1993)
		Leptus sp.	USA	Townsend et al. (2006)
	L. longipes	*L. indianensis*	USA	Fain et al. (1987)
	L. nigripes	*L. nearcticus*		
	L. speciosum	*L. indianensis*	USA	Fain et al. (1987)
	L. ventricosum			
	L. vittatum	*L. nearcticus*	USA	Fain et al. (1987)
	Prionostemma sp.	*Leptus* sp.	Trinidad	Townsend et al. (2008)
	Trachyrhinus marmoratus	*Leptus* sp.	USA	MacKay, Grimsley, and Cokendolpher (1992)
Family Phalangiidae	*Lacinius ephippiatus*	*L. holmiae*	Slovakia	Stašiov (2003)
	Lophopilio palpinalis	*L. holmiae*	Slovakia	Stašiov (2003)
		L. ignotus	Poland	Haitlinger (1987) (as *Odiellus palpinalis*)
		L. phalangii	Poland	Gabrys (1991) (as *Odiellus palpinalis*)
	Megabunus diadema	*L. beroni*	France	Fain and Amico (1997)
	Mitopus morio	*L. beroni*	Belgium	Fain (1991a)
			France	Fain and Amico (1997)

(*Continued*)

Table 1. (Continued)

Host	Host species	*Leptus* species	Country	References
		L. holmiae	Denmark	Southcott (1992)
			Iceland	
			Ireland	
			Poland	
		L. ignotus	Bulgaria	Beron (1975)
		L. kalaallus	Greenland	Southcott (1992)
		Leptus spp.	Norway	Åbro (1988)
		L. holmiae	Poland	Haitlinger (1991)
	Oligolophus hansenii	*L. beroni*	France	Fain and Amico (1997)
	Oligophorus tridens	*L. holmiae*	Poland	Haitlinger (1991)
			Slovakia	Stašiov (2003)
	Opilio canestrinii	*L. holmiae*	Denmark	Southcott (1992)
	O. pentaspinulatus	*L. hidakai*	Japan	Kawashima (1958)
	O. ruzickai	*L. ignotus*	Bulgaria	Beron (1975)
	Opilio sp.	*L. holmiae*	Sweden	Southcott (1992)
		L. ignotus	Sweden	Oudemans (1912)
	Phalangium opilio	*L. holmiae*	UK	Southcott (1992)
		L. ignotus	Poland	Haitlinger (1987)
			France	Fain and Amico (1997)
		L. mariae	Poland	
		L. molochinus	Poland	
		L. phalangii	Poland	Gabrys (1991)
		L. phalangii	UK	Evans (1910)
		Leptus spp.	Norway	Åbro (1988)
	P. partietinum	*L. ignotus*	Netherlands	Oudemans (1912)
	Phalangium spp.	*L. ignotus*	France	Bruyant (1911)
	Rilaena triangularis	*L. holmiae*	UK	Southcott (1992) (as *Platybunus triangularis*)
		L. ignotus	Poland	Haitlinger (1987) (as *Platybunus triangularis*)
		L. phalangii	Poland	Gabrys (1991) (as *Platybunus triangularis*)
	Rhamsinitis fissidens	*L. rwandae*	Rwanda	Fain and Jocqué (1996)
Family Samoidae	*Maracaynatum trinidadense*	*Leptus* sp.	Trinidad	Townsend et al. (2008)
	Pellobunus longipalpus	*Leptus* sp.	Trinidad	Townsend et al. (2008)
Family Cranaidae	*Phareicranaus calcariferus*	*Leptus* sp.	Trinidad	Townsend et al. (2008)
	Santinezia serratotibilis	*Leptus* sp.	Trinidad	Townsend et al. (2008)

(Continued)

Table 1. (Continued)

Host	Host species	*Leptus* species	Country	References
Family Manaosbiidae	*Rhopalocranaus albilineatus*	*Leptus* sp.	Trinidad	Townsend et al. (2008)
Family Stygnidae	*Stygnoplus clavotibialis*	*Leptus* sp.	Trinidad	Townsend et al. (2008)
Family Agoristenidae	*Avima* sp.	*Leptus* sp.	Trinidad	Townsend et al. (2008) (as *Trinella* sp.)
Undetermined		*L. bicristatus*	Malawi	Fain and Elsen (1987)
		L. jocquei		
		L. puylaerti		
		L. polythrix		
		L. stieglmayri	Brazil	Oudemans (1905)
		L. ignotus	Poland	Haitlinger (1991)
		Leptus sp.	USA	Welbourn (1983)
Order Pseudoscorpiones				
Family Neobisiidae	*Neobisium* sp.	*Leptus* sp.	France	Judson and Mąkol (2011)
Undetermined	-	*L. chelonethus*	Australia	Womersley (1934)
Order Scorpiones				
Family Buthidae	*Buthus occitanus*	*L. pyrenaeus*	France	Andre (1953)
	Centruroides vittatus	*Leptus* sp.	USA	Welbourn (1983)
	Hemilychas alexandrinus	*L. waldockae*	Australia	Fain (1991b) (as *Lychas alexandrines*)
	Lychas sp.	*L. korematus*	Australia	Southcott (1999)
Family Bothriuridae	*Cercophonius squama*	*L. charon*	Australia	Southcott (1999)
			Tasmania	Seeman and Miller (2002)
Family Scorpionidae	*Urodacus manicatus*	*Leptus* sp.	Australia	Southcott (1955) (as *Urodacus abruptus*)
		L. baudini	Australia	Southcott (1999)
		L. urodaci	Australia	Southcott (1999)
		L. smithi	Australia	Southcott (1999)
		L. pistoris	Australia	Southcott (1999)
		L. carduus	Australia	Southcott (1999)
	U. armatus	*Leptus* sp.	Australia	Fain (1991b)
	U. hoplurus			
	U. yaschenkoi			
	U. hartmeyeri	*L. korematus*	Australia	Southcott (1999)
	U. varians	*L. korematus*	Australia	Southcott (1999)
	Urodacus cf. *yaschenkoi*	*L. barmeedius*	Australia	Southcott (1999)
Undetermined	-	*Leptus* sp.	Mexico	Welbourn (1983)
		Leptus sp.	Costa Rica	Welbourn (1983)

Source: Modified from Baker and Selden (1997).

Little is known about the impact of *Leptus* larva feeding on their host upon the survival, locomotion, or reproductive capacity of their spider hosts, but it is known that *Leptus* larvae are able to transmit *Spiroplasma* bacteria which can be mutualistic or pathogentic (DiBlasi et al., 2011). Although specific associations between deutonymphs of Astigmata and Heterostigmata mites, and larvae of Prostigmata mites, and spiders are well documented, little is known about the spider mite associations in Brazil and the implications for the host.

This paper revealed an unrecorded association between trapdoor *Actinopus* spider and a *Leptus* mite for Brazil, which indicated that similar interactions (parasitic and non-parasitic) are likely to be far more diverse. Additional field and laboratory studies of the life history and ecology of parasite and host species are required.

Acknowledgments
We thank Dr José Albertino Rafael for lab support on realization of photos and measurements of the spider at Diptera laboratory, National Institute for Amazonian Research (INPA).

Funding
The authors received no direct funding for this research.

Competing Interests
The author declares no competing interest.

Author details
Lidianne Salvatierra[1]
E-mail: lidiannetrigueiro@gmail.com
ORCID ID: http://orcid.org/0000-0002-2479-9924
Marlus Q. Almeida[2]
E-mail: marlusqazoo@gmail.com
ORCID ID: http://orcid.org/0000-0001-5580-088X
[1] Instituto Nacional de Pesquisas da Amazônia, Laboratório de Ecologia e Sistemática de Invertebrados do Solo, Av. André Araújo, 2.936, CEP 69067-375, Manaus, Amazonas, Brazil.
[2] Laboratório de Ecologia Terrestre, Universidade Federal do Amazonas, Av. Gal. Rodrigo Otávio Jordão Ramos 3000, CEP 69067-000, Manaus, Amazonas, Brazil.

References

Åbro, A. (1988). The mode of attachment of mite larvae (*Leptus* spp.) to harvestmen (Opiliones). *Journal of Natural History, 22*, 123–130.
Andre, M. (1953). Une espèce nouvelle de Leptus (Acarien) parasite de scorpions. *Bulletin du Museum National d'Histoire Naturelle, Paris, 2e serie, 25*, 150–154.
Armas, L. F., & Trueba, D. P. (2003). Primer registro de ácaros parasitos de amblipígidos (Arachnida: Amblypygi). *Revista Ibérica de Aracnología, 7*, 133–134.
Baker, A. S., & Selden, P. A. (1997). New morphological and host data for the ectoparasitic larva of *Leptus hidakai* Kawashima (Acari, Acariformes, Erythraeidae). *Systematic Parasitology, 36*, 183–191. http://dx.doi.org/10.1023/A:1005757014689
Beron, P. (1975). Erythraeidae (Acariformes) larvaires de Bulgarie. *Acta Zoologica Bulgarica, 1*, 45–75.
Brescovit, A. D., Bonaldo, A. B., Bertani, R., & Rheims, C. A. (2002). Araneae. In J. Adis (Ed.), *Amazonian Arachnida and Myriapoda: Identification keys to all classes, orders, families, some genera, and lists of known terrestrial species* (pp. 303–343). Sofia: Pensoft Publishers.
Bruyant, L. (1911). Nouvelles notes sur des larves d'Acariens Prostigmata. *Zoologischer Anzeiger, 37*, 257–262.
Cokendolpher, J. C. (1993). Pathogens and parasites of Opiliones (Arthropoda: Arachnida). *Journal of Arachnology, 21*, 120–146.
Cokendolpher, J. C., Horner, N. V., & Jennings, D. T. (1979). Crab spiders of north-central Texas (Araneae: Philodromidae and Thomisidae). *Journal of the Kansas Entomological Society, 52*, 723–734.
Coyle, F. A., Goloboff, P. A., & Samson, R. A. (1990). *Actinopus* trapdoor spiders (Araneae, Actinopodidae) killed by the fungus, *Nomuraea atypicola* (Deuteromycotina). *Acta Zoologica Fennica, 190*, 89–93.
DiBlasi, E., Morse, S., Mayberry, J. R., Avila, L. J., Morando, M., & Dittmar, K. (2011). New *Spiroplasma* in parasitic *Leptus* mites and their Agathemera walking stick hosts from Argentina. *Journal of Invertebrate Pathology, 107*, 225–228. http://dx.doi.org/10.1016/j.jip.2011.05.013
Ebermann, E., & Goloboff, P. A. (2002). Association between Neotropical burrowing spiders (Araneae: Nemesiidae) and mites (Acari: Heterostigmata, Scutacaridae). *Acarologia, 42*, 173–184.
Evans, W. (1910). Notes on "*Leptus phalangii*" and "*Leptus autumnalis*"; and their parent earth mites. Proceedings of the Royal Physical Society of Edinburgh, 18, 100–101.
Fain, A. (1991a). Two new larvae of the genus *Leptus* latreille, 1796 (Acari, Erythraeidae) from Belgium. *International Journal of Acarology, 17*, 107–111. http://dx.doi.org/10.1080/01647959108683890
Fain, A. (1991b). Notes on mites parasitic or phoretic on Australian centipedes, spiders and scorpion. *Record of the Western Australian Museum, 15*, 69–82.
Fain, A., & Amico, F. D. (1997). Observations on larval mites (Acari) parasitic on Opiliones from the French Pyrenees. *International Journal of Acarology*, 23(1). doi:10.1080/01647959708684118
Fain, A., & Elsen, P. (1987). Observations sur les larves du genre *Leptus* Latreille, 1795 (Acari, Erythraeidae) d'Afrique centrale. *Revue Zoologique Africaine, 101*, 103–123.
Fain, A., & Jocqué, R. (1996). A new larva of the genus *Leptus* latreille, 1796 (Acari: Erythraeidae) parasitic on a spider from rwanda. *International Journal of Acarology, 22*, 101–108. http://dx.doi.org/10.1080/01647959608684084
Fain, A., Gummer, S. L., & Whitaker, J. J. O. (1987). Two new species of *Leptus* Latreille, 1796 (Acari, Erythraeidae) from the USA. *International Journal of Acarology, 13*, 135–140. doi:10.1080/01647958708683493
Flechtmann, C. H. W. (1980). Dois ácaros associados à abelha (*Apis mellifera* L.) no Perú. *Anais da Escola Superior de Agricultura Luiz de Queiroz, 37*, 737–741. http://dx.doi.org/10.1590/S0071-12761980000200010

Gabrys, G. (1991). New data on the distribution and hosts of larvae of Erythraeidae (Acari, Actinedida) in Poland. *Wiadomosci Parazytologiczne, 37*, 103–105.

Haitlinger, R. (1987). The genus *Leptus* Latreille, 1796 and *Charletonia* Oudemans, 1910 (Acari, Prostigmata, Erythraeidae) in Poland (larvae). *Polskie Pismo Entomologiczne, 57*, 339–349.

Haitlinger, R. (1991). New data on distribution of larvae from the genus *Leptus* Latreille, 1796 (Acari, Prostigmata, Erythraeidae) in Poland with the description of *Leptus miromiri* n. sp. *Wiadomości parazytologiczne, 37*, 499–506.

Haitlinger, R. (2004). Three new species of *Leptus* Latreille, 1796 and the first record of *Leptus onnae* Haitlinger, 2000 (Acari: Prostigmata: Erythraeidae) from Brazil. *Systematic and Applied Acarology, 9*, 147–156. http://dx.doi.org/10.11158/saa.9.1

Judson, M. L. I., & Mąkol, J. (2011). Pseudoscorpions (Chelonethi: Neobisiidae) parasitized by mites (Acari: Trombidiidae, Erythraeidae). *Journal of Arachnology, 39*, 345–348. doi:10.1636/CHa10-69.1

Kamran, M., Afzal, M., Bashir, M. H., & Raza, A. B. M. (2009). A new species of the genus *Leptus* Latreille (Acari: Erythraeidae) parasitising aphids in Pakistan. *Pakistan Journal of Zoology, 41*, 17–20.

Kawashima, K. (1958). Studies on larval erythraeid mites parasitic on arthropods from Japan (Acarina: Erythraeidae). *Kyushu Journal of Medical Science, 9*, 190–211.

Lawrence, R. F. (1940). New larval forms of South African mites from arthropod hosts. *Annals of the Natal Museum, 9*, 401–408.

MacKay, W. P., Grimsley, C., & Cokendolpher, J. C. (1992). Seasonal changes in a population of desert harvestmen, *Trachyrhinus marmoratus* (Arachnida: Opiliones), from Western Texas. *Psyche: A Journal of Entomology, 99*, 207–213. http://dx.doi.org/10.1155/1992/90348

Mąkol, J., & Wohltmann, A. (2012). An annotated checklist of terrestrial *Parasitengona* (Actinotrichida: Prostigmata) of the World, excluding *Trombiculidae* and *Walchiidae*. *Annales Zoologici, 62*, 359–562. http://dx.doi.org/10.3161/000345412X656671

Masan, P., Simpson, C., Perotti, M. A., & Braig, H. R. (2012). Mites parasitic on Australasian and African spiders found in the pet trade; a redescription of *Ljunghia pulleinei* Womersley. *PLoS ONE, 7*, e39019. http://dx.doi.org/10.1371/journal.pone.0039019

Miglio, L. T. (2009). *Taxonomia das espécies Brasileiras de Actinopus Perty, 1833 (Araneae, Mygalomorphae, Actinopodidae)* (p. 107). Belém: Universidade Federal do Pará.

Mohamed, M. I., & Mohamed, A. A. (2011). Natural infestation of *Pimeliaphilus joshuae* on scorpion species from Egypt. *Experimental and Applied Acarology, 55*, 77–84. doi:10.1007/s10493-011-9452-6

Norton, R. A., Welbourn, W. C., & Cave, R. D. (1988). First records of Erythraeidae parasitic on oribatid mites (Acari, Prostigmata: Acari, Oribatida). *Proceedings of the Entomological Society of Washington, 90*, 407–410.

Oudemans, A. C. (1902). Acarologische Aanteekeningen III. *Entomologische Berichten, 6*, 36–39.

Oudemans, A. C. (1905). Acarologische aanteekenigen XVIII. *Entomologische Berichten, Amsterdam, 1*, 236–241.

Oudemans, A. C. (1910a). Acarologische aanteekenigen XXXI. *Entomologische Berichten, Amsterdam, 3*, 47–51.

Oudemans, A. C. (1910b). Acarologische aanteekenigen XXXII. *Entomologische Berichten, Amsterdam, 3*, 67–74.

Oudemans, A. C. (1912). Die bis jetzt bekannten Larven von Thrombidiidae und Erythraeidae mit besonderer Berucksichtigung der fur den Menschen sch adlichen Arten. *Zoologischer Jahrbucher, Supplement, 14*, 1–230.

Parker, J. R. (1962). Ectoparasitic mites on spiders. *Entomologist's Monthly Magazine, 98*, 264.

Penney, D., & Green, D. I. (2011). *Fossils in Amber: Remarkable snapshots of prehistoric forest life* (p. 226). Manchester, NH: Siri Scientific Press.

Raven, R. J. (1985). The spider infraorder Mygalomorphae (Araneae): Cladistics and systematics. *Bulletin of the American Museum of Natural History, 182*, 1–180.

Reillo, P. R. (1989). Mite parasitism of the polymorphic spider, *Enoplognatha ovata* (Araneae, Theridiidae), from coastal Maine. *Journal of Arachnology, 17*, 246–249.

Ríos, T. D. (2014). A new species of the genus *Actinopus* (Mygalomorphae: Actinopodidae) from Argentina. *Acta Arachnologica, Tokyo, 63*, 73–77.

Seeman, O., & Miller, A. (2002). Parasitism of scorpions by mites. *Tasmanian Naturalist, 124*, 49–55.

Sorkin, L. N. (1982). Parasites of *Pardosa* wolf spiders (Acarina, Erythraeidae; Insecta, Hymenoptera; Araneae, Lycosidae). *American Arachnology, 26*, 6.

Southcott, R. V. (1992). Revision of the larvae of *Leptus* Latreille (Acarina: Erythraeidae) of Europe and North America, with descriptions of post-larval instars. *Zoological Journal of the Linnean Society, 105*, 1–153. http://dx.doi.org/10.1111/zoj.1992.105.issue-1

Southcott, R. V. (1994). Two new larval Erythraeidae (Acarina) from Thailand, with keys to the larvae of *Leptus* for Asia and New Guinea, and world larvae of *Hauptmannia*. *Steenstrupia, 20*, 165–176.

Southcott, R. V. (1955). Some observations on the biology, including mating and other behaviour of the Australian scorpion *Urodacus abruptus* Pocock. *Transactions of the Royal Society of South Australia, 78*, 145–154.

Southcott, R. V. (1999). Larvae of *Leptus* (Acarina : Erythraeidae), free-living or ectoparasitic on arachnids and lower insects of Australia and Papua New Guinea, with descriptions of reared post-larval instars. *Zoological Journal of the Linnean Society, 127*, 113–276. http://dx.doi.org/10.1111/j.1096-3642.1999.tb00677.x

Stašiov, S. (2003). *Leptus holmiae* Southcott 1992 (Acarina, Erythraeidae) associated with harvestmen (Opilionida). *Ekologia Bratislava, 22*, 23–27.

Teixeira, E. W. (2011). Larvas de *Leptus* sp. Latreille 1796 (Acarina: Erythraeidae) em abelhas africanizadas *A. mellifera* Linnaeus 1758 (Hymenoptera: Apidae), no Brasil. *Pesquisa & Tecnologia, 8*, 27.

Townsend, V. R., Mulholland, K. A., Bradford, J. O., Proud, D. N., & Parent, K. M. (2006). Seasonal variation in parasitism by *Leptus* mites (Acari, Erythraeidae) upon the harvestman, *Leiobunum formosum* (Opiliones, Sclerosomatidae). *Journal of Arachnology, 34*, 492–494. doi:10.1636/T05-44.1

Townsend, V. R., Proud, D. N., Moore, M. K., Tibbetts, J. A., Burns, J. A., Hunter, R. K., ... Felgenhauer, B. E. (2008). Parasitic and phoretic mites associated with neotropical harvestmen from Trinidad, West Indies. *Annals of the Entomological Society of America, 101*, 1026–1032. doi:10.1603/0013-8746-101.6.1026

Turk, F. A. (1945). Studies of Acari. V. Notes on and descriptions of new and little-known British Acari. *Journal of Natural History Series 11, 12*, 785–820. http://dx.doi.org/10.1080/00222934508654785

Welbourn, W. C. (1983). Potential use of trombidioid and erythraeoid mites as biological control agents of insect pests. In M. A. Hoy, G. L. Cunningham, & L. Knutson (Eds.), *Biological control of pests by mites* (pp. 103–140). Berkeley: Agricultural Experiment Station, Division of Agriculture & Natural Resources, University of California, Special Publication 3304.

Welbourn, W., & Young, O. P. (1988). Mites parasitic on spiders, with a description of a new species of Eutrombidium (Acari, *Eutrombidiidae*). *Journal of Arachnology, 16*, 373–385.

Wendt, F. E., Olomski, R., Leimann, J., & Wohltmann, A. (1992). Parasitism, life cycle and phenology of *Leptus trimaculatus* (Hermann, 1804) (Acari: Parasitengonae: Erythraeidae) including a description of the larva. *Acarologia, 33*, 55–68.

Wilson, W. T., Rubink, W. L., & Collins, A. M. (1990). A larval species of erythraeid mite (*Leptus* sp., Acarina: Erythraeidae) ectoparasitic on adult honey bees (*Apis mellifera* L.) in south Texas. *BeeScience, 1*, 18–22.

Wilson, W. T., Wooley, T. A., Nunamaker, R. A., & Rubink, W. L. (1987). An erythraeid mite externally parasitic on honey bees (*Apis mellifera*). *American Bee Journal, 127*, 853–854.

Womersley, H. (1934). A revision of the trombid and erythraeid mites of Australia with descriptions of new genera and species. *Record of the South Australian Museum, 5*, 179–254.

6

Acute and sub-acute toxicity studies on the effect of *Senna alata* in Swiss Albino mice

S. Roy[1], B. Ukil[1] and L.M. Lyndem[1*]

*Corresponding author: Larisha M. Lyndem, Parasitology Research Laboratory, Department of Zoology, Visva-Bharati University, Santiniketan 731235, West Bengal, India

E-mail: lyndemlarisha@gmail.com

Reviewing editor: Hani El-Nezami, University of Eastern Finland, Finland

Additional information is available at the end of the article

Abstract: *Senna alata* has attracted the attention of many researchers due to its numerous medicinal properties. This study aims to test the acute and sub-acute toxicity of its leaf extracts in Swiss albino mice. Studies were carried out with a fixed dose of 1,000, 2,000, and 3,000 mg/kg body weight through oral administration daily. Signs of toxicity in terms of behavior and mortality were noted after every two hours till 24 h of administration for acute toxicity and further administration of extracts till 15 days to analyze the physical, biochemical, hematological parameters, and histopathological studies in liver, kidney, and spleen for sub-acute study. The highest dose administered did not produce mortality or changes in the general behavior of the test animals. All parameters were unaltered throughout the study. The present study revealed no obvious toxicity in mice treated with *S. alata*. These results indicate the safety of the oral administration of leaf extract.

Subjects: Pharmaceutical Science; Pharmacy; Drug Design & Development; Natural Products

Keywords: toxicity; *Senna alata*; extracts; non-toxic; mice

ABOUT THE AUTHOR

Larisha Mawkhlieng Lyndem PhD is an associate professor in the Department of Zoology, Visva-Bharati University, West Bengal, India. She teaches Parasitology and related courses for graduate and postgraduate students. Among other areas, her research interest includes the development of anthelmintic agents from natural sources mainly plant source. Together with her research team members Lyndem has extensive publications on the anthelmintic efficacy of medicinal plants widely used in traditional medicine. Saptarshi Roy and Bidisha Ukil are active members of Lyndem's team of research.

PUBLIC INTEREST STATEMENT

Helminthiasis is a neglected disease and has shown resistance to some available marketed drugs. Scientists continue the search for new anthelmintic especially from natural sources agents and plants have proved to be a potential source for this purpose. There are few species of *Senna* plant that showed to have medicinal properties, three species viz. *S. alexandrina*, *S alata* and *S. occidentalis* leaf extracts have been reported for the first time from our laboratory to have cestocidal property. Amongst the three plants, *S. alata* leaf extracts showed to have more anthelmintic efficacy and is apparently believed to be nontoxic, but detail pre-clinical toxicological evaluation in animals have not been evaluated. Moreover, not all medicinal plants are safe for consumption in the crude form. Some level of toxicity arises from the potent toxic compounds present in it and nontoxic compounds can also behave like a toxic compound even at a lower dose, and can produce an adverse effect in human or animal cells. Thus, it is required to examine the toxicity profile of *S. alata* leaf extract, given its widespread consumption by human.

1. Introduction

According to World Health Organization (2003), more than 80% of the world's population rely on traditional medicine for their primary health care and more than 30% of the plant species have been used for this purpose. A major population of the world are attracted to this type of traditional medicine due to scarcity and high costs of available drugs (Hudaib et al., 2008) and an easy access to these plants in some regions of the world (Humber, 2002). Besides, a large amount of evidences has shown immense potential of medicinal plants for prevention, diagnosis, and treatment of various diseases (Abera, 2014; Ghosh, Sahoo, Das, Duley, & Palhy, 2014). In India, over 3,000 plants were officially recognized for their medicinal value, but it is generally estimated that over 6,000 plants are being used in traditional, folklore, and herbal medicines. Although the scientific study of some medicinal plants clearly validates the effectiveness and reliability of ethno-medical knowledge and traditional use in managing diseases, however herbal medicines are complex mixtures of many bioactive phytochemicals which may differ in different mechanisms (Sengupta, Sharma, & Chakraborty, 2011). Some level of toxicity arises from the potent toxic compounds present in it, and nontoxic compounds can also behave like a toxic compound even at a lower dose, and can produce an adverse effect by interacting with human or animal cells. Therefore, not all medicinal plants are safe for consumption in the crude form. Thus, such plants should be investigated to better understand their properties, safety and efficiency.

Senna alata Linn. (Family Fabaceae) has been recognized in traditional medicine for its medicinal activities (Lim, 2013; Karthika, Manivannan, & Mohamed, 2016) and was also recently reported to have anthelmintic activity (Kundu, Roy, & Lyndem, 2012, 2014). Though this plant is apparently believed to be nontoxic, but detail pre-clinical toxicological evaluation in animals have not been evaluated. Thus, it is required to examine the toxicity profile of *S. alata* leaf extract given its widespread consumption by human.

2. Results and discussion

During the 15-day period of toxicity study, mice showed no signs of behavioral distress or change in skin color, no changes in the eyelids, sleep, food, and water intake, and no observable toxicity symptoms or death. The experimented mice survived till the completion of the experimental duration at all levels of treatment (Table 1). Thus, this indicates there was no disturbance in carbohydrate, protein, or fat metabolism (Klaassen, 2001).

Though the body weight gradually increased in control and treated groups, there was no significant difference in mean body weight amongst the different treated groups and the control (Table 2) which indicated that the extract has negligible levels of toxicity on the growth of the animals as also observed by Mir, Sexena, and Malla (2013) and Rajalakshmi, Jayachitra, Gopal, and Krithiga (2014). There were no significant changes in organ weight and relative organ weight of liver, kidney, and spleen with respect to the body weight as well (Table 3). Kluwe (1981), documented that the increase in organ weight had been observed to be a relative sensitive indicator of nephrotoxicity. Thus, *S. alata* did not induce any toxic effect on the kidneys and the other organs going by this indicator.

Hematological tests showed no significant differences in hemoglobin, RBC, WBC (total and differential), and platelet count in all doses as compared to control. In WBC differential count, lymphocytes showed slight variation in the dose of 2,000 and 3,000 mg/kg body weight compared to 1,000 mg/kg body weight and also to control, while eosinophil count showed no significant differences at all dose levels compared to control (Table 4). According to Onyeyilli, Iwuoha, and Akinniyi (1998), administration of an agent can result in loss of blood cells and/or inhibition of blood cell

Table 1. Acute toxicity study of ethanol extract of *Senna alata* in mice

Observations	Response			
	Control	1,000 mg/kg body wt.	2,000 mg/kg body wt.	3,000 mg/kg body wt.
Consciousness	+	+	+	+
Grooming	–	–	–	–
Touch response	+	+	+	+
Sleeping duration	+	+	+	+
Movement	+	+	+	+
Gripping strength	+	+	+	+
Righting reflex	+	+	+	+
Food intake	+	+	+	+
Water consumption	+	+	+	+
Tremors	–	–	–	–
Diarrhea	–	–	–	–
Hyper activity	–	–	–	–
Pinna reflex	+	+	+	+
Corneal reflex	+	+	+	+
Salivation	+	+	+	+
Skin color	+	+	+	+
Lethargy	–	–	–	–
Convulsion	–	–	–	–
Morbidity	–	–	–	–
Sound response	+	+	+	+

Notes: + = normal, – = absent.

Table 2. Body weight of mice during sub-acute toxicity study after administration of *S. alata* extract

	Body weight			
	Initial day	After 5 days	After 10 days	After 15 days
Control	26.3 ± 0.35	27.45 ± 0.5	29.95 ± 1.51	34.34 ± 4.33
1,000 mg/kg	24.67 ± 0.3	26.20 ± 0.54	28.14 ± 0.41	30.87 ± 2.38
2,000 mg/kg	24.63 ± 0.25	26.59 ± 0.35	27.51 ± 0.35	29.32 ± 0.66
3,000 mg/kg	25.02 ± 0.32	26.94 ± 0.51	28.11 ± 0.5	30.44 ± 1.79

Note: All values are expressed as mean ± SD of 6 animals.

Table 3. Absolute organ weight (g) and relative organ weight (g) during sub-acute toxicity study of *S. alata* extract

	Liver		Kidney		Spleen	
	Absolute organ weight	Relative organ weight	Absolute organ weight	Relative organ weight	Absolute organ weight	Relative organ weight
Control	1.58 ± 0.11^{a}	4.59 ± 0.23^{b}	$0.25 \pm .014^{c}$	0.72 ± 0.03^{d}	0.23 ± 0.02^{e}	0.66 ± 0.05^{f}
1,000 mg/kg	1.62 ± 0.02^{a}	5.23 ± 0.28^{b}	0.25 ± 0.02^{c}	0.79 ± 0.03^{d}	0.20 ± 0.03^{e}	0.61 ± 0.06^{f}
2,000 mg/kg	1.57 ± 0.09^{a}	5.33 ± 0.20^{b}	0.23 ± 0.02^{c}	0.79 ± 0.10^{d}	0.21 ± 0.03^{e}	0.72 ± 0.10^{f}
3,000 mg/kg	1.50 ± 0.14^{a}	5.00 ± 0.54^{b}	$0.22 \pm 0.03_{c}$	0.73 ± 0.05^{d}	0.20 ± 0.02^{e}	0.67 ± 0.08^{f}

Notes: Values are expressed as mean ± SD of 6 animals. A rows means followed by a common superscript are not significantly at 5% by using DMRT.

Table 4. Hematological parameters of sub-acute toxicity study of *S. alata* extract in mice

Hematological parameters	Group 1 control	Group 2 1,000 mg/kg body wt.	Group 3 2,000 mg/kg body wt.	Group 4 3,000 mg/kg body wt.	Unit of values
Hemoglobin	14.03 ± 0.45ᵃ	13.7 ± 0.41ᵃ	14 ± 0.43ᵃ	14.16 ± 0.67ᵃ	gm/dl
RBC count	7.07 ± 0.81ᵇ	7.07 ± 0.31ᵇ	7 ± 0.57ᵇ	7.1 ± 0.51ᵇ	Million/Cu mm
Total lymphocytes	11.71 ± 1ᶜ	11.68 ± 0.4ᶜ	11.57 ± 0.23ᶜ	11.53 ± 0.5ᶜ	10^3/µl
Neutrophils	25.5 ± 3.78ᵈ	24.83 ± 3.82ᵈ	25.17 ± 3.31ᵈ	25.17 ± 2.56ᵈ	%
Lymphocytes	66.55 ± 2.91ᵉ	65.1 ± 3.89ᵉ	64.79 ± 2.35ᵉ	64.45 ± 3.34ᵉ	%
Eosinophils	1.33 ± 0.52ᶠ	1.33 ± 0.52ᶠ	1.33 ± 0.52ᶠ	1.33 ± 0.52ᶠ	%
Monocytes	0	0	0	0	%
Basophils	0	0	0	0	%
Platelet count	273 ± 7.35ᵍ	268.17 ± 11.34 ᵍ	268.83 ± 6.33ᵍ	271.67 ± 25.57 ᵍ	10^3/µl

Notes: Values are expressed as mean ± SD of 6 animals. A column means followed by a common superscript are not significant at 5% by using DMRT.

synthesis and decrease in such hematological parameters in experimental animals has been associated with anemia. The above results suggest the nontoxicity of *S. alata* in mice. A similar observation was reported by Ping, Darah, Chen, Sreeramanan, and Sasidharan (2013), after oral administration of *Euphorbia hirta, Carica papaya, Petroselinum crispum*, and *Lygodium flexuosum*.

The serum analyses showed no significant difference in calcium and chloride level between the control and experimental group. However, there is a less significant level of difference in phosphorus at 1,000 and 2,000 mg/kg body weight compared to control group (Table 5). Similarly, levels of creatinine and uric acid were not significantly different between the control and the experimental group of mice (Table 6). This observation was also made by Ping et al. (2013). Moreover, levels of total protein, albumin, SGPT, total bilirubin, direct bilirubin, cholesterol, triglyceride, alkaline phosphatase (ALP), and aspartate transaminase (AST) also showed no significant difference, however slight variation in the latter two was observed at 1,000 mg/kg body weight of the control group (Table 7). Liver injury is characterized as hepatocellular when there is predominant elevation of the ALT, while AST is a mitochondria enzyme whose increased activity reflects severe tissue injuries (Martin, 2006). Hypo-proteinaemia, a common finding in liver damage (Larrey, 2002), was also not observed in the present study. This indicates that the extract did not cause any overt liver damage at the dose levels studied. Further, there was a low level of glucose in the treated group as compared to the control group (Table 7) which may be due to inadequate insulin secretion that indicates normal functioning of the liver. Similar observations were made by Rajalakshmi et al. (2014), Nabukenya

Table 5. Sub-acute toxicity study on the electrolytes of mice after treatment with *S. alata* alcoholic leaf extracts

Serum biochemical parameter	Group 1 control	Group 2 1,000 mg/kg body wt.	Group 3 2,000 mg/kg body wt.	Group 4 3,000 mg/kg body wt.	Unit of values
Phosphorus	5.5 ± 0.5ᵃ	5.1 ± 0.83ᵃ	5.0 ± 0.4ᵃ	5.4 ± 0.3ᵃ	mg/dl
Chloride	102.82 ± 3.94ᵇ	102.77 ± 3.36ᵇ	102.4 ± 1.6ᵇ	100.13 ± 1.88ᵇ	mmol/lit
Calcium	10.47 ± 0.52ᶜ	10 ± 0.01ᶜ	10.06 ± 0.2ᶜ	10.1 ± 0.5ᶜ	mg/dl

Notes: Values are expressed as mean ± SD of 6 animals. A column means followed by a common superscript are not significant at 5% by using DMRT.

Table 6. Sub-acute toxicity study on biomarkers of kidney malfunction in mice treated with *S. alata*

Serum biochemical parameter	Group 1 control	Group 2 1,000 mg/kg body wt.	Group 3 2,000 mg/kg body wt.	Group 4 3,000 mg/kg body wt.
Creatinine (mg/dl)	0.87 ± 0.12^{d}	0.84 ± 0.18^{d}	0.87 ± 0.18^{d}	0.86 ± 0.12^{d}
Uric acid (mg/dl)	4.45 ± 0.38^{e}	4.41 ± 0.51^{e}	4.25 ± 0.79^{e}	4.21 ± 0.16^{e}

Notes: Values are expressed as mean ± SD of 6 animals. A column means followed by a common superscript are not significant at 5% using DMRT.

Table 7. Sub-acute toxicity study of biomarkers of liver malfunction in mice treated with *S. alata*

Serum biochemical parameter	Group 1 control	Group 2 1,000 mg/kg body wt.	Group 3 2,000 mg/kg body wt.	Group 4 3,000 mg/kg body wt.	Unit of values
Glucose	103.78 ± 9.3^{a}	99.61 ± 7.6^{a}	103.88 ± 1.81^{a}	86.3 ± 1.41^{a}	mg/dl
Total protein	6.55 ± 1.15^{b}	6.33 ± 0.1.15^{b}	6.21 ± 0.42^{b}	6 ± 0.44^{b}	g/dl
Albumin	3.95 ± 0.1^{c}	3.86 ± 0.15^{c}	3.87 ± 0.18^{c}	3.8 ± 0.43^{c}	g/dl
Alkaline phosphatase	99.44 ± 8.65^{d}	95.6 ± 12.33^{d}	95.2 ± 15.04^{d}	97 ± 6.33^{d}	IU/L
Aspartate transaminase	102.28 ± 27.29^{e}	93.41 ± 10.92^{e}	97.91 ± 12.23^{e}	97.83 ± 8.57^{e}	IU/L
SGPT	38 ± 7.6^{f}	36.27 ± 6.4^{f}	36 ± 11.8^{f}	38.05 ± 6^{f}	IU/L
Direct bilirubin	0.173 ± 0.03^{g}	0.156 ± 0.04^{g}	0.165 ± 0.05^{g}	0.165 ± 0.06^{g}	mg/dl
Total bilirubin	0.87 ± 0.06^{h}	0.85 ± 0.13^{h}	0.86 ± 0.09^{h}	0.86 ± 0.06^{h}	mg/dl
Cholesterol	196.64 ± 20.69^{i}	193.85 ± 18.16^{i}	196.4 ± 20.9^{i}	195.15 ± 8.85^{i}	mg/dl
Triglycerides	131.1 ± 10.08^{j}	130.45 ± 9.1^{j}	129.66 ± 9.53^{j}	130 ± 7.41^{j}	mg/dl

Notes: Values are expressed as mean ± SD of 6 animals. A column means followed by a common superscript are not significant at 5% by using DMRT.

et al. (2014), Priyadarshini, Mazumder, and Choudhury (2014) and Bello et al. (2016). Cholesterol and triglyceride levels have no significant difference in treated animals which concurs that this plant extract does not present any risk of hypercholesterolemia or artherosclerosis at a high level of doses (Bello et al., 2016).

Histological studies revealed no abnormalities in liver, kidney and spleen tissue in treated mice. The liver tissue displayed normal hepatocytes without any enlargement in sinusoidal vein, central vein, and portal triad in all treated groups compared to control (Figure 1). Similar type of observation was also seen by Bello et al. (2016) in rat liver. Kidney micrograph revealed normal architecture of glomerulus and Bowman's capsules with no degeneration, necrosis, or inflammation (Figure 2), which are comparable with the study made by Ping et al. (2013) and Nabukenya et al. (2014). Histological features of spleen showed normal splenocytes with prominent nucleus in both treated and control groups (Figure 3). These observations agreed with that of Ping et al. (2013) in rat model that have been treated with *Euphorbia hirta*. Thus, histopathological evaluation indicated that the extract did not have any adverse effect on morphology of the tissues and these observations supported the biochemical results mentioned above. Therefore, it is concluded that *S. alata* did not produce any toxic effect in male albino mice.

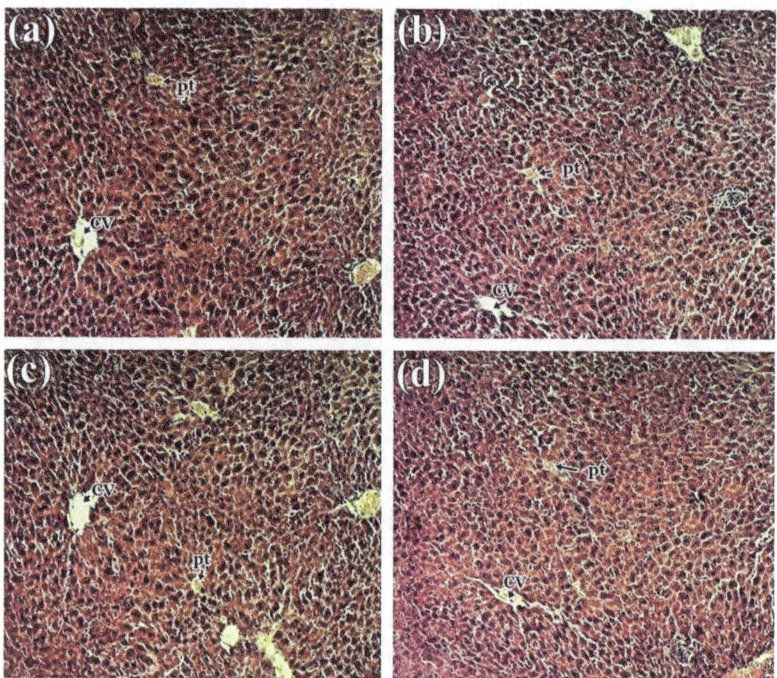

Figure 1. Histology study of liver from of mice: (a) control group; (b) 1,000 mg/kg; (c) 2,000 mg/kg and (d) 3,000 of *S. alata* leaf extract in a 15-day sub-acute toxicity.

Notes: No significant damage was detected in any treatment group Indicators: Bowman's capsule (BW), glomerulus (G), proximal collecting tubule (P), distal collecting tubule (D).

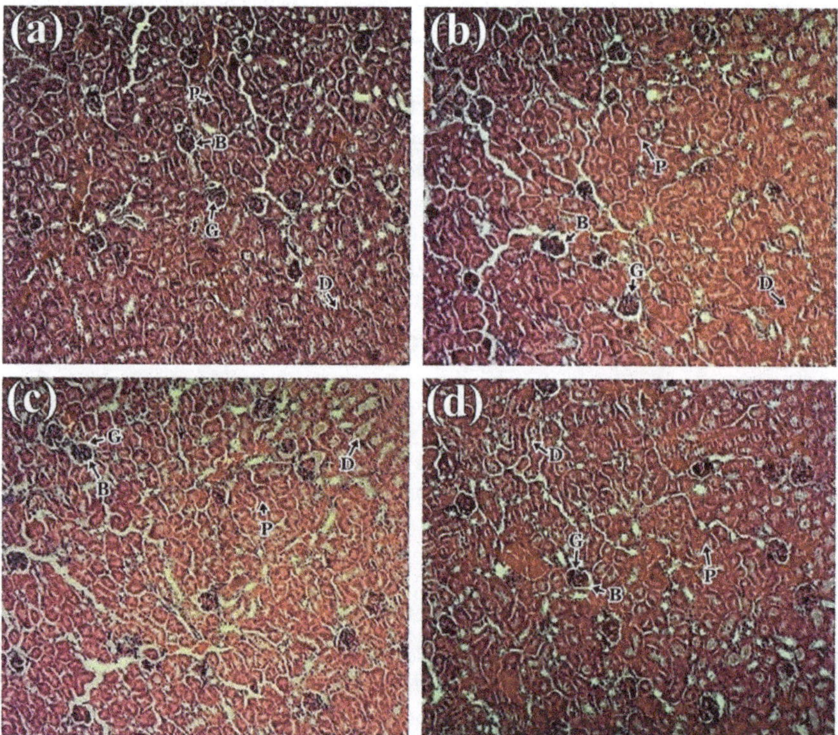

Figure 2. Histology study of kidney of mice: (a) control group; (b) 1,000 mg/kg; (c) 2,000 mg/kg and (d) 3,000 of *S. alata* leaf extract in a 15-day sub-acute toxicity.

Notes: No significant damage was detected in any treatment group. Indicators: Portal Triad (pt); Central Vein (CV).

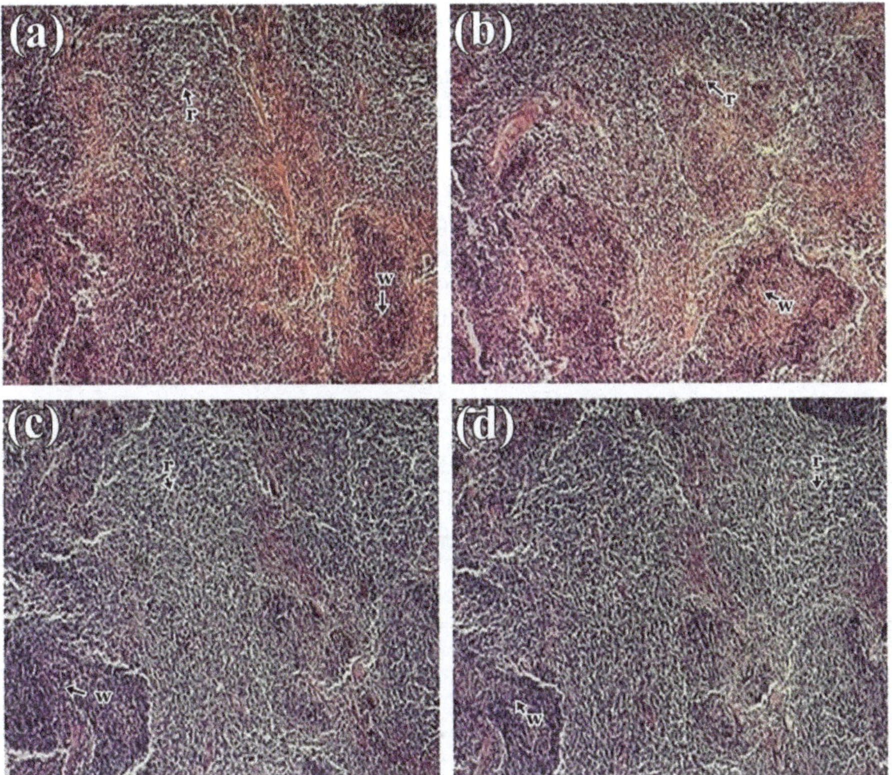

Figure 3. Histology study of spleen of mice showing normal splenocytes with defined red pulp and white pulp in control and treated group: (a) Control; (b) 1,000 mg/kg; (c) 2,000 mg/kg and (d) 3,000 mg/kg body weight of *S.alata* leaf extract.

Notes: No significant damage was detected in any treatment group Indicators: Red pulp (r), White pulp (w).

3. Materials and methods

3.1. Preparation of plant extract

S. alata leaves were collected from in and around the University campus of Visva-Bharati, Santiniketan. Young leaves were washed with distilled water, allowed to dry in an oven at 50°C, and crushed to powder. About 250 g of the powdered form was extracted with 1 L of ethanol (90%) in a Soxhlet apparatus for 7–8 h, and the final crude extract was recovered using rotary evaporator and stored at 4°C until further use.

3.2. Expremental designs

Twenty-four Swiss albino male mice weighed 24–28 g were divided into four groups of six animals each (Group 1–4). Group 1 is control group, fed daily with only normal laboratory diet and water. Group 2–4 were treated with a dose of 1,000, 2,000, and 3,000 mg/kg body weight, respectively, for 15 days through an oral needle following a period of 10-h fasting. All animals were maintained on standard laboratory diets with water *ad libitum*. The experimental protocol and procedures used in this study were approved by the Institutional Animal Ethical Committee (IAEC), Visva-Bharati, Santiniketan.

3.3. Acute oral toxicity study

After administration of the extract, animals were monitored continuously for every two hours for a day to detect acute changes in morphological and behavioral responses, spontaneous activity, irritability, corneal reflex, tremors, convulsion, salivation, diarrhea, lethargy if any, and also monitored for any mortality during the course of toxicity study.

3.4. Sub- acute oral toxicity study

Body weight of each animal was recorded every five days interval till the last day of experiment. After the 15th day, all animals were sacrificed after light chloroform inhalation of anesthesia and different hematological and biochemical studies were performed.

3.4.1. Hematological assay

About 1.5–2 ml of blood was drawn directly with a hypodermic syringe to minimize the damage of serum contamination through a cardiac puncture (Jochems, Valk, Stafleu, & Baumans, 2002). About 100 µl of the collected blood sample was used for the determination of hematological parameters like hemoglobin concentration, total RBC count, total WBC count, WBC differential count, and total platelet count following the methods of Smith (1995) and Kjeldsberg (1998).

3.4.2. Analysis of serum biochemical parameters

The rest of the collected blood sample was prepared for serum isolation according to the method of Singh and Rana (2007). In brief, blood was kept for 20 min at room temperature of 30°C and then centrifuged at 2,500 rpm for 5 min at 4°C. The serum obtained as supernatant, was collected in an eppendorf tube and kept at 4°C till use. Determination of glucose, calcium, phosphorus, chloride, total protein, albumin, ALP, AST, SGPT, total bilirubin, direct bilirubin, creatinine, cholesterol, triglyceride, and uric acid were analyzed by different assay kits following the manufacturer protocol.

3.4.3. Measurement of relative organ weight

Liver, kidneys, and spleen were carefully dissected out and weighed separately. The relative organ weight of each animal was then calculated as follows:

$$\text{Relative organ weight} = \frac{\text{Absolute organ weight (g)}}{\text{Body weight of rat on sacrifice day}}\,(\text{g}) \times 100$$

3.4.4. Histological examination

Each selected organs were cut into small pieces and kept in Bouin's fixative for 24 h, and processed for histological study following methods of Mayer (1896) with slight modification and later observed under a light microscope.

3.5. Chemicals

All the chemicals used were of analytical grade. Ethanol was supplied by Bengal Chemicals, Kolkata. Stains and fixatives were purchased from Sigma-Aldrich. All biochemical assay kits were purchased from Coral clinical system company, Goa, India, and all other reagents were obtained from Merck Life Science Pvt. Ltd., Merck India.

3.6. Statistical analysis

Data are expressed as a mean ± SD. Total variations present in a set of data were estimated by one way Analysis of Variance (ANOVA) comparisons were made between the treated groups. All data were analyzed using Duncan's Multiple Range Test (DMRT). $p < 0.05$ was considered as the level statistical significance.

4. Conclusions

The absence of gross and histopathological lesions in the organs as well as no significant differences in hematological and biochemical test in the treated groups from the control could suggest the level of safety of the leaf alcoholic extract on the animals. In conclusion, to our knowledge, this is the first investigation of the various parameters of toxicity studies made on the *S. alata* at higher dose. This study has shown that sub-acute administration of the alcoholic leaf extract of *S.alata* may be safe and thereby provide a support to the use of *S.alata* leaves as an alternative system of medicine.

Acknowledgments
We also wish to thank the Department of Zoology, Centre for Advanced Studies, Visva-Bharati for providing infrastructural support.

Funding
The authors' gratefully acknowledge the University Grants Commission (UGC), New Delhi for providing financial assistance through a major research project (No: UGC/SR/40-385/2011) sanctioned to Larisha M. Lyndem.

Competing Interests
The authors declare no competing interest

Author details
S. Roy[1]
E-mail: roysaptarshi89@gmail.com
B. Ukil[1]
E-mail: ukilbidisha@gmail.com
L.M. Lyndem[1]
E-mail: lyndemlarisha@gmail.com
[1] Parasitology Research Laboratory, Department of Zoology, Visva-Bharati University, Santiniketan 731235, West Bengal, India.

Cover image
Source: Author.

References

Abera, B. (2014). Medicinal plants used in traditional medicine by Oromo people, Ghimbi District, Southwest Ethiopia. *Journal of Ethnobiology and Ethnomedicine, 10*, 40. http://dx.doi.org/10.1186/1746-4269-10-40

Bello, I., Bakkouri, A. S., Tabana, Y. M., Hindi, B. A., Mansoub, M. A. A., Mahmud, R., & Asmawi, M. Z. (2016). Acute and sub-acute toxicity evaluation of the methanolic extract of *Alstonia scholaris* stem bark. *Medical Sciences, 4*, 4.

Ghosh, G., Sahoo, S., Das, D., Duley, D., & Palhy, R. N. (2014). Antibacterial and antioxidant activities of methanol extract and fractions of *Clerodendrum viscosum* vent. leaves. *Indian Journal of Natural Products and Resources, 5*, 134–142.

Hudaib, M., Mohammad, M., Bustanji, Y., Tayyem, R., Yousef, M., Aburjaie, M. & Aburjai, T. (2008). Ethnopharmacological survey of medicinal plants in Jordan, Mujib Nature Reserve and surrounding area. *Journal of Ethnopharmacology, 120*, 63–71. http://dx.doi.org/10.1016/j.jep.2008.07.031

Humber, J. M. (2002). The role of complementary and alternative medicine: Accommodating pluralism. *JAMA: The Journal of the American Medical Association, 288*, 1655–1656.

Jochems, C. E. A., Valk, J. B. F., Stafleu, F. R., & Baumans, V. (2002). Ethical or scientific problem? *ATIA, 30*, 219–227.

Karthika, C., Manivannan, S., & Mohamed, R. K. (2016). Phytochemical analysis of *Ruellia patula* using gas chromatography-mass spectrometry. *Asian Journal of Pharmaceutical and Clinical Research, 9*, 253–257.

Kjeldsberg, C. R. (1998). Principios de examehematologico. In G. R. Lee, T. C. Bithell, J. Forester, J. W. Athens, & J. N. Lukens (Eds.), *Wintrobehematol.clinica*. (pp. 7–42). São Paulo: Manole.

Klaassen, C. D. (2001). Principles of toxicology. In *Casarett and Doull's toxicology: The basic science of poisons* (5th ed., p. 13). New York, NY: McGraw-Hill.

Kluwe, W. M. (1981). Renal function tests as indicators of kidney injury in subacute toxicity studies. *Toxicology and Applied Pharmacology, 57*, 414–424. http://dx.doi.org/10.1016/0041-008X(81)90239-8

Kundu, S., Roy, S., & Lyndem, L. M. (2012). *Cassia alata* L.: Potential role as anthelmintic agent against *Hymenolepis diminuta*. *Parasitology Research, 111*, 1187–1192. http://dx.doi.org/10.1007/s00436-012-2950-6

Kundu, S., Roy, S., & Lyndem, L. M. (2014). Broad spectrum anthelmintic potential of Cassia plants. *Asian Pacific Journal of Tropical Biomedicine, 4*, S436–S441. http://dx.doi.org/10.12980/APJTB.4.2014C1252

Larrey, D. (2002). Epidemiology and individual susceptibility to adverse drug reactions affecting the liver. *Seminars in Liver Disease, 22*, 145–156. http://dx.doi.org/10.1055/s-2002-30101

Lim, T. K. (2013). *Edible medicinal and non-medicinal plants* (pp. 841–859). Dordrecht: Springer Science Business Media.

Martin, A. C. (2006). *Clinical chemistry and metabolic medicine* (7th ed., pp. 7–15). London: Edward Arnold.

Mayer, P. (1896). *Mitt. Zool. Stn* (12th ed., p. 303). Neapel.

Mir, A. H., Sexena, M., & Malla, M. Y. (2013). An acute oral toxicity study of methanolic extract from *Tridex procumbens* in Sprague Dawley's Rats as per OECD guidelines 423. *Asian Journal of Plant Science, 3*, 16–20.

Nabukenya, I., Rubaire-Akiiki, C., Mugizi, D., Kateregga, J., Olila, D., & Hoglund, J. (2014). Sub-acute toxicity of aqueous extracts of *Tephrosia vogelii*, *Vernonia amygdalina* and *Senna occidentalis* in rats. *Natural Products Chemistry & Research, 2*, 143. doi:10.4172/2329-6836.1000143

Onyeyilli, P. A., Iwuoha, C. L., & Akinniyi, J. A. (1998). Chronic toxicity study of *Ficus platyphtlla* blume in rats. *West African Journal of Pharmacology and Drug Research, 14*, 27–30.

Ping, K. Y., Darah, I., Chen, Y., Sreeramanan, S., & Sasidharan, S. (2013). Acute and subchronic toxicity study of *Euphorbia hirta* L. methanol extract in rats. *BioMed Research International, 182064*. doi:10.1155/2013/182064

Priyadarshini, L., Mazumder, P. B., & Choudhury, M. D. (2014). Acute toxicity and oral glucosetolerance test of ethanol and methanol extracts of antihyperglycaemic plant *Cassia alata* Linn. *Journal of Pharmacy and Biological Sciences, 9*, 43–46.

Rajalakshmi, A., Jayachitra, A., Gopal, P., & Krithiga, N. (2014). Toxicity analysis of different medicinal plant extracts in swiss albino mice. *Pharmacology and Toxicology, 1*(2), 1–6.

Sengupta, M., Sharma, G. D., & Chakraborty, B. (2011). Hepatoprotective and immunomodulatory properties of aqueous extract of *Curcuma longa* in carbon tetra chloride intoxicated Swiss albino mice. *Asian Pacific Journal of Tropical Biomedicine, 1*, 193–199. http://dx.doi.org/10.1016/S2221-1691(11)60026-9

Singh, S., & Rana, S. V. S. (2007). Amelioration of arsenic toxicity by L-Ascorbic acid in laboratory rat. *Journal of Environmental Biology, 28*, 377–384.

Smith, J. E. (1995). Comparative hematology. In E. Beutler, M. A. Lichtman, B. S. Coller, & T. J. Kipps (Eds.), *Williams hematology* (5th ed., pp. 77–85). New York, NY: McGraw-Hill.

WHO. (2003). Retrieved from http://www.who.int/mediacentre/factsheets/2003/fs134/en/

7

Etomidate decreases adrenal gland apoptosis and necrosis associated with hemorrhagic shock in a rat model (*Rattus norvegicus*)

Nuno M. Félix[1*], Isabelle Goy-Thollot[2], Ronald S. Walton[3], Pedro M. Borralho[4], Hugo Pissara[1], Ana S. Matos[5], Cecília M.P. Rodrigues[4] and Maria M.R.E. Niza[1]

*Corresponding author: Nuno M. Félix, CIISA, Faculty of Veterinary Medicine, ULisboa, Avenida da Universidade Técnica, 1300-477 Lisboa, Portugal
E-mail: nuno.felixgrey@gmail.com

Reviewing editor: Hsu Tsai-Ching, Chung Shan Medical University, Taiwan, Province Of China
Additional information is available at the end of the article

Abstract: *Purpose*: Evaluate if etomidate modulates adrenal apoptosis and if this influences the development of critical illness-related corticosteroid insufficiency (CIRCI) in hemorrhagic shock (HS). *Material and methods*: Four groups of 16 male Wistar rats: G0 (control group anesthetized with isoflurane and mechanical ventilation), G1 (like G0, but with buprenorphine), G2 (like G1 with HS), and G3 (like G2 with etomidate 1 mg/kg, IV, before HS). HS induced by collecting 30% of blood volume. Resuscitation performed 90 min later with the collected blood and normal saline. Hemodynamic parameters, blood gas analysis, adrenocorticotropic hormone (ACTH), corticosterone (CS), and TNF-α, IL6, IL10 were determined at 0, 90, 150, and 240 min post-HS induction (at the corresponding time points in G0 and G1). Apoptosis and necrosis were determined by TUNEL and caspase-3 immunofluorescence and a necrosis score, respectively. *Results*: HS groups had significantly higher levels of apoptosis and necrosis than G1 and G0. Compared with G2, etomidate-treated animals had significantly lower levels of CS (compatible with CIRCI), PO_2, PO_2/FiO_2, BE, HCO_3, apoptosis, and necrosis and significantly higher cytokine levels. *Conclusions*: Etomidate was associated with CIRCI. HS was associated with adrenal gland

ABOUT THE AUTHOR

Nuno M. Félix was licensed as Doctor in Veterinary Medicine in 1999 by the Lisbon Faculty of Veterinary Medicine, completed his master of science (Neuroscience) in 2004 by the Lisbon Faculty of Medicine, became human medical doctor by the Lisbon Faculty of Medicine in 2011, and finished his PhD about hemorrhagic shock in 2017, by the Lisbon Faculty of Veterinary Medicine. He worked as assistant professor of Small Animal Internal Medicine in the same faculty from 2001 to 2014 and is currently a human doctor, resident in Pediatric Medicine. The study here reported was related with his PhD research project, which aimed to evaluate endocrine changes associated to hemorrhagic shock. He and his research group continue to research several critical illnesses in small animals and how these can be translated in the Human field, with clinical benefits for all species.

PUBLIC INTEREST STATEMENT

Hemorrhagic shock (HS) is still a significant Health problem, being responsible for more than a third of deaths associated to trauma. Etomidate is an anesthetic commonly used for intubation in HS. However, its use has been recently questioned in critically ill patients because it affects adrenal function. Apoptosis is a type of cell death which has been associated to development of organ dysfunction in critical illness. In this study, the effects of etomidate on adrenal apoptosis and necrosis in a rat model of HS were evaluated. The study found that etomidate attenuated significantly the development of adrenal apoptosis and necrosis. Etomidate also increased morbidity and was associated to significant adrenal dysfunction. Thus, etomidate´s effects in HS are complex and can be beneficial and deleterious. Until these effects are more thoroughly evaluated and its clinical significance determined, etomidate should be used cautiously in HS and only in specific circumstances.

apoptosis and necrosis. The latter were decreased by etomidate, possibly by both direct and indirect mechanisms.

Subjects: Bioscience; Health and Social Care; Medicine, Dentistry, Nursing & Allied Health

Keywords: hemorrhagic; shock; apoptosis; necrosis; adrenal; etomidate; cytokine; rat; CIRCI

1. Introduction

Etomidate is an anesthetic commonly used in critical conditions, such as Hemorrhagic Shock (HS), due to its favorable hemodynamic profile (Forman, 2011). However, it also directly inhibits adrenal gland steroidogenesis, even after single bolus administration (Allolio et al., 1985; Wagner, White, Kan, Rosenthal, & Feldman, 1984). This "pharmacologic adrenal suppression" has been associated to development of critical illness-related corticosteroid insufficiency (CIRCI) in several conditions, including sepsis and septic shock (Cherfan et al., 2011; den Brinker et al., 2008), trauma (Cotton et al., 2008) and burns (Mosier, Lasinski, & Gamelli, 2015). CIRCI is defined as the inadequate glucocorticoid intracellular anti-inflammatory activity for the severity of the patient's illness (Marik, 2009). The clinical significance of etomidate-associated CIRCI has been extensively debated and it is still incompletely understood. The most recently available meta-analysis, however, have reached the conclusion that etomidate administration to critically ill patients is associated with increased risk of Multi-Organ Failure (Bruder, Ball, Ridi, Pickett, & Hohl, 2015).

The development of CIRCI following HS has been demonstrated in both clinical (Cotton et al., 2008; Hoen et al., 2002; Stein et al., 2013) and experimental models (Rushing, Britt, & Britt, 2006; Wang et al., 1999). How HS induces CIRCI is still incompletely understood although several mechanisms have been suggested. These include the following: adrenal necrosis caused by HS-associated ischemia (Kajihara, Malliwah, Matsumura, Taguchi, & Iijima, 1983; Rushing et al., 2006); reduced adrenal levels of corticosterone (CS) and cyclic adenosine monophosphate; reduced CS release following corticotropin stimulation; decreased adrenocorticotropic hormone (ACTH) secretion (Wang et al., 1999) and adrenal dysfunction due to HS-associated inflammation (Hoen et al., 2002). Etomidate administration has also been suggested as a possible mechanism by some but not all authors. For example, Hoen et al. (2002) did not found any relation between CIRCI's occurrence and etomidate administration when studying trauma patients with HS. In contrast, and also in trauma with HS, another study found that etomidate was the only modifiable factor associated to CIRCI (Cotton et al., 2008).

In recent years, experimental studies have shown that increased adrenal apoptosis can cause adrenal dysfunction in several types of critical illnesses, including acute necrotizing pancreatitis (ANP) and sepsis (Liu et al., 2016; Polito et al., 2010; Yu et al., 2012, 2016). To our knowledge only one study described the development of adrenal apoptosis in HS, and this did not report any occurrence of CIRCI (Rushing & Britt, 2007).

Etomidate has been shown to modulate apoptosis in several cell types (Wu et al., 2011; Xu, Chen, Luo, & Firoj, 2014). This has also been found in the adrenal gland of both normal and septic animals, where etomidate's modulation of apoptosis seems to be a dose- and time-dependent (Liu et al., 2016; Liu, Zhang, Han, Lv, & Xiong, 2015; Zhang et al., 2015). The influence of etomidate in apoptosis can be both direct, through direct interference in the intra-signaling pathways and proteins associated to the apoptotic process and indirect, through modulation of oxidative and nitrosative stress and inflammation (Liu et al., 2015, 2016; Zhang et al., 2015).

To our knowledge, the contribution of adrenal apoptosis to HS-associated CIRCI and the effects of etomidate in this process have not been described. Therefore, we conducted an experimental rat model of HS to evaluate several hypotheses: that HS is associated to development of adrenal gland apoptosis contributing to CIRCI; that etomidate significantly modulates adrenal apoptosis in HS; that etomidate will exert mainly an anti-apoptotic effect in the adrenal.

2. Clinical significance

- Etomidate is associated with development of CIRCI in critically ill patients although the clinical significance of this is still being debated.
- Few studies have addressed the development of etomidate-associated CIRCI in HS and none in an experimental model.
- This study demonstrated significant morbidity associated with etomidate-associated CIRCI in experimental HS.
- In addition, the study demonstrated for the first time that HS-associated adrenal apoptosis and necrosis were decreased in etomidate-treated groups.
- This finding launches the possibility of using etomidate therapeutically to decrease adrenal apoptosis and necrosis in HS.

3. Methods

All experiments were performed accordingly with the ethical standards of Faculdade de Medicina Veterinária da Universidade de Lisboa, in compliance with the Portuguese legislation for the use of animals for experimental purposes (Decreto-Lei nº 129/92 and Portaria nº 1005/92, DR nº 245, série I-B, 4930-42), and with the European Union legislation (EU Directive 2010/63/EU).

3.1. Experimental animals

Twelve-week-old male Wistar rats (*Rattus norvegicus*) (Charles Rivers, Barcelona, Spain), weighing 250–450 g were used. Animals were housed (*n* = 3 per cage) in a climate-controlled room under standard conditions (20–24°C; 12 h light/dark cycle). Water was provided *ad libitum* and food consisted of rat chow (Harlan ®). All animals were acclimatized for seven days before the experiments.

3.2. Study groups

Rats were randomly and blindly allocated to one of four body weight-matched groups: G0 (*n* = 16), G1 (*n* = 16), G2 (*n* = 16), and G3 (*n* = 16). Rats from G0 (the control group) were subjected to general anesthesia (GA), mechanical ventilation, and surgical intervention. Rats from G1 had the same procedures as G0, but also received buprenorphine. Rats from G2 had the same procedures as G1 but were subjected to HS. Rats from G3 had the same procedures as G2, but also received etomidate.

3.3. Experimental procedures

The experimental protocol is displayed in Figure 1.

3.4. Anesthesia and surgical procedure

Rats from all experimental groups except G0 were pre-medicated with 0.05 mg/kg buprenorphine (Budale®, Dechra, UK) to induce pre-emptive analgesia. Buprenorphine was injected subcutaneously (SC), 20 min prior induction of GA. GA was initiated by placing the animals in an induction chamber (World Precision Instruments, UK, Europe) previously saturated with 100% oxygen and 5% isoflurane (IsoFLo®, Abbott, USA). Once anesthetized, rats were moved and placed in dorsal recumbency over a water-based heated pad. Isoflurane anesthesia was maintained through face mask until a tracheostomy tube was placed. ECG was registered continuously through lead wire probes (ECG; ML136 Animal Bio Amp, ADInstruments, UK). A respiratory sensor was placed over the thoracic wall to measure respiratory rate. A rectal probe (MLT1403, ADInstruments, UK) and an oximetry tail sensor (ADInstruments, UK,) were placed to continuously record rectal temperature (kept between 35 and 38°C) and pulse oximetry, respectively. The pedal withdrawal reflex was used to help in

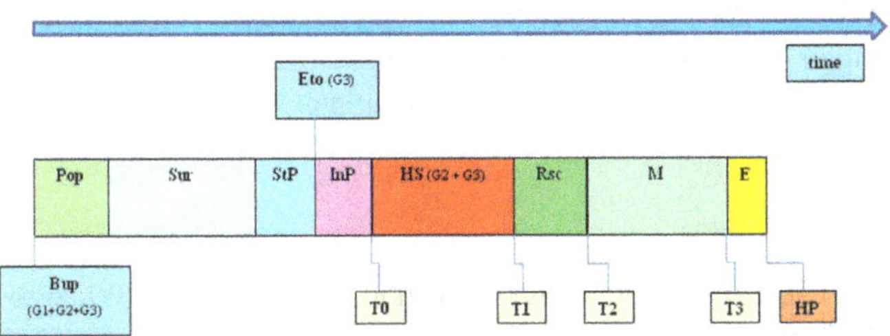

Figure 1. Experimental time line.

Notes: G0: control animals with general anesthesia, mechanical ventilation; G1: similar to G0 but with buprenorphine; G2: similar to G1 but subjected to shock; G3: similar to G2 but with etomidate administration. Pop-preoperative period; Sur, surgical intervention; StP, stabilization period; InP, intermediate period (in G3 only, lasting 5 min); HS, hemorrhagic shock; Rsc, resuscitation; M, maintenance stage; E, euthanasia; HP-histopathological analysis; Bup, buprenorphine; Eto, etomidate; Eto was only administered in G3. In G0 and G1, after StP, the animals went directly from M to E because they did not had the HS and Rsc stages.

assessing the depth of anesthesia. The anesthetic plane was considered adequate if a toe pinch did not induce the withdrawal reflex. An intravenous catheter (Introcan®, 26 Gauge, B. Braun Medical, Portugal) was placed in the left femoral vein for fluid and drug administration. Another catheter was placed in the right external carotid artery and connected to a three-way stopcock for blood sampling and arterial blood pressure measurement. The carotid catheter and the three-way stopcock were connected to a fluid-filled pressure transducer (MLT844, ADInstruments, UK) which in turn was connected to blood pressure amplifiers (ML221 Bridge Amp, ADInstruments, UK). Arterial blood pressure [systolic, diastolic and mean arterial pressure (MAP)], ECG, rectal temperature, pulse oximetry, and respiratory frequency data were transmitted to a data acquisition unit (PowerLab®, ADInstruments, UK), to be analyzed by a specific software (LabChart Pro®, ADInstruments, UK).

After placement of a carotid catheter, the tracheostomy tube was placed. Volume-controlled ventilation was initiated with a small animal ventilator (CWE Small Animal Ventilator, World Precision Instruments, UK). General anesthesia was maintained with isoflurane at concentrations of 1.5–2%, administered through the tracheostomy tube. Oxygen was also delivered by the tracheostomy tube. The inspiratory fraction of O_2 (FiO_2) was kept at 100% until the end of the experiment. Initial ventilator settings consisted in a tidal volume of 10 ml/kg and a respiratory rate of 110–125 breaths/min. These parameters were adjusted as the experiment progressed, accordingly to the results of arterial blood gas analysis, with the aim of maintaining normocapnia (35–45 mm Hg) and normoxemia [arterial pressure of O_2 (PO_2)>85 mm Hg]. Once the surgical procedure was completed, a stabilization period of 10 min (15 min in G2) was allowed to happen before inducing HS in G2 and G3. The purpose of the stabilization period was to permit the recovery of the hemodynamic instability caused by initiation of mechanical ventilation. In G3, the stabilization period was followed by the intermediate period of 5 min in order that HS was not initiated immediately after etomidate injection. At T3 and after blood sampling, euthanasia was performed by pentobarbital administration (Eutasil®, Ceva, Portugal) at 100–150 $mg.kg^{-1}$ through the femoral catheter.

3.5. Sample collection and anesthesia monitoring

At four time points (T0, T1, T2, and T3), a sample of blood (500 μl) was collected for analysis. To prevent activation of the hypothalamic-pituitary axis following blood collection 1 ml of normal saline was administered IV following each blood sampling, except at T0 in G2 and G3, when blood collection coincided with HS induction. At T0, T1, T2, and T3, MAP, heart rate (HR), temperature, ECG, ventilator parameters, and depth of anesthesia were recorded. T0, the initial time point, was set at the end of the stabilization period (intermediate period in G3). T1, T2, and T3 were set at 90, 150, and 240 min post-T0, respectively. HR and MAP collected at each time point were used for analysis and constituted the hemodynamic variables.

3.6. Immunological and hormonal variables

To determine hormone and cytokine levels, 200 μl of the collected blood were placed in sterile heparin-coated tubes (FactorMed, Portugal) and centrifuged (12,000 rpm for 15 min) to obtain plasma. Plasma was then stored at −20°C until further analysis. ACTH, CS (rat stress hormone panel Millipore, Arium Laboratórios, Portugal) TNF-α, IL6, and IL10 (Rat Cytokine/Chemokine, Millipore, Arium Laboratórios, Portugal) plasma levels were determined by Multiplex/Luminex technology (Vignali, 2000).

3.7. Metabolic variables

At all time points, 100 μl of blood was used to measure metabolic variables, including pH, PO_2, arterial pressure of CO_2 (PCO_2), BE, HCO_3, and lactate, through a portable blood gas analyzer (I-STAT Analyzer 300, I-STAT Corporation, Abbot, USA). The arterial pressure of O_2/Inspiratory Fraction of O_2 (PO_2/FiO_2) ratio was calculated as indicator of lung dysfunction.

3.8. HS induction and resuscitation

In G2 and G3, at T0 and after blood sampling, HS was induced by collecting 30% of blood volume (estimated as 54 ml/kg) (Rushing et al., 2006) from the carotid artery. Blood collection was divided into two stages. In the first stage, 15% of blood volume was collected in 10 min. The remaining 15% was collected on the following 20 min. We divided blood collection in two stages of different velocities, to induce a more natural model of hypovolemia. Hemorrhage was stopped when 30% of the estimated blood volume was removed and/or when MAP reached and stabilized at 45 mm Hg. The blood collected was placed into 1 ml sterile aliquots previously filled with 0.12 ml of sodium citrate until it was used in the resuscitation phase.

Resuscitation was initiated at T1, after blood sampling for analysis. It was performed by administrating the blood which was removed to induce HS added to normal saline in a 1:3 ratio, respectively. Both fluids were warmed at body temperature and administered by a syringe pump (Perfusor® fm (MFC), B. Braun Medical, Portugal) through the femoral vein catheter. Resuscitation lasted 30 min and once completed, a constant-rate infusion of warmed normal saline was administered at maintenance rate (2 ml/kg/h) until the end of the experiment.

3.9. Etomidate administration

In G3, etomidate (Etomidate®, Lipuro, 2 mg/ml, B. Braun Medical, Portugal) was administered by bolus at 1 mg/kg, IV, 5 min before T0. The total volume of anesthetic solution was 0.5 ml/kg for each animal. In the other groups, the same amount of normal saline was given instead of etomidate.

3.10. Histopathological variables

3.10.1. Adrenal gland necrosis assessment

After euthanasia both adrenals were collected and placed in 1% formaldehyde. Several 10 μm-thick tissue sections were than obtained and stained with hematoxylin and eosin. Two slides of each adrenal gland were evaluated by a pathologist blinded to the study, which looked for changes such as tissue edema, lipid vacuolization, and inflammatory cell infiltration. The degree of adrenal necrosis was quantified in a score adapted from a score used in other study (Rushing et al., 2006). In our scoring system, necrosis was quantified in five categories depending on the percentage of gland affected. Thus, score 0 represented a complete absence of necrosis and scores 1, 2, 3, and 4 corresponded to 25%, 25–50%, 50–75%, and more than 75% of the gland being affected by necrosis, respectively. The necrosis score was first determined in each of the two original slides of the adrenal gland. This first score was used to determine the average necrosis score of each gland. The scores of the two glands were used to calculate the average necrosis score of each animal. A final average necrosis score for each group calculated the necrosis scores of all animals belonging to the correspondent group.

3.10.2. TUNEL assay

TUNEL staining was performed according to the manufacturer's instructions (ApopTag® Plus Peroxidase In Situ Apoptosis Kit, #S7101, Merck KGaA, Darmstadt, Germany). Tissue specimens were examined using an Axioskop bright-field microscope (Carl Zeiss GmbH, Gena, Germany). Tissue sections 10 μm thick were deparaffinized in three changes of xylene, and next hydrated with two changes of absolute ethanol, followed by three consecutive washes in 95%, in 70% ethanol, and in phosphate buffer saline (PBS), 5 min per wash. Then, samples were incubated with freshly diluted proteinase K (20 μg/ml) in PBS, for 15 min at room temperature, followed by two washes in ddH2O, for 2 min also at room temperature. In the following step samples were treated with 3% hydrogen peroxide to quench endogenous peroxidase activity. After adding the equilibration buffer, sections were treated with terminal deoxynucleotidyltransferase for 60 min at 37°C. This was followed by incubation with anti-digoxigenin peroxidase conjugate for 30 min at 37°C, staining with 3, 3′-diaminobenzidine substrate, and counterstaining with 0.5% methyl green. Finally, slides were rinsed, dehydrated, and mounted. A negative control was prepared by omitting the TdT enzyme to control for non-specific incorporation of nucleotides or binding of enzyme-conjugate.

The slides were observed and photo documented using an Olympus BH-2 microscope. To quantify TUNEL-positive nuclei, a total of 12 high-power fields (x400) were randomly examined for each animal, six from each adrenal gland (three fields from the adrenal cortex and three from the adrenal medulla). The apoptotic index (in percentage) was obtained from the ratio between the number of apoptotic and total number of cells counted in each field, as described by others (Yu et al., 2012). The cell count in each field was determined using Image J software (NIH, http://rsb.info.nih.gov/ij/), especially adapted for this purpose. The apoptotic index obtained in each field was used to calculate the average cortical and medullary apoptotic index of each adrenal gland. The average cortical and medullary apoptotic indexes of each animal were subsequently obtained from the results of the two adrenal glands . Finally, the average cortical and medullary apoptotic index for each group was calculated from the values of all animals belonging to each specific group. These were then compared through statistically methods.

3.10.3. Caspase-3 immunofluorescence microscopy analysis

Apoptosis was confirmed by determining the presence of active caspase-3 (Porter & Jänicke, 1999; Yu et al., 2012). For this assessment, 10 μm thick paraffin-embedded adrenal sections were deparaffinized, rehydrated, and boiled three times in 10 mM citrate buffer, pH 6. Sections were then incubated for 60 min in blocking buffer, containing 10% (v/v) normal donkey serum (Jackson ImmunoResearch Laboratories Inc., West Grove, PA, USA) in PBS with 0.1% (v/v) Triton X-100 (Sigma-Aldrich), and subsequently in diluted primary antibodies overnight at 4°C. After rinsing, the primary antibody was developed by incubating with DyLight 488- (Jackson ImmunoResearch) or Alexa Fluor 594- (Invitrogen, Grand Island, NY, USA) conjugated secondary antibodies against the corresponding species, for 2 h at room temperature. To confirm the apoptotic phenotype, we used a rabbit polyclonal active caspase-3 antibody (R&D Systems, Lille, France; 1:100). No staining was observed in control sections, where primary antibody was replaced by blocking buffer. Immunofluorescence analysis was performed using an epifluorescence microscope (Leica DM R HC model, Wetzlar, Germany). The data-sets were acquired by Adobe Photoshop CS5 software (Adobe Systems, Inc., San Jose, USA) and images were subsequently processed with Image J open source software (version 1.46r).

3.11. Statistical Analysis

Statistical analyses were performed with the Statistica software, version 8.0 (Statsoft Ibérica, Lisboa, Portugal), using a mixed linear model. Data were expressed as mean and standard error. For each variable, a two-way ANOVA was used where variance components were obtained through the Variance Estimation and Precision (VEPAC) Statistica module. Variance components in the model were estimated by Restricted Maximum Likelihood (REML) estimation. Least Squares means and standard errors for Least square means were computed from the solution of the mixed model equations. Correlation analysis between variables was performed with the Spearman rank correlation, considering the complete set of four time points. In all analysis, statistical significance was set at a $p < 0.05$.

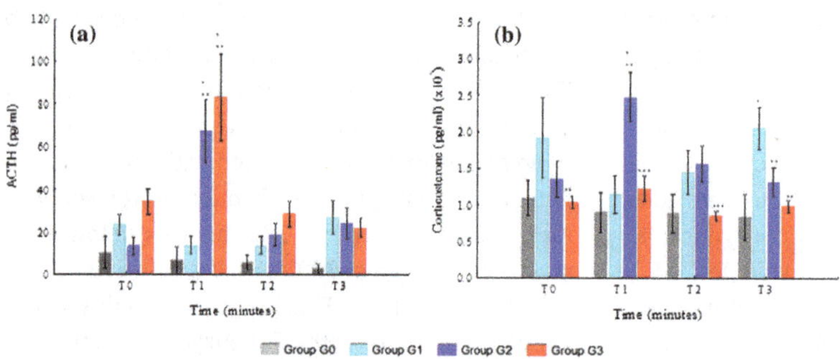

Figure 2. Variation of hormonal variables from T0 to T3 in G0, G1, G2, and G3: (a) ACTH and (b) CS.

Notes: G0: control animals with general anesthesia and mechanical ventilation only; G1: similar to G0 but with buprenorphine; G2: similar to G1 but subjected to shock; G3: similar to G2 but with etomidate administration. Data expressed as mean ± SE. * indicates statistically significant differences between G1, G2, or G3 with indicates statistically significant differences between G2 and G3 with G1; *** indicates statistically significant differences between G3 and G2. Statistical significance was established at *p*-value < 0.05. ACTH, adrenocorticotropic hormone; CS, corticosterone.

4. Results

Data were obtained in all animals of G0 and G1. Three animals of G2 and two animals of G3 were excluded due to insufficient data.

4.1. Hormonal variables

The statistical analysis is shown in Figure 2 and Supplementary Table 1. In both HS groups, ACTH levels reached its highest peak at T1. The G3's value of ACTH at T1 was the highest of all groups and time points. ACTH levels were always higher in G3 than in G2 except at T3. ACTH levels were higher in G2 than in G1 at T1 and T2, but not at T0 and T3. G0's ACTH levels were the lowest of all groups. The statistically significant differences between groups were between G2 and G3 with G1 at T1 (G2 with G1, $p = 0.001$; G3 with G1, $p = 0.000$) and between G2 and G3 with G0 at T1 (G2 with G0, $p = 0.001$; G3 with G0, $p = 0.000$). The interaction between ACTH and time was also statistically significant ($p = 0.006$).

The highest levels of CS were observed in G2 at T1. CS levels were lower in G3 than in G2 at all time points, although this difference was only statistically significant at T1 ($p = 0.000$) and T2 ($p = 0.031$). CS levels were also lower in G3 than in G1 at T0, T2, and T3. However, this was only statistically significant at T0 ($p = 0.003$) and T3 ($p = 0.001$). The levels of CS in G3 and G0 were similar and not statistically different. In G3, CS levels were lower than10000 pg/ml at T0, T1, and T2.

The levels of CS were higher in G2 than in G1 at T1 and T2, but not at T0 and T3. These differences were only statistically significant at T1 ($p = 0.000$) and T3 ($p = 0.046$). The levels of CS in G2 were always higher than in G0, although none of the differences reached statistically significance. The levels of CS in G1 were always higher than in G0, but only at T3 the difference was statistically significant ($p = 0.002$). The interaction between CS and time was statistically significant ($p = 0.015$).

4.2. Immunological variables

Immunological variables are displayed in Figure 3 and Supplementary Table 2. TNF-α was only detectable in G3. In this group, it increased from T0 to T2 and then decreased at T3. The differences between G3 with other groups were only statistically significant at T2 (G3 with G0, $p = 0.017$; G3 with G1, $p = 0.008$; G3 with G2, $p = 0.012$) and T3 (G3 with G0, $p = 0.040$; G3 with G1, $p = 0.022$; G3 with G2, $p = 0.030$). The differences between G0, G1, and G2 were not statistically significant.

G3 had always higher levels of IL6 than other groups. These differences were only statistically significant at T3 (G3 with G0, $p = 0.004$; G3 with G1, $p = 0.000$; G3 with G2, $p = 0.002$). In G3, the levels of IL6 increased from T0 to T3, which contrasts with the other groups, where they remained stable. The levels of IL6 of G2 were always higher than those of G0 and G1. However none of the differences between G2, G1, and G0 were found to be statistically significant.

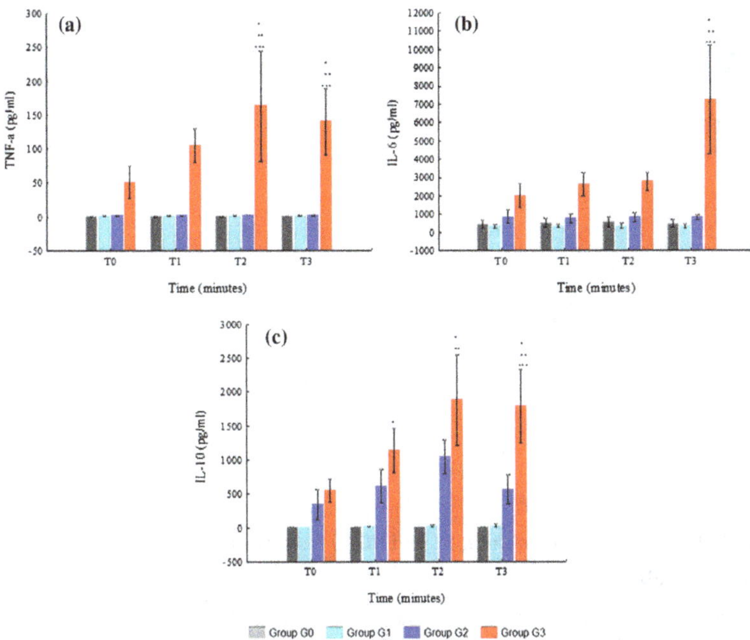

Figure 3. Immunological variables: (a) TNF-α; (b) IL6 and (c) IL10.

Notes: G0: control animals with general anesthesia and mechanical ventilation only; G1: similar to G0 but with buprenorphine; G2: similar to G1 but subjected to shock; G3: similar to G2 but with etomidate administration. Data expressed as mean ± SE. * indicates statistically significant differences between G1, G2, or G3 with G0; ** indicates statistically significant differences between G2 and G3 with G1; *** indicates statistically significant differences between G3 and G2.Statistical significance was est *p*-value < 0.05. TNF-α, Tumor Necrosis Factor-α; IL6, Interleukin 6; IL10, Interleukin 10.

IL10 levels were only detectable in G2 and G3. Its highest level was observed in G3 at T2. The levels of IL10 were always higher in G3 than in G2, although the only difference which was statistically significant was at T3 ($p = 0.004$). The levels of IL10 in G3 also differed significantly from those in G1 at T2 ($p = 0.001$) and T3 ($p = 0.002$) and from those in G0 at T1 ($p = 0.051$), T2 ($p = 0.003$), and T3 ($p = 0.005$). The differences between G2, G1, and G0 were not statistically significant.

ablished at

4.3. Metabolic variables

In metabolic variables the groups differed significantly at several time points (Figure 4 and 5 and Supplementary Table 3).

The pH of G2 and G3 was always lower than G1 and G0 at all time points. G3 and G1 were the group with the lowest and highest pH levels, respectively.

The levels of lactate increased after hemorrhage, decreased after resuscitation and increased from T2 to T3 in both G2 and G3. Lactate levels were always higher in G3 than in G2 although the differences were not statistically significant. The levels of lactate in G3 were also higher than in G1 and G0 at T1, T2, and T3.

BE levels decreased markedly with hemorrhage, then improved slightly with resuscitation and finally decreased from T2 to T3, in both G2 and G3. G1 was the group which presented the highest BE levels. BE was always lower in G3 than in G2 and G1. The interaction between time and BE was statistically significant ($p = 0.016$).The results of HCO_3 were similar to BE. G1 had the highest HCO_3 levels of all groups. HCO_3 levels were lower in G3 than in G2 at all time points.

G3 animals had the lowest PO_2 levels of all groups. In G3, PO_2 levels did not significantly change with hemorrhage, but then decreased continuously after resuscitation until T3. G2's PO_2 levels

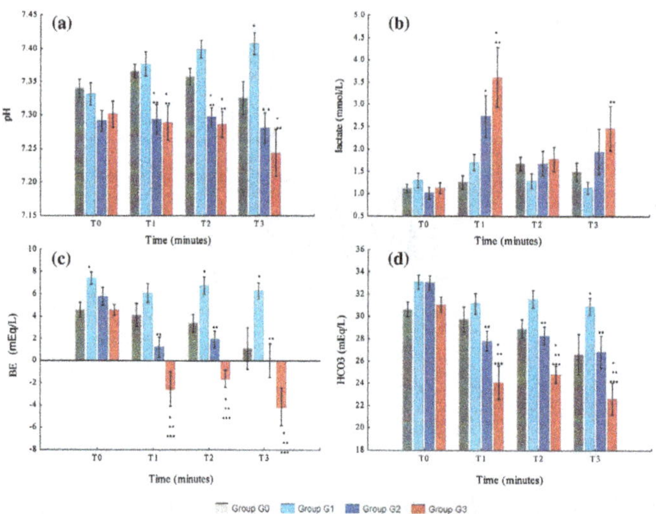

Figure 4. Variation of metabolic variables (pH, lactate, BE and HCO_3) from T0 to T3 in G0, G1, G2, and G3: (a) pH; (b) lactate; (c) BE and (d) HCO_3.

Notes: G0: control animals with general anesthesia and mechanical ventilation only; similar to G0 but with buprenorphine; G2: similar to G1 but subjected to shock; G3: similar to G2 but with etomidate administration. Data expressed as mean ± SE. * indicates statistically significant differences between G1, G2, or G3 with G0; ** indicates statistically significant differences between G2 and G3 with G1; *** indicates statistically significant differences between G3 and G2.Statistical significance was established at p-value < 0.05. BE, base excess.

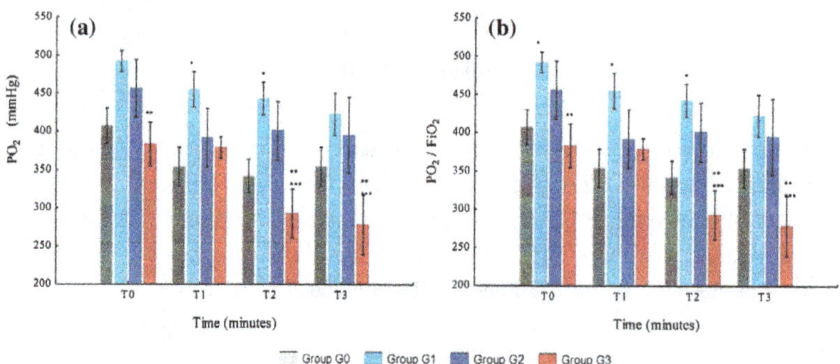

Figure 5. Variation of metabolic variables (PO_2 and PO_2/FiO_2) from T0 to T3 in G0, G1, G2, and G3: (a) PO_2 and (b) PO_2/FiO_2.

Notes: G0: control animals with general anesthesia and mechanical ventilation only; similar to G0 but with buprenorphine; G2: similar to G1 but subjected to shock; G3: similar to G2 but with etomidate administration. Data expressed as mean ± SE. * indicates statistically significant differences between G1, G2, or G3 with G0; ** indicates statistically significant differences between G2 and G3 with G1; *** indicates statistically significant differences between G3 and G2.Statistical significance was established at p-value < 0.05.

decreased especially from T0 to T1, but contrarily to G3, they remained stable from T1 to T3. The PO_2/FiO_2 ratio in G3 decreased progressively from T0 to T3 and reached levels below 300 at T2 and T3.

4.4. Hemodynamic variables

Hemodynamic variables are shown in Figure 6 and in Supplementary Table 4.

G3 animals had always lower HR than other groups. In contrast, G2 was the group with the highest HR levels. In G3, HR was stable from T0 to T1 and then decreased continuously until T3. In G2, HR was highest at T0 and then decreased continuously until T3. None of the differences regarding HR between the studied groups were statistically significant, except the ones between G3 and G2 at T2 (p = 0.0.037) and T3 (p = 0.003).

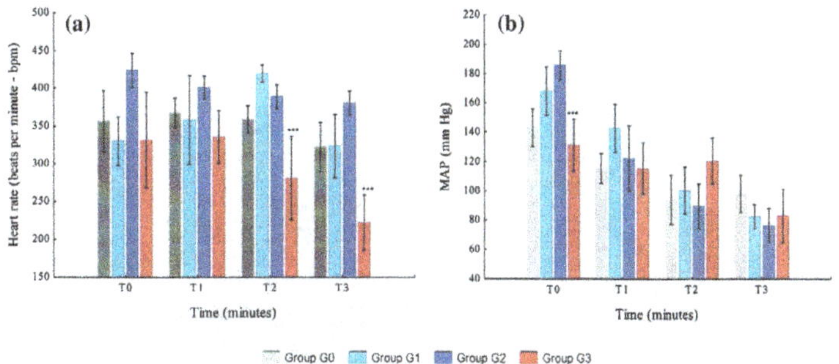

Figure 6. Variation of hemodynamic variables from T0 to T3 in G0, G1, G2, and G3: (a) HR and (b) MAP.

Notes: G0: control animals with general anesthesia and mechanical ventilation only; similar to G0 but with buprenorphine; G2: similar to G1 but subjected to shock; G3: similar to G2 but with etomidate administration. Data expressed as mean ± SE. * indicates statistically significant differences between G1, G2, or G3 with G0; ** indicates statistically significant differences between G2 and G3 with G1; *** indicates statistically significant differences between G3 and G2.Statistical significance was established at p-value < 0.05. HR, heart rate; MAP, mean arterial pressure.

G3's MAP decreased slightly with hemorrhage, increased mildly at T2 and decreased from T2 to T3. At T2, G3 was the group which presented the highest MAP levels. In both G1 and G2, MAP decreased continuously from T0 to T3. The difference between G2 and G3 at T0 was the only one between all groups which was statistically significant ($p = 0.014$).

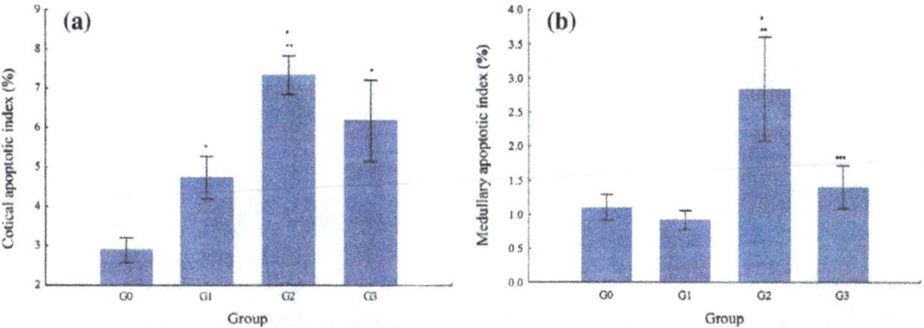

Figure 7. Variation of the apoptotic index G0, G1, G2, and G3. Apoptosis score in G0, G1, G2, and G3: (a) apoptosis score in the adrenal cortex and (b) apoptosis score in the adrenal medulla.

Notes: G0: control animals with general anesthesia and mechanical ventilation only; similar to G0 but with buprenorphine; G2: similar to G1 but subjected to shock; G3: similar to G2 but with etomidate administration. Data expressed as mean ± SE. * indicates statistically significant differences between G1, G2, or G3 with G0; ** indicates statistically significant differences between G2 and G3 with G1; *** indicates statistically significant differences between G3 and G2.Statistical significance was established at p-value < 0.05.

4.5. Histopathological variables

4.5.1. Apoptosis

Cortical and medullary apoptosis statistical analysis is described in Supplementary Table 5 and Figures 7. Figures 8–11 display the presence of apoptosis in the adrenal cortex and medulla. In the cortex, TUNEL positive nuclei were more frequently found in *Zona reticularis* than in *Zona fasciculata* and were infrequently found in Zona *glomerulosa* (Figure 8–10).The statistical analysis of the necrosis score is described in Supplementary Table 5 and Figure 12.

In all groups, apoptotic index was higher in the cortex than in medulla. G2 had the highest cortical apoptotic index, followed by G3, G1, and G0. G2 had also the highest medullary apoptotic index followed by G3, G0, and G1.

Regarding the statistical analysis of cortical apoptotic index, the differences between G0 with G1, G2, and G3 were statistically significant, with the level of significance being $p = 0.039$, $p = 0.000$, and

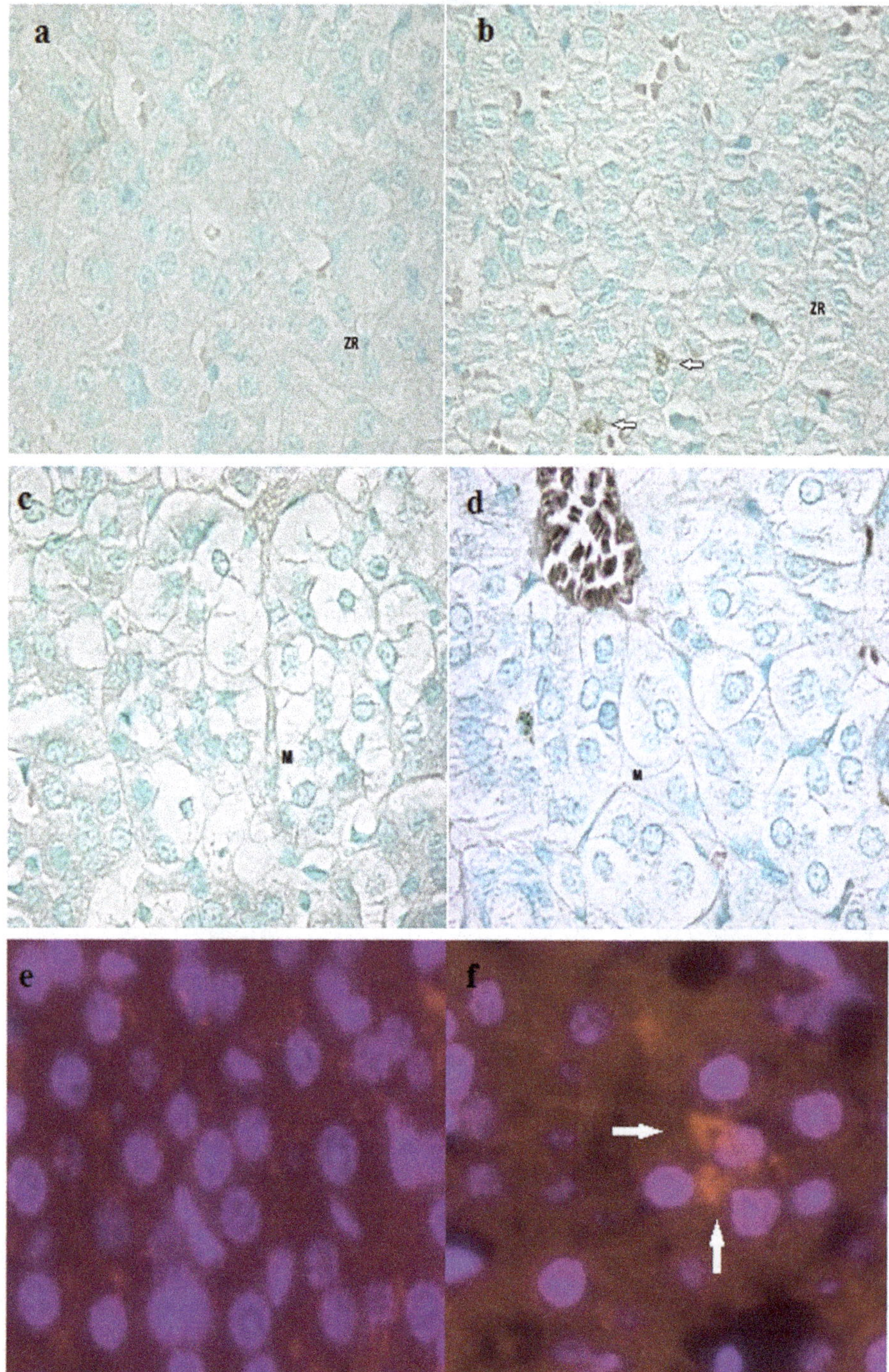

Figure 8. Pictures of TUNEL positive nuclei in the adrenal cortex and medulla and caspase-3 immunofluorescence analysis in G0 and G1.

Notes: First row corresponds to sections of adrenal cortex; second row corresponds to sections of adrenal medulla; third row corresponds to pictures of caspase-3 immunofluorescence analysis in the adrenal cortex. The column of the left corresponds to G0 and the column of the right to G1. Pictures were taken at 400x magnification. (a) TUNEL in adrenal cortex of G0; (b) TUNEL in adrenal cortex of G1; (c) TUNEL in adrenal medulla of G0; (d) TUNEL in adrenal medulla of G1; (e) caspase-3 immunofluorescent analysis in G0; (f) caspase-3 immunofluorescent analysis in G1. Small white arrows in b indicate adrenal cortical cells from G1which stained TUNEL positive. Small white arrows in f indicate caspase-3 positive areas in adrenal cortical cells from G1, which appear with bright red. ZR, *zona reticularis*; M, adrenal medulla.

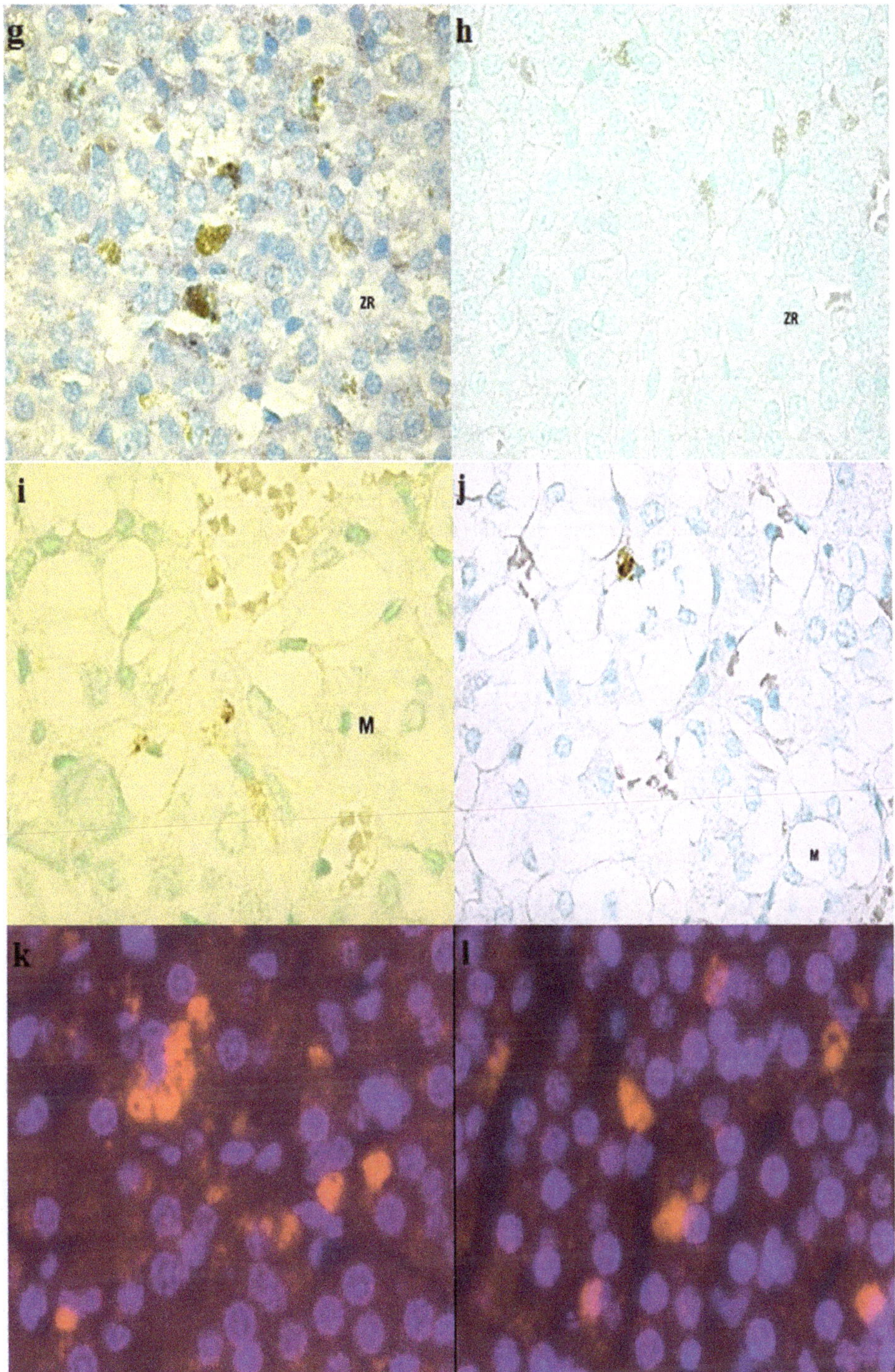

Figure 9. Pictures of TUNEL positive nuclei in the adrenal cortex and medulla and caspase-3 immunofluorescence analysis in G2 and G3.

Notes: First row corresponds to sections of adrenal cortex; second row corresponds to sections of adrenal medulla; third row corresponds to pictures of caspase-3 immunofluorescence analysis in the adrenal cortex. The column of the left corresponds to G2 and the column of the right to G3. Pictures were taken at 400x magnification. (g) TUNEL in adrenal cortex of G2; (h) TUNEL in adrenal cortex of G3; (i)TUNEL in adrenal medulla of G2; (j) TUNEL in adrenal medulla of G3; (k) caspase-3 immunofluorescent analysis in G2; (l) caspase-3 immunofluorescent analysis in G3. ZR, *zona reticularis*; M, medulla.

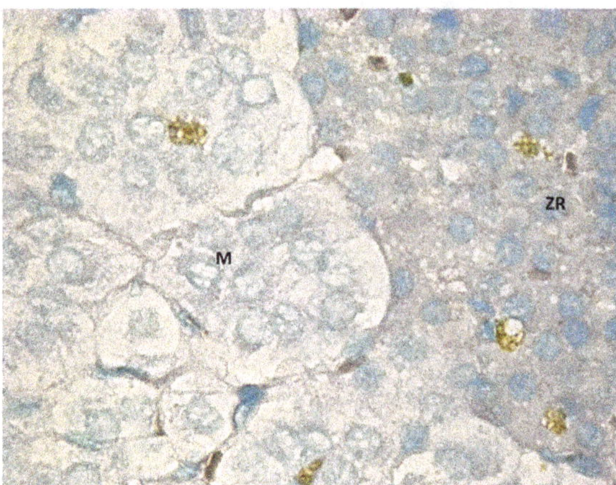

Figure 10. TUNEL positive cells in both adrenal cortex (*zona reticularis*) and medulla from an animal of G2.

Notes: The picture shows that the adrenal cortex had a higher number of TUNEL positive cells than the medulla. Picture take at a 400x magnification ZR, *zona reticularis*; M, medulla.

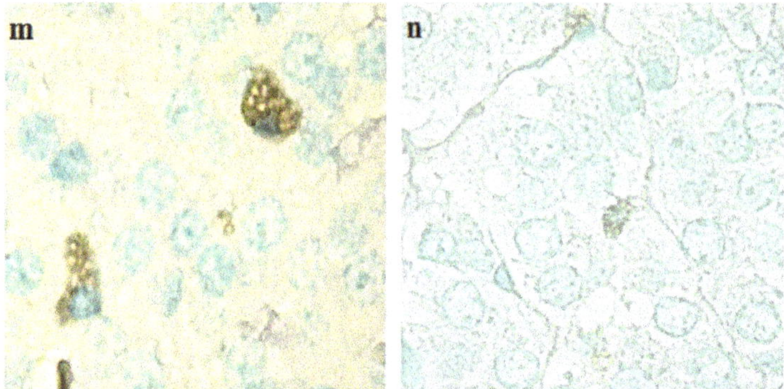

Figure 11. TUNEL positive adrenal cells in high resolution. TUNEL positive adrenal cells in high resolution; (m) *zona reticularis* of an animal of G2 and (n) adrenal medulla from an animal of G3.

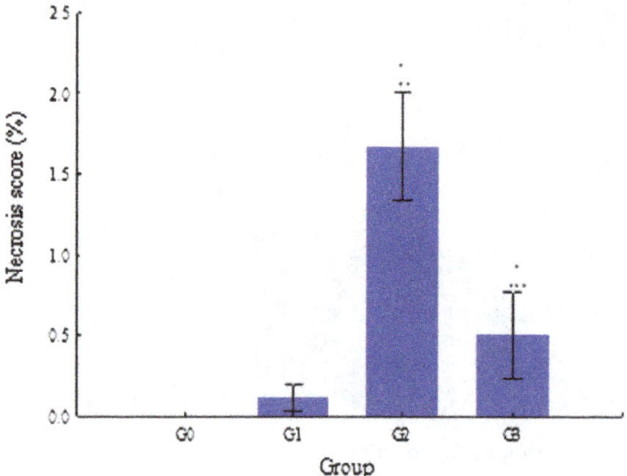

Figure 12. Variation of the necrosis score in G0, G1, G2, and G3.

Notes: G0: control animals with general anesthesia and mechanical ventilation only; similar to G0 but with buprenorphine; G2: similar to G1 but subjected to shock; G3: similar to G2 but with etomidate administration. Data expressed as mean ± SE. * indicates statistically significant differences between G1, G2, or G3 with G0; ** indicates statistically significant differences between G2 and G3 with G1; *** indicates statistically significant differences between G3 and G2.Statistical significance was established at *p*-value < 0.05.

$p = 0.000$, respectively. G2 also differed significantly from G1 ($p = 0.008$). The difference between G2 and G3 was not statistically significant.

The statistical analysis of the medullary apoptotic index revealed that the one of G2 was significantly different from the one of G0 ($p = 0.000$), G1 ($p = 0.000$), and G3 ($p = 0.000$). The differences between G0, G1, and G3 were not statistically significant.

G2 and G3 were the only groups where adrenal necrosis was found (Figure 12). Necrosis was highest in G2, where it reached an average score of 2. The necrosis score of G2 differed significantly from the one of G0 ($p = 0.000$), G1 ($p = 0.000$), and G3 ($p = 0.000$). The necrosis score also differed significantly between G3 and G0 ($p = 0.034$).

5. Discussion

5.1. Buprenorphine was associated to increased apoptotic rate

In this study, animals of G1 had a higher rate of adrenocortical apoptosis than animals of G0. To our knowledge, no study has described a pro-apoptotic effect of buprenorphine (and of any other opioid) in adrenocortical cells. The underlying mechanisms of buprenorphine's pro-apoptotic effects could not be determined in this study. However, based in what has been described in nerve and cancer cells, buprenorphine could have promoted apoptosis by increasing mitochondrial permeability and cytochrome c release, by increasing caspase-3 activity and/or by promoting the expression of genes associated with the polyubiquitination of apoptotic proteins (Kugawa, Arae, Ueno, & Aoki, 1998; Kugawa, Nakamura, Ueno, & Aoki, 2004). In addition, because G1 animals had increased CS levels, buprenorphine could have increased apoptosis indirectly by increasing glucocorticoids levels as these are known to increase adrenal cell apoptosis (Almeida, Matos, & Neves, 2007).

5.2. HS increases adrenal gland apoptosis

As expected, HS was associated to increased development of adrenal apoptosis. An increase in adrenal apoptosis has been described in several critical illnesses including trauma (Didenko, Wang, Yang, & Hornsby, 1996), ANP (Yu et al., 2012, 2016), and sepsis (Flierl et al., 2008; Kanczkowski, Chatzigeorgiou, Grossklaus, et al., 2013, Kanczkowski, Chatzigeorgiou, Samus, et al., 2013; Liu et al., 2016; Wang et al., 2015). In rat models of ANP (Yu et al., 2012, 2016) and sepsis (Kanczkowski, Chatzigeorgiou, Grossklaus, et al. 2013; Liu et al., 2016), adrenal apoptosis was also associated to decreased adrenal function. In contrast, although increased adrenal apoptosis was described in HS this was not associated to adrenal dysfunction (Rushing & Britt, 2007).

G2 and G3 had several conditions which could have increased the risk of adrenal apoptosis. This include the pro-inflammatory state associated to HS as observed in our study: higher cytokine levels; combination of decreased PO_2 and decreased blood pressure, higher lactate and lower BE, HCO_3, and pH levels (decreased tissue oxygen delivery); increased risk of developing ischemic/reperfusion (I/R) injury due to resuscitation and, in the case of G2, the increased CS levels (Papathanassoglou, Moynihan, & Ackerman, 2000).

In this study, adrenal apoptosis was apparent just 4 h after the induction of a mild state of HS. These results suggest that in this pathologic condition, adrenal apoptosis can develop very early in the disease process and even after a mild insult. If this reflects a specific vulnerability of the adrenal gland to the pro-apoptotic effects of HS or any other reason remains unknown, and to our knowledge, it has never been reported before.

Etomidate-treated animals had a lower apoptotic index than G2, despite having higher TNF-α levels, lower PO_2, lower blood pressure, and worse indicators of tissue perfusion following HS. The reasons for these findings were not determined. We hypothesize that this could have been due to two possible mechanisms: a direct anti-apoptotic effect of etomidate; an indirect effect, where the decreased rate of apoptosis was due to etomidate's effects in other mediators, such as cytokines and hormones.

5.3. Etomidate's direct anti-apoptotic effects

Studies have shown that etomidate can be both pro- and anti-apoptotic (Roesslein et al., 2008; Wang et al., 2014; Wu et al., 2011; Xu et al., 2014; Zhang, Xiong, Jiao, Wang, & Zuo, 2010). In addition, recent studies demonstrated that etomidate influences the development of adrenal apoptosis in both normal and septic animals in a time- and dose-dependent manner (Liu et al., 2015, 2016; Zhang et al., 2015). In septic animals when etomidate was administered by constant-rate infusion (2 mg/kg/h) for 2 h or by IV bolus (2 mg/kg), it increased adrenal apoptosis (Liu et al., 2016; Zhang et al., 2015). In contrast, when administered by IV bolus at 0.6 mg/kg, immediately after the induction of sepsis, it prevented apoptosis (Liu et al., 2016; Zhang et al., 2015). The results in normal animals paralleled those of septic animals. For example, when etomidate was administrated by constant-rate infusion (1.2 mg/kg/h) for 6 h, it increased adrenal apoptosis in normal rats (Liu et al., 2015). However, if the constant-rate infusion of etomidate was preceded by its administration as a bolus of 0.6 mg/kg, than apoptosis was prevented. The results of these studies seem to suggest that etomidate's anti-apoptotic effects are manifest when the drug is administered at lower doses and before or very shortly after the beginning of the pathologic process. This is compatible with some sort of preconditioning effect induced by etomidate. In fact a preconditioning effect by etomidate to prevent apoptosis has been reported before and seems to involve the activation of the mitochondrial ATP-sensitive potassium channel (Zhang et al., 2010).

Etomidate was shown to directly affect apoptosis by several mechanisms. It can interfere directly with cytochrome c levels, increase the opening of the mitochondrial permeability transition pore and modulate the activity of several intracellular transcription factors involved in apoptosis (Liu et al., 2015; Wu et al., 2011; Zhang et al., 2010).

Alternatively etomidate can also affect apoptosis by its direct pro-oxidant effects (Wang et al., 2014; Yagmurdur et al., 2004). It also has direct activity upon intracellular anti-oxidant systems such as superoxide dismutase, catalase, and glutathione-peroxidase and modulates directly the inflammatory cascade (Liu et al., 2016; Zhang et al., 2010, 2015), including by a direct interference in the NF-kB pathway (Zhang et al., 2015). Studies have shown that etomidate's effects upon oxidative stress and inflammation are also dose- and time-dependent and can also be pro or anti-apoptotic (Liu et al., 2015, 2016; Zhang et al., 2015). Furthermore, by attenuating I/R injury (Ergün, Darendeli, Imrek, Kilinç, & Oksüz, 2010; Yu et al., 2010), etomidate can also prevent apoptosis.

We believe that using etomidate at the doses of 1 mg/kg, immediately before inducing HS, we were able to prevent adrenal apoptosis in a similar manner to what was described in septic models. We hypothesize that the mechanisms involved could have included a direct preconditioning effect of etomidate or its effects in oxidative stress, I/R injury, and inflammation-related intracellular pathways. However, because the pathophysiology of HS is different from sepsis and our dose was higher than the one used in septic models, this hypothesis needs to be confirmed in future studies.

5.4. Etomidate's indirect effects in adrenal apoptosis

5.4.1. Could have decreased CS levels contributed to the lower apoptotic rate of etomidate-treated animals?

As expected, etomidate-treated animals had lower levels of CS, because etomidate inhibits glucocorticoid adrenal production (Crozier, Beck, Schlaeger, Wuttke, & Kettler, 1987). We hypothesize that the decreased CS levels might have contributed to the decrease in adrenal apoptosis. In fact it has been reported that glucocorticoids augment adrenal apoptosis by increasing the levels of Bcl-2 and caspase-3 (Almeida et al., 2007) and especially in *zona reticularis*, which was the mostly affected by apoptosis in our study. Besides increased glucocorticoid levels were associated to increased parenchymal cell apoptosis in burns (Fukuzuka et al., 1999) and sepsis (Kamiyama et al., 2008).

5.4.2. Did increased ACTH levels contributed to the lower apoptotic rate of etomidate-treated animals?

ACTH levels were higher in etomidate-treated animals, possibly due to the decreased CS levels (Preziosi & Vacca, 1982), increased TNF-α and IL6 levels (John & Buckingham, 2003), and the higher degree of hypotension at T0 and T1 (Turnbull & Rivier, 1999). We believe that the higher ACTH levels could have contributed to the lower apoptotic rate of etomidate-treated animals. It is known that ACTH has anti-apoptotic effects in adrenal cells (Carsia, Macdonald, Gibney, Tilly, & Tilly, 1996; Keramidas, Feige, & Thomas, 2004), which are especially manifest in *zona reticularis* and *zona fasciculata* (Almeida, Matos, Ferreira, & Neves, 2006; Almeida et al., 2007; Carsia et al., 1996).

How ACTH prevents apoptosis is poorly understood. In part it results from its activation of the JAK/ERK/STAT pathway after ligating to the melanocortin receptors (Ottani et al., 2013; Si et al., 2013). In addition, ACTH activates the vagus nerve-mediated brain cholinergic anti-inflammatory pathway (Guarini et al., 2004). Through this way ACTH prevents vascular dysfunction (Bertuglia & Giusti, 2004), decreases leukocyte adhesion and its infiltration into tissues (Bertuglia & Giusti, 2004; Guarini et al., 1996; Squadrito et al., 1999) and inhibits the inflammatory cascade (Bazzani, Bertolini, & Guarini, 1997; Guarini et al., 2004; Squadrito et al., 1999) and ROS production (Guarini et al., 1996).

5.4.3. Did increased IL6 and IL10 levels contributed to the lower apoptotic rate of etomidate-treated animals?

Etomidate-treated animals had higher levels of TNF-α, IL6, and IL10 than other groups. Increased cytokine levels following etomidate administration have been previously described in rats (Félix et al., 2016; Pejo et al., 2012) and Humans (den Brinker et al., 2005; Jameson, Desborough, Bryant, & Hall, 1997) and they most likely result from etomidate's suppression of CS production (Félix et al., 2016; Pejo et al., 2012). It is believed that inappropriately low serum and tissue levels of glucocorticoids can lead to an exaggeration of the immunological response to injury in etomidate-treated animals (Hermoso & Cidlowski, 2003; Pejo et al., 2012).

The influence of cytokines in apoptosis depends from several factors, including the type of cytokine and the target cell and/or tissue. TNF-α is known to be mainly pro-apoptotic in adrenal cells (Mikhaylova, Kuulasmaa, Jääskeläinen, & Voutilainen, 2007; Tkachenko, Jääskeläinen, Jääskeläinen, Palvimo, & Voutilainen, 2011). In contrast, to our knowledge, the effects of IL6 and IL10 in adrenal apoptosis have not yet been studied. Nevertheless, their preventive role against apoptosis has been well recognized in other occasions (Dhingra, Bagchi, Ludke, Sharma, & Singal, 2011; Moran et al., 2009; Rollwagen, Yu, Li, & Pacheco, 1998; Thacker, Robinson, Abel, & Tweardy, 2013; Thompson, Zurko, Hanna, Hellenbrand, & Hanna, 2013). IL6 prevents apoptosis by several mechanisms: by activating STAT3, a transcription factor which down-regulates the expression of pro-apoptotic genes (Hirano, Ishihara, & Hibi, 2000; Moran et al., 2009; Rollwagen et al., 1998; Thacker et al., 2013); by increasing the activity of DNA-repair systems (Centurione & Aiello, 2016); by suppressing cytosolic Ca^{2+} increase and oxidative stress; and by preventing mitochondrial dysfunction (Smart et al., 2006). Likewise, IL10 also activates STAT3 (Moore, de Waal Malefyt, Coffman, & O'Garra, 2001) and prevents apoptosis indirectly, by decreasing TNF-α, ILβ1, IFNγ (Fiorentino et al., 1991; Fuchs et al., 1996), glucocorticoid (Koldzic-Zivanovic et al., 2006), and ROS production (Thompson et al., 2013).

Based in these studies, we hypothesize that the increased IL6 and IL10 levels combined with the already cited etomidate's direct anti-apoptotic effects, and associated increased ACTH and decreased CS levels counteracted the pro-apoptotic effects of HS, including those promoted by increased TNF-levels.

5.5. Apoptotic index was different accordingly to the different areas

The distribution of apoptosis was heterogeneous across the adrenal gland, being more severe in the cortex than in the medulla. In addition, apoptosis was also heterogeneously distributed across the adrenal cortex itself, being particularly predominant in *zona reticularis*. It was less present in the inner half of *zona fasciculata* and was infrequently found in *zona glomerulosa*. The heterogeneous distribution of apoptosis in the adrenal gland has also been described in other studies (Bozzo et al., 2006; Wolkersdörfer et al., 1996). For instances in rats, each adrenocortical zone has its specific and independent pattern of apoptosis (Bozzo et al., 2006) and this zone-specificity is maintained in the presence of exogenous pro-apoptotic stimulus. Two of these are dexamethasone administration and decreased ACTH levels (for example, by hypophysectomy) and rats submitted to these stimuli have increased rate of adrenocortical apoptosis which mainly affects *zona reticularis* (Almeida et al., 2007; Carsia et al., 1996).

The mechanisms behind the heterogeneous pattern of apoptosis in the adrenal gland are only partially understood. It may result from different paracrine and/or autocrine regulatory mechanisms which are specific to each adrenal zone (Bozzo et al., 2006). In addition, the adrenal distribution of proteins involved in the apoptotic process is not uniform (Vinson, 2016). Examples of this include superoxide dismutase (Vinson, 2016), the IL6 receptor, (mainly found in *zona reticularis* and inner *zona fasciculata*) (Päth, Bornstein, Ehrhart-Bornstein, & Scherbaum, 1997), thrombospondin-2 (absent from *zona reticularis*) (Feige, Keramidas, & Chambaz, 1998), and Bak (only found in adrenal cortex) (Krajewski, Krajewska, & Reed, 1996). In addition, more than 90% of the arterial blood supply to the adrenal cortex reaches first *zona glomerulosa*, and only after this, it arrives to *zona reticularis* and *zona fasciculata*. This pattern of arterial blood supply increases the risk of ischemia development in *zona reticularis* and *zona fasciculata* (Sparrow & Coupland, 1987), especially during low-flow states such as HS (Hamaji et al., 1986; Jasper, McDermott, Gann, & Engeland, 1990).

Two additional factors can predispose the inner adrenocortical zones to develop increased rates of apoptosis. The first is that *zona reticularis and fasciculata* have capillary walls with larger fenestrations than *zona glomerulosa* (Motta, Muto, & Fujita, 1979).This factor can facilitate the diffusion of pro-apoptotic molecules from the plasma to the adrenal interstitium. The second is that adrenal gland macrophages are especially abundant in the inner cortical regions (Almeida, Ferreira, & Neves, 2004). Due to pivotal role of macrophages in intra-adrenal inflammation and apoptosis (Wang et al., 2015) it is possible that this heterogeneous distribution also contributes to the higher apoptotic rate of the inner cortical regions.

5.6. HS increased adrenal gland necrosis

In our study, the degree and location of adrenal necrosis (preferentially affecting *zona fasciculata* and *reticularis*) were similar to what was found in previous experimental studies involving rats and dogs (Kajihara, Hirata, & Miyoshi, 1977; Kajihara et al., 1983; Rushing et al., 2006). What caused adrenal necrosis in our model remains undetermined. HS-associated adrenal ischemia is one major possibility as necrosis was especially predominant in the inner cortical areas (Rushing et al., 2006), as described before. Other possibilities include oxidative stress, I/R injury, and increased systemic and adrenal inflammation. Increased cytokines levels, in particular TNF-α, have been associated with increased cellular necrosis (Galluzzi et al., 2012; Vanlangenakker, Berghe, Krysko, Festjens, & Vandenabeele, 2008). Furthermore, an experimental study in dogs submitted to HS showed that adrenal necrosis was mainly apparent after fluid resuscitation (Kajihara et al., 1983) which may also suggest I/R injury as a contributing factor.

Adrenal necrosis could have also been due to necroptosis, an active form of necrosis which has been recognized in several critical illnesses (Duprez et al., 2011; Sharma, Matsuo, Yang, Wang, & Wang, 2014). This can be triggered by cytokines such as TNF-α, pathogen-associated molecular patterns, damage-associated molecular patterns, I/R injury, Ca^{2+} overload, hypoxia, infectious agents, DNA damage, and oxidative and nitrosative stress (Vanlangenakker et al., 2008). The presence of necroptosis was not investigated in our setting.

Etomidate-treated animals had a lower necrosis score than G2. The reason for this finding remains unclear. To our knowledge, no study has reported a protective effect of etomidate against cell necrosis. We hypothesize that etomidate might have prevented adrenal necrosis by the same mechanisms that prevented apoptosis, because apoptosis and necrosis share many pathophysiological mechanisms (Formigli et al., 2000).

5.7. Did adrenal gland apoptosis or necrosis cause CIRCI in our model?

In experimental septic and ANP models (Liu et al., 2016; Polito et al., 2010; Yu et al., 2012, 2016), adrenal apoptosis was associated to development of CIRCI. In our study, the results suggest that G2 animals could have also developed adrenal dysfunction at some degree. For instances, their CS levels were similar to those found in rats considered to have an inadequate adrenal response to HS (Rushing et al., 2006; Stein et al., 2013). In addition, in G2, the rise in ACTH levels from T2 to T3 was not associated to a correspondent increase of CS levels, suggesting a possible decreased adrenal responsiveness to ACTH's stimulatory effects. In addition, adrenal apoptosis was mainly concentrated in the areas responsible for glucocorticoid production in rats (Vinson, 2016). We hypothesize based on other models (Liu et al., 2016; Polito et al., 2010; Rushing et al., 2006; Yu et al., 2012, 2016) that the development of adrenal apoptosis (associated to development of adrenal necrosis) could in part explain this adrenal dysfunction,. However, contrarily to what was described in septic and ANP models, the decrease in CS levels was not sufficiently low to classify the adrenal dysfunction as CIRCI, accordingly to current guidelines (Marik, 2009). The reason why our results differ from the septic and ANP studies remains unknown but it could be related with differences in experimental design and methodology. Besides, we cannot rule out that maybe the pathophysiology of adrenal apoptosis is different between HS and sepsis or ANP.

In contrast, etomidate-treated animals had CS levels compatible with the diagnosis of CIRCI in Humans (Marik, 2009). In G3, we believe that the decreased CS levels were mainly due to the effects of etomidate in CS production. However previous studies have also shown that even in rats treated with etomidate, CIRCI resulted in part from adrenal apoptosis (Liu et al., 2015, 2016). Thus, we cannot exclude that adrenal apoptosis contributed to CIRCI in G3's animals. However, if it did, we believe that it played a minor role, because etomidate-treated animals had a lower apoptotic rate.

5.7.1. Implications

Our findings regarding etomidate effects upon adrenal apoptosis and necrosis open the possibility of this drug being used to modulate the adrenal apoptotic and necrotic processes in HS. This modulation can be advantageous to clarify their role in CIRCI development during HS and eventually clinically beneficial. However, before we consider using etomidate therapeutically for this purpose, one should consider that the anesthetic was associated to significant morbidity, including a higher degree of hypoperfusion, hypoxemia, and lung injury, as described more in detail in a previous study by our group (Félix et al., 2016). The latter study and others performed in septic models (Pejo et al., 2012) suggested that this increased morbidity was partially associated to the significant increase in cytokine levels. This in turn might have resulted from the lower levels of glucocorticoids caused by etomidate, which allowed the immune system to act without restraint. Thus, the literature suggests a cautious approach for the use of etomidate in HS. If the balance between the desired and adverse effects of etomidate administration in HS is time- and dose-dependent as described in septic models can only be answered by future studies.

5.8. Limitations of the study

This study had some limitations. The first was the use of a limited number of animals. In addition the animals were exclusively young male Wistar rats, limiting the translation of our results to other species. Besides, the response to HS is strongly influenced by sex, strain, and age of animals (Klemcke et al., 2011; Mees et al., 2008). We also induced a mild state of HS. It is possible that a severer and/or a longer model of HS can lead to a higher degree of apoptosis and adrenal dysfunction.

How etomidate modulated apoptosis was not defined. Future studies should address this issue, possibly using different doses of etomidate, its administration at different time points, and eventually by manipulating factors such as ACTH and cytokine levels.

In addition, although the methods chosen to identify the presence of apoptosis and necrosis are commonly used for this purpose, they do not provide information regarding its underlying mechanisms. The quantification of gene expression of *bax and bcl2* by PCR, the subsequent calculation of the *bax/bcl2* ratio, and the determination of intracellular factors involved in the apoptotic and necrotic pathways could be useful in this context and should be considered in future studies.

6. Conclusion

In this study, a mild form of HS was associated to development of adrenal gland apoptosis and necrosis. Apoptosis was heterogeneously distributed, affecting mostly the adrenal cortex and in particular, *zona reticularis* and *zona fasciculata.* Etomidate was associated to decrease rate of adrenal apoptosis, necrosis, and CS levels (compatible with CIRCI). The development of adrenal apoptosis and necrosis was not clearly associated to CIRCI. Etomidate could have modulated apoptosis and necrosis directly and/or indirectly, by decreasing CS levels and by contributing to increased IL6, IL10, and ACTH levels. Future studies should be performed to confirm and clarify the mechanisms of these findings, and eventually to explore the use of etomidate in preventing adrenal apoptosis and necrosis in HS.

Acknowledgements

The authors would like to thank Dr Jeanne Kehren for her assistance and technical advice to prepare the final version of the manuscript and Prof Solange Gil, Prof Luísa Mateus, Dr Sofia Gomes, and Dr Tiago Rafael for their contributions to conduct the experimental procedure.

Funding

The authors received no direct funding for this research.

Author details

Nuno M. Félix[1]
E-mail: nuno.felixgrey@gmail.com
Isabelle Goy-Thollot[2]
E-mail: Isabelle.goy-thollot@vetagro-sup.fr
Ronald S. Walton[3]
E-mail: sailblue2@gmail.com
Pedro M. Borralho[4]
E-mail: borralho@ff.ul.pt
Hugo Pissara[1]
E-mail: hpissara@fmv.ulisboa.pt
Ana S. Matos[5]
E-mail: asvm@fct.unl.pt
Cecília M.P. Rodrigues[4]
E-mail: cmprodrigues@ff.ul.pt
Maria M.R.E. Niza[1]
E-mail: necas@fmv.ulisboa.pt

[1] Faculty of Veterinary Medicine, CIISA, ULisboa, Avenida da Universidade Técnica, 1300-477 Lisboa, Portugal.

[2] VetAgro Sup - Hémostase, Inflammation et Sepsis, SIAMU, Université de Lyon, F-69280 Marcy L'Etoile, France.

[3] Speciality and Emergency of Tacoma, 5608 S. Durango Street,Tacoma, Washington 98409, USA.

[4] Faculty of Pharmacy, Research Institute for Medicines (iMed. ULisboa), Universidade de Lisboa, Avenida Professor Gama Pinto, 1649-003 Lisboa, Portugal.

[5] Faculdade de Ciências e Tecnologia, Departamento de Engenharia Mecânica e Industrial, Universidade Nova de Lisboa (UNIDEMI), 2829-516 Caparica, Portugal.

References

Allolio, B., Dörr, H., Stuttmann, R., Knorr, D., Engelhardt, D., & Winkelmann, W. (1985). Effect of a single bolus of etomidate upon eight major corticosteroid hormones and plasma ACTH. *Clinical Endocrinology, 22*(3), 281–286. https://doi.org/10.1111/j.1365-2265.1985.tb03241.x

Almeida, H., Ferreira, J., & Neves, D. (2004). Macrophages of the adrenal cortex: A morphological study of the effects of aging and dexamethasone administration. *Annals of the New York Academy of Sciences, 1019*, 135–140. https://doi.org/10.1196/annals.1297.024

Almeida, H., Matos, L., Ferreira, J., & Neves, D. (2006). Age-related effects of dexamethasone administration in adrenal zona reticularis. *Annals of the New York Academy of Sciences, 1067*, 354–360. https://doi.org/10.1196/annals.1354.050

Almeida, H., Matos, L., & Neves, D. (2007). Caspase-3 and Bcl-2 expression in aging in adrenal zona reticularis after dexamethasone administration. *Annals of the New York Academy of Sciences, 1119*, 190–195. https://doi.org/10.1196/annals.1404.021

Bazzani, C., Bertolini, A., & Guarini, S. (1997). Inhibition of nitric oxide synthases enhances the effect of ACTH in hemorrhagic shock. *Life Sciences, 61*(19), 1889–1897. https://doi.org/10.1016/S0024-3205(97)00828-X

Bertuglia, S., & Giusti, A. (2004). Influence of ACTH-(1-24) and plasma hyperviscosity on free radical production and capillary perfusion after hemorrhagic shock. *Microcirculation, 11*(3), 227–238. https://doi.org/10.1080/10739680490425930

Bozzo, A., Soñez, C. A., Mugnaini, M. T., Pastorino, I. C., Rolando, A. N., Romanini, M. C., & Gauna, H. F. (2006). Chronic stress effects on the apoptotic index of the adrenal cortex of pregnant rats. *Biocell, 30*(3), 439–445.

Bruder, E. A., Ball, I. M., Ridi, S., Pickett, W., & Hohl, C. (2015). Single induction dose of etomidate versus other induction agents for endotracheal intubation in critically ill patients. *The Cochrane database of systematic reviews, 1*, CD010225.

Carsia, R. V., Macdonald, G. J., Gibney, J. A., Tilly, K. I., & Tilly, J. L. (1996). Apoptotic cell death in the rat adrenal gland: An *in vivo* and *in vitro* investigation. *Cell Tissue Research, 283*, 247–254. https://doi.org/10.1007/s004410050535

Centurione, L., & Aiello, F. B. (2016). DNA repair and cytokines: TGF-β, IL-6, and thrombopoietin as different biomarkers of radioresistance. *Frontiers Oncology, 6*, 175.

Cherfan, A. J., Tamim, H. M., AlJumah, A., Rishu, A. H., Al-Abdulkareem, A., Al Knawy, B. A., ... Arabi, Y. M. (2011). Etomidate and mortality in cirrhotic patients with septic shock. *BMC Clinical Pharmacology, 11*, 22. https://doi.org/10.1186/1472-6904-11-22

Cotton, B. A., Guillamondegui, O. D., Fleming, S. B., Carpenter, R. O., Patel, S. H., Morris, Jr., J. A., & Arbogast, P. G. (2008). Increased risk of adrenal insufficiency following etomidate exposure in critically injured patients. *Archives of Surgery, 143*, 62–67. https://doi.org/10.1001/archsurg.143.1.62

Crozier, T. A., Beck, D., Schlaeger, M., Wuttke, W., & Kettler, D. (1987). Endocrinological changes following etomidate, midazolam, or methohexital for minor surgery. *Anesthesiology, 66*, 628–635. https://doi.org/10.1097/00000542-198705000-00006

den Brinker, M., Joosten, K. F., Liem, O., de Jong, F. H., Hop, W. C., Hazelzet, J. A., ... Hokken-Koelega, A. C. (2005). Adrenal insufficiency in meningococcal sepsis: Bioavailable cortisol levels and impact of interleukin-6 levels and intubation with etomidate on adrenal function and mortality. *The Journal of Clinical Endocrinology & Metabolism, 90*, 5110–5117. https://doi.org/10.1210/jc.2005-1107

den Brinker, M., Hokken-Koelega, A. C., Hazelzet, J. A., de Jong, F. H., Hop, W. C., & Joosten, K. F. (2008). One single dose of etomidate negatively influences adrenocortical performance for at least 24 h in children with meningococcal sepsis. *Intensive Care Medicine, 34*, 163–168. https://doi.org/10.1007/s00134-007-0836-3

Dhingra, S., Bagchi, A. K., Ludke, A. L., Sharma, A. K., & Singal, P. K. (2011). Akt regulates IL-10 mediated suppression of TNFα-induced cardiomyocyte apoptosis by upregulating Stat3 phosphorylation. *PLoS ONE, 6*, e25009. https://doi.org/10.1371/journal.pone.0025009

Didenko, V. V., Wang, X., Yang, L., & Hornsby, P. J. (1996). Expression of p21 (WAF1/CIP1/SDI1) and p53 in apoptotic cells in the adrenal cortex and induction by ischemia/reperfusion injury. *Journal of Clinical Investigation, 97*, 1723–1731. https://doi.org/10.1172/JCI118599

Duprez, L., Takahashi, N., Van Hauwermeiren, F., Vandendriessche, B., Goossens, V., Vanden Berghe, T., ... Vandenabeele, P. (2011). RIP kinase-dependent necrosis drives lethal systemic inflammatory response syndrome. *Immunity, 35*(6), 908–918. https://doi.org/10.1016/j.immuni.2011.09.020

Ergün, Y., Darendeli, S., Imrek, S., Kilinç, M., & Oksüz, H. (2010). The comparison of the effects of anesthetic doses of ketamine, propofol, and etomidate on ischemia-reperfusion injury in skeletal muscle. *Fundamentals of Clinical Pharmacology, 24*(2), 215–222.

Feige, J. J., Keramidas, M., & Chambaz, E. M. (1998). Hormonally regulated components of the adrenocortical cell environment and the control of adrenal cortex homeostasis. *Hormone and Metabolic Research, 30*, 421–425. https://doi.org/10.1055/s-2007-978908

Félix, N. M., Goy-Thollot, I., Walton, R. S., Gil, S. A., Mateus, L. M., Matos, A. S., & Niza, M. M. R. E. (2016). Effects of etomidate in the adrenal and cytokine responses to hemorrhagic shock in rats. *European Journal of Inflammation, 14*(3), 147–161. doi:10.1177/172172 7X16677604

Fiorentino, D. F., Zlotnik, A., Vieira, P., Mosmann, T. R., Howard, M., Moore, K. W., & O'Garra, A. (1991). IL-10 acts on the antigen-presenting cell to inhibit cytokine production by Th1 cells. *Journal of Immunology, 146*, 3444–3451.

Flierl, M. A., Rittirsch, D., Chen, A. J., Nadeau, B. A., Day, D. E., Sarma, J. V., ... Ward, P. A. (2008). The complement anaphylatoxin C5a induces apoptosis in adrenomedullary cells during experimental sepsis. *PLoS ONE, 3*, e2560. doi:10.1371/journal.pone.0002560

Forman, S. A. (2011). Clinical and molecular pharmacology of etomidate. *Anesthesiology, 114*, 695–707. https://doi.org/10.1097/ALN.0b013e3181ff72b5

Formigli, L., Papucci, L., Tani, A., Schiavone, N., Tempestini, A., Orlandini, G. E., ... Orlandini, S. Z. (2000). Aponecrosis: Morphological and biochemical exploration of a syncretic process of cell death sharing apoptosis and necrosis. *Journal of Cellular Physiology, 182*, 41–49. https://doi.org/10.1002/(ISSN)1097-4652

Fuchs, A. C., Granowitz, E. V., Shapiro, L., Vannier, E., Lonnemann, G., Angel, J. B., ... Dinarello, C. A. (1996). Clinical, hematologic, and immunologic effects of interleukin-10 in humans. *Journal of Clinical Immunology, 16*, 291–303. https://doi.org/10.1007/BF01541395

Fukuzuka, K., Rosenberg, J. J., Gaines, G. C., Edwards, C. K., 3rd, Clare-Salzler, M., MacKay, S. L., ... Mozingo, D. W. (1999). Caspase-3-dependent organ apoptosis early after burn injury. *Annals of Surgery, 229*, 851–858. https://doi.org/10.1097/00000658-199906000-00012

Galluzzi, L., Vitale, I., Abrams, J. M., Alnemri, E. S., Baehrecke, E. H., Blagosklonny, M. V., ... Kroemer, G. (2012). Molecular definitions of cell death subroutines: Recommendations of the Nomenclature Committee on Cell Death 2012. *Cell Death and Differentiation, 19*, 107–120. https://doi.org/10.1038/cdd.2011.96

Guarini, S., Bazzani, C., Ricigliano, G. M., Bini, A., Tomasi, A., & Bertolini, A. (1996). Influence of ACTH-(1-24) on free radical levels in the blood of haemorrhage-shocked rats: Direct ex vivo detection by electron spin resonance spectrometry. *British Journal of Pharmacology, 119*(1), 29–34. https://doi.org/10.1111/bph.1996.119.issue-1

Guarini, S., Cainazzo, M. M., Giuliani, D., Mioni, C., Altavilla, D., Marini, H., ... Bertolini, A. (2004). Adrenocorticotropin reverses hemorrhagic shock in anesthetized rats through the rapid activation of a vagal anti-inflammatory pathway. *Cardiovascular Research, 63*(2), 357–365. https://doi.org/10.1016/j.cardiores.2004.03.029

Hamaji, M., Nakamura, M., Izukura, M., Nakaba, H., Hashimoto, T., Tanaka, Y., ... Harrison, T. S. (1986). Autoregulation and regional blood flow of the dog during hemorrhagic shock. *Circulatory Shock, 19*(3), 245–255.

Hermoso, M. A., & Cidlowski, J. A. (2003). Putting the brake on inflammatory responses: The role of glucocorticoids. *IUBMB Life (International Union of Biochemistry and Molecular Biology: Life), 55*, 497–504. https://doi.org/10.1080/15216540310001642072

Hirano, T., Ishihara, K., & Hibi, M. (2000). Roles of STAT3 in mediating the cell growth, differentiation and survival signals relayed through the IL-6 family of cytokine receptors. *Oncogene, 19*, 2548–2556. https://doi.org/10.1038/sj.onc.1203551

Hoen, S., Asehnoune, K., Brailly-Tabard, S., Mazoit, J. X., Benhamou, D., Moine, P., & Edouard, A. R. (2002). Cortisol response to corticotropin stimulation in trauma patients: Influence of hemorrhagic shock. *Anesthesiology, 97*, 807–813. https://doi.org/10.1097/00000542-200210000-00010

Jameson, P., Desborough, J. P., Bryant, A. E., & Hall, G. M. (1997). The effect of cortisol suppression on interleukin-6 and white blood cell responses to surgery. *Acta Anaesthesiologica Scandinavica, 41*, 304–308. https://doi.org/10.1111/aas.1997.41.issue-2

Jasper, M. S., McDermott, P., Gann, D. S., & Engeland, W. C. (1990). Measurement of blood flow to the adrenal capsule, cortex and medulla in dogs after hemorrhage by fluorescent microspheres. *Journal of the Autonomic Nervous System, 30*(2), 159–167. https://doi.org/10.1016/0165-1838(90)90140-E

John, C. D., & Buckingham, J. C. (2003). Cytokines: Regulation of the hypothalamo-pituitary-adrenocortical axis. *Current Opinion in Pharmacology, 3*(1), 78–84. https://doi.org/10.1016/S1471-4892(02)00009-7

Kajihara, H., Hirata, S., & Miyoshi, N. (1977). Changes in blood catecholamine levels and ultrastructure of dog adrenal medullary cells during hemorrhagic shock. *Virchows Archives B Cell Pathology, 23*(1), 1–16.

Kajihara, H., Malliwah, J. A., Matsumura, M., Taguchi, K., & Iijima, S. (1983). Changes in blood cortisol and aldosterone levels and ultrastructure of the adrenal cortex during hemorrhagic shock. *Pathology, Research and Practice, 176*, 324–340. https://doi.org/10.1016/S0344-0338(83)80022-3

Kamiyama, K., Matsuda, N., Yamamoto, S., Takano, K., Takano, Y., Yamazaki, H., ... Hattori, Y. (2008). Modulation of glucocorticoid receptor expression, inflammation, and cell apoptosis in septic guinea pig lungs using methylprednisolone. *AJP: Lung Cellular and Molecular Physiology, 295*, 998–1006. https://doi.org/10.1152/ajplung.00459.2007

Kanczkowski, W., Chatzigeorgiou, A., Grossklaus, S., Sprott, D., Bornstein, S. R., & Chavakis, T. (2013). Role of the endothelial-derived endogenous anti-inflammatory factor Del-1 in inflammation-mediated adrenal gland dysfunction. *Endocrinology, 154*(3), 1181–1189. https://doi.org/10.1210/en.2012-1617

Kanczkowski, W., Chatzigeorgiou, A., Samus, M., Tran, N., Zacharowski, K., Chavakis, T., & Bornstein, S. R. (2013). Characterization of the LPS-induced inflammation of the adrenal gland in mice. *Molecular and Cellular Endocrinology, 371*, 228–235. https://doi.org/10.1016/j.mce.2012.12.020

Keramidas, M., Feige, J. J., & Thomas, M. (2004). Coordinated regression of adrenocortical endocrine and endothelial compartments under adrenocorticotropin deprivation. *Endocrine Research, 30*(4), 543–549. https://doi.org/10.1081/ERC-200043623

Klemcke, H. G., Joe, B., Calderon, M. L., Rose, R., Oh, T., Aden, J., & Ryan, K. L. (2011). Genetic influences on survival time after severe haemorrhage in inbred rat strains. *Physiological Genomics, 43*, 758–765. https://doi.org/10.1152/physiolgenomics.00245.2010

Koldzic-Zivanovic, N., Tu, H., Juelich, T. L., Rady, P. L., Tyring, S. K., Hudnall, S. D., ... Hughes, T. K. (2006). Regulation of adrenal glucocorticoid synthesis by interleukin-10: A preponderance of IL-10 receptor in the adrenal zona fasciculata. *Brain, Behavior, and Immunity, 20*(5), 460–468. https://doi.org/10.1016/j.bbi.2005.09.003

Krajewski, S., Krajewska, M., & Reed, J. C. (1996). Immunohistochemical analysis of *in vivo* patterns of Bak expression, a proapoptotic member of the Bcl-2 protein family. *Cancer Research, 56*, 2849–2855.

Kugawa, F., Arae, K., Ueno, A., & Aoki, M. (1998). Buprenorphine hydrochloride induces apoptosis in NG108-15 nerve cells. *European Journal of Pharmacology, 347*, 105–112. https://doi.org/10.1016/S0014-2999(98)00080-6

Kugawa, F., Nakamura, M., Ueno, A., & Aoki, M. (2004). Over-expressed Bcl-2 cannot suppress apoptosis via the mitochondria in buprenorphine hydrochloride-treated NG108-15 cells. *Biological & Pharmaceutical Bulletin, 27*, 1340–1347. https://doi.org/10.1248/bpb.27.1340

Liu, N., Zhang, Y., Xiong, J. Y., Liu, S., Zhu, J., & Lv, S. (2016). The pituitary adenylate cyclase-activating polypeptide (PACAP) protects adrenal function in septic rats administered etomidate. *Neuropeptides, 58*, 53–59. https://doi.org/10.1016/j.npep.2016.03.005

Liu, S., Zhang, X. P., Han, N. N., Lv, S., & Xiong, J. Y. (2015). Pretreatment with low dose etomidate prevents etomidate-induced rat adrenal insufficiency by regulating oxidative stress-related MAPKs and apoptosis. *Environmental Toxicology and Pharmacology, 39*(3), 1212–1220. https://doi.org/10.1016/j.etap.2015.04.014

Marik, P. E. (2009). Critical illness-related corticosteroid insufficiency. *Chest, 135*, 181–193. https://doi.org/10.1378/chest.08-1149

Mees, S. T., Gwinner, M., Marx, K., Faendrich, F., Schroeder, J., Haier, J., & Kahlke, V. (2008). Influence of sex and age on morphological organ damage after hemorrhagic shock. *Shock, 29*, 670–674.

Mikhaylova, I. V., Kuulasmaa, T., Jääskeläinen, J., & Voutilainen, R. (2007). Tumor necrosis factor-alpha regulates steroidogenesis, apoptosis, and cell viability in the human adrenocortical cell line NCI-H295R. *Endocrinology, 148*(1), 386–392. https://doi.org/10.1210/en.2006-0726

Moore, K. W., de Waal Malefyt, R., Coffman, R. L., & O'Garra, A. (2001). Interleukin-10 and the interleukin-10 receptor. *Annual Review of Immunology, 19*, 683–765. https://doi.org/10.1146/annurev.immunol.19.1.683

Moran, A., Tsimelzon, A. I., Mastrangelo, M. A., Wu, Y., Yu, B., Hilsenbeck, S. G., ... Tweardy, D. J. (2009). Prevention of trauma/hemorrhagic shock-induced lung apoptosis by IL-6-mediated activation of Stat3. *Clinical and Translational Science, 2*, 41–49. https://doi.org/10.1111/cts.2009.2.issue-1

Mosier, M. J., Lasinski, A. M., & Gamelli, R. L. (2015). Suspected adrenal insufficiency in critically ill burned patients: Etomidate-induced or critical illness-related corticosteroid insufficiency?-A review of the literature. *Journal of Burn & Care Research, 36*(2), 272–278. https://doi.org/10.1097/BCR.0000000000000099

Motta, P., Muto, M., & Fujita, T. (1979). Three-dimensional organization of mammalian adrenal cortex. A scanning electron microscopic study. *Cell Tissue Research, 30*, 23–38.

Ottani, A., Galantucci, M., Ardimento, E., Neri, L., Canalini, F., Calevro, A., ... Guarini, S. (2013). Modulation of the JAK/ERK/STAT signaling in melanocortin-induced inhibition of local and systemic responses to myocardial ischemia/reperfusion. *Pharmacological Research, 72*, 1–8. https://doi.org/10.1016/j.phrs.2013.03.005

Papathanassoglou, E. D., Moynihan, J. A., & Ackerman, M. H. (2000). Does programmed cell death (apoptosis) play a role in the development of multiple organ dysfunction in critically ill patients? A review and a theoretical framework. *Critical Care Medicine, 28*, 537–549. https://doi.org/10.1097/00003246-200002000-00042

Päth, G., Bornstein, S. R., Ehrhart-Bornstein, M., & Scherbaum, W. A. (1997). Interleukin-6 and the interleukin-6 receptor in the human adrenal gland: Expression and effects on steroidogenesis. *Journal of Clinical Endocrinology and Metabolism, 82*(7), 2343–2349.

Pejo, E., Feng, Y., Chao, W., Cotten, J. F., Le Ge, R., & Raines, D. E. (2012). Differential effects of etomidate and its pyrrole analogue carboetomidate on the adrenocortical and cytokine responses to endotoxemia. *Critical Care Medicine, 40*, 187–192. https://doi.org/10.1097/CCM.0b013e31822d7924

Polito, A., Lorin de la Grandmaison, G., Mansart, A., Louiset, E., Lefebvre, H., Sharshar, T., & Annane, D. (2010). Human and experimental septic shock are characterized by depletion of lipid droplets in the adrenals. *Intensive Care Medicine, 36*, 1852–1858. https://doi.org/10.1007/s00134-010-1987-1

Porter, A. G., & Jänicke, R. U. (1999). Emerging roles of caspase-3 in apoptosis. *Cell Death Differentiation, 6*, 99–104. https://doi.org/10.1038/sj.cdd.4400476

Preziosi, P., & Vacca, M. (1982). Etomidate and corticotrophic axis. *Archives Internationales de Pharmacodynamie et de Therapie, 256*, 308–310.

Roesslein, M., Schibilsky, D., Muller, L., Goebel, U., Schwer, C., Humar, M., ... Loop, T. (2008). Thiopental protects human T lymphocytes from apoptosis in vitro via the expression of heat shock protein 70. *Journal of Pharmacology and Experimental Therapeutics, 325*, 217–225. https://doi.org/10.1124/jpet.107.133108

Rollwagen, F. M., Yu, Z. Y., Li, Y. Y., & Pacheco, N. D. (1998). IL-6 rescues enterocytes from hemorrhage induced apoptosis in vivo and in vitro by a bcl-2 mediated mechanism. *Clinical Immunology Immunopathology, 89*, 205–213. https://doi.org/10.1006/clin.1998.4600

Rushing, G. D., & Britt, L. D. (2007). Inhibition of NF-KB does not induce c-Jun N-Terminal kinase-mediated apoptosis in reperfusion injury. *Journal of the American College of Surgeons, 204*(5), 964–967. https://doi.org/10.1016/j.jamcollsurg.2007.01.029

Rushing, G. D., Britt, R. C., & Britt, L. D. (2006). Effects of hemorrhagic shock on adrenal response in a rat model. *Annals of Surgery, 243*, 652–656. https://doi.org/10.1097/01.sla.0000216759.36819.1b

Sharma, A., Matsuo, S., Yang, W. L., Wang, Z., & Wang, P. (2014). Receptor-interacting protein kinase 3 deficiency inhibits immune cell infiltration and attenuates organ injury in sepsis. *Critical Care, 18*(4), R142. https://doi.org/10.1186/cc13970

Si, J., Ge, Y., Zhuang, S., Wang, L.J., Chen, S., & Gong, R. (2013). Adrenocorticotropic hormone ameliorates acute kidney injury by steroidogenic-dependent and -independent mechanisms. *Kidney International, 83*, 635–646. https://doi.org/10.1038/ki.2012.447

Smart, N., Mojet, M. H., Latchman, D. S., Marber, M. S., Duchen, M. R., & Heads, R. J. (2006). IL-6 induces PI 3-kinase and nitric oxide-dependent protection and preserves mitochondrial function in cardiomyocytes. *Cardiovascular Research, 69*, 164–177. https://doi.org/10.1016/j.cardiores.2005.08.017

Sparrow, R. A., & Coupland, R. E. (1987). Blood flow to the adrenal gland of the rat: Its distribution between the cortex and the medulla before and after haemorrhage. *Journal Anatomy, 155*, 51–61.

Squadrito, F., Guarini, S., Altavilla, D., Squadrito, G., Campo, G. M., Arlotta, M., ... Caputi, A. P. (1999). Adrenocorticotropin reverses vascular dysfunction and protects against splanchnic artery occlusion shock. *British Journal of Pharmacology, 128*(3), 816–822. https://doi.org/10.1038/sj.bjp.0702848

Stein, D. M., Jessie, E. M., Crane, S., Kufera, J. A., Timmons, T., Rodriguez, C. J., ... Scalea, T. M. (2013). Hyperacute adrenal insufficiency after hemorrhagic shock exists and is associated with poor outcomes. *Journal of Trauma Acute Care Surgery, 74*(2), 363–370. https://doi.org/10.1097/TA.0b013e31827e2aaf

Thacker, S. A., Robinson, P., Abel, A., & Tweardy, D. J. (2013). Modulation of the unfolded protein response during hepatocyte and cardiomyocyte apoptosis in trauma/hemorrhagic shock. *Scientific Reports, 3*, 1. https://doi.org/10.1038/srep01187

Thompson, C. D., Zurko, J. C., Hanna, B. F., Hellenbrand, D. J., & Hanna, A. (2013). The therapeutic role of interleukin-10 after spinal cord injury. *Journal of Neurotrauma, 30*(15), 1311–1324. https://doi.org/10.1089/neu.2012.2651

Tkachenko, I. V., Jääskeläinen, T., Jääskeläinen, J., Palvimo, J. J., & Voutilainen, R. (2011). Interleukins 1α and 1β as regulators of steroidogenesis in human NCI-H295R adrenocortical cells. *Steroids, 76*(10-11), 1103–1115. https://doi.org/10.1016/j.steroids.2011.04.018

Turnbull, A. V., & Rivier, C. L. (1999). Regulation of the hypothalamic-pituitary-adrenal axis by cytokines: Actions and mechanisms of action. *Physiological Reviews, 79*(1), 1–71.

Vanlangenakker, N., Berghe, T. V., Krysko, D. V., Festjens, N., & Vandenabeele, P. (2008). Molecular mechanisms and pathophysiology of necrotic cell death. *Current Molecular Medicine, 8*, 207–220. https://doi.org/10.2174/156652408784221306

Vignali, D.A. (2000). Multiplexed particle-based flow cytometric assays. *Journal of Immunological Methods, 243*, 243–255. https://doi.org/10.1016/S0022-1759(00)00238-6

Vinson, G. P. (2016). Functional zonation of the adult mammalian adrenal cortex. *Frontiers Neuroscience, 10*, 238.

Wagner, R. L., White, P. F., Kan, P. B., Rosenthal, M. H., & Feldman, D. (1984). Inhibition of adrenal steroidogenesis by the anesthetic etomidate. *New England Journal of Medicine, 310*(22), 1415–1421. https://doi.org/10.1056/NEJM198405313102202

Wang, P., Ba, Z. F., Jarrar, D., Cioffi, W.G., Bland, K. I., & Chaudry, I. H. (1999). Mechanism of adrenal insufficiency following trauma and severe hemorrhage: Role of hepatic 11beta-hydroxysteroid dehydrogenase. *Archives of Surgery, 134*, 394–401. https://doi.org/10.1001/archsurg.134.4.394

Wang, C. N., Duan, G. L., Liu, Y. J., Yu, Q., Tang, X. L., Zhao, W., ... Ni, X. (2015). Overproduction of nitric oxide by endothelial cells and macrophages contributes to mitochondrial oxidative stress in adrenocortical cells and adrenal insufficiency during endotoxemia. *Free Radical Biology & Medicine, 83*, 31–40. https://doi.org/10.1016/j.freeradbiomed.2015.02.024

Wang, X., Jin, A., An, M., Ding, Y., Tuo, Y., & Qiu, Y. (2014). Etomidate deteriorates the toxicity of advanced glycation end products to human endothelial Eahy926 cells. *The Journal of Toxicological Sciences, 39*(6), 887–896. https://doi.org/10.2131/jts.39.887

Wolkersdörfer, G. W., Ehrhart-Bornstein, M., Brauer, S., Marx, C., Scherbaum, W. A., & Bornstein, S. R. (1996). Differential regulation of apoptosis in the normal human adrenal gland. *Journal Clinical Endocrinology Metabolism, 81*, 4129–4136.

Wu, R. S., Wu, K. C., Yang, J. S., Chiou, S. M., Yu, C. S., Chang, S. J., ... Chung, J. G. (2011). Etomidate induces cytotoxic effects and gene expression in a murine leukemia macrophage cell line (RAW264.7). *Anticancer Research, 31*, 2203–2208.

Xu, H., Chen, S., Luo, W., & Firoj, K. M. (2014). Effect of stress on myocardial apoptosis in ischemic preconditioning in rabbit hearts. *Zhong Nan da Xue Xue Bao Yi Xue Ban, 39*, 477–482.

Yagmurdur, H., Cakan, T., Bayrak, A., Arslan, M., Baltaci, B., Inan, N., & Kilinc, K. (2004). The effects of etomidate, thiopental and propofol in induction on hypoperfusion-reperfsuion phenomenon during laparoscopic cholecystectomy. *Acta Anaesthesiologica Scandinavica, 48*(6), 772–777. https://doi.org/10.1111/aas.2004.48.issue-6

Yu, J., Xu, S., Wang, W. X., Deng, W. H., Jin, H., Chen, X. Y., ... Sun, H. T. (2012). Changes of inflammation and apoptosis in adrenal gland after experimental injury in rats with acute necrotizing pancreatitis. *Inflammation, 35*, 11–22. https://doi.org/10.1007/s10753-010-9284-2

Yu, J., Zuo, T., Deng, W., Shi, Q., Ma, P., Chen, C., ... Wang, W. (2016). Poly(ADP-ribose) polymerase inhibition suppresses inflammation and promotes recovery from adrenal injury in a rat model of acute necrotizing pancreatitis. *BMC Gastroenterology, 16*(1), 11. https://doi.org/10.1186/s12876-016-0493-5

Yu, Q., Zhou, Q., Huang, H., Wang, Y., Tian, S., & Duan, D. (2010). Protective effect of etomidate on spinal cord ischemia-reperfusion injury induced by aortic occlusion in rabbits. *Annals of Vascular Surgery, 24*(2), 225–232. https://doi.org/10.1016/j.avsg.2009.06.023

Zhang, X., Xiong, J., Jiao, Y., Wang, G., & Zuo, Z. (2010). Involvement of mitochondrial ATP-sensitive potassium channels in etomidate preconditioning-induced protection in human myeloid HL-60 cells. *Environmental Toxicology and Pharmacology, 29*(3), 320–322. https://doi.org/10.1016/j.etap.2010.02.003

Zhang, Y., Li, R., Zhu, J., Wang, Z., Lv, S., & Xiong, J. Y. (2015). Etomidate increases mortality in septic rats through inhibition of nuclear factor kappa-B rather than by causing adrenal insufficiency. *Journal of Surgical Research, 193*(1), 399–406. https://doi.org/10.1016/j.jss.2014.07.001

8

Hepatoprotective activity of *Ganoderma lucidium* (Curtis) P. Karst against cyclophosphamide-induced liver injury in mice

Hong Ngoc Pham[1]*, Le Son Hoang[1] and Van Trung Phung[2]

*Corresponding author: Hong Ngoc Pham, Department of Biochemistry, School of Biotechnology, International University - Vietnam National University, HCMC, Vietnam

E-mail: phamngocst@gmail.com

Reviewing editor: Tsai-Ching Hsu, Chung Shan Medical University, Taiwan

Additional information is available at the end of the article

Abstract: This study aims to investigate the hepatoprotective activity of *Ganoderma lucidium* mushroom growing on the dead ironwood tree in the Central Highlands of Vietnam. The total extract was orally administrated to experimental mice with hepatotoxicity induced by cyclophosphamide at the dose of 150 mg/kg (intraperitoneal injection). The liver malondialdehyde (MDA) content and the level of endogenous antioxidant glutathione (GSH) were measured. The results revealed that the total extract at the oral doses of 330, 230, and 120 mg/kg body weight which were equivalent to 5, 10, and 15 g/kg of dry material alleviated the increase of heaptic MDA content and restored the decrease of GSH level significantly which almost have the same potent as reference drug (silymarin). Biochemical observations were also supported with histopathological examination of liver. This finding demonstrated that *G. lucidium* in Vietnam could represent a promising approach for effective liver protective agents.

Subjects: Pharmaceutical Science; Pharmacology; Biology; Health & Society; Medicine

Keywords: *Ganoderma lucidium*; hepatoprotection; lipidperoxidation; cyclophosphamide

1. Introduction

Asian countries have a long convention of handling mushrooms as medicine, whereas in Western this has been used since last decade (Lindequist, Niedermeyer, & Julich, 2005). Mushrooms have been known to be a potential source of antioxidants as well as hepatoprotective agents and capable of strong inhibition of lipid peroxidation (Acharya, Chatterjee, & Ghosh, 2011; Acharya, Yonzone, Rai, & Acharya, 2005; Pal, Ganguly, Tahsin, & Acharya, 2000). *Ganoderma lucidium* (reishi mushroom, Ling Zhi) has been an economically crucial species, particularly in the Far East countries. It is broadly

ABOUT THE AUTHORS

Our research group is interested in exploring the bioactive and novel compounds from Vietnamese medicinal materials. Recent interest studies have concerned about hepatoprotective activity of some valuable mushrooms in Vietnam including *Tramates versicolor, and Ganoderma colossum* collected in nature. This study had focused on Lim Xanh (*Ganoderma lucidium*) as a potential source for elucidating the target hepatoprotective agents.

PUBLIC INTEREST STATEMENT

Medicinal mushrooms have a long history of ultilization in traditional therapy; and fungal metabolite are increasingly operated not to be considered as food, some of them could have biological properties including antioxidant, antitumor, and hepatoprotective activity. In Vietnam, *Ganoderma lucidium* has been found growing on the dead ironwood tree in the Central of Highlands. It is boardly presented to a commercial scale or harvested in nature and commonly purchased for its medicinal values. This study has provided the scientific evidence to validate the usage of this mushroom in folk medicine as the treatment of liver disorders and the prevention of hepatic injury.

grown on a commercial scale or harvested in nature and commonly purchased for its medicinal and spiritual properties. In a broad review about the hepatoprotective property of *G. lucidium,* Gao et al. (2003) collected evidence to suggest possible molecular mechanisms to explain its hepatoprotective action.

This annual mushroom grows on a great variety of dead or dying tree, e.g. deciduous trees, especially oak, maple, elm, willow, sweet gum, magnolia, and locust. *G. lucidium* grows originally on plum trees and found on stumps near the soil surface (Wasser, 2005). It had been found growing in the dead ironwood tree in the Central of Highlands of Vietnam. Due to the growth of this mushroom in the green ironwood tree, it was a so-called "Green Ironwood mushroom." Local people harvested the fruiting body of *G. lucidium* and then dried under shade. It has been often sold to patients suffering cancer or even disorders such as cirrhosis. *G. lucidium* is also believed that the regular consumption of this mushroom may preserve human vitality and to promote the longevity. In general, the mushroom has been traditionally used for the prevention or treatment of numerous diseases related to liver although whether it was effective or not is presumably unknown. The aim of this study is to validate the use of this mushroom in folk medicine by evaluating the hepatoprotective activity of *G. lucidium* which grows on the dead ironwood tree against the cyclophosphamide (CP)-induced liver toxicity. The lipid peroxidation and antioxidant parameters MDA and GSH were observed in liver homogenates and histopathological examination of liver sections was conducted to confirm the activity. The potent hepatoprotective effect of natural material exhibit an appealing advance for liver protective agents, especially at the moment there is an urgent need concerning on new innovative drugs.

2. Materials and methods

2.1. Materials

The fruiting body of *G. lucidium* was harvested from Central Highlands of Vietnam in August 2014. The mushroom was identified by Associate Prof. Dr Tran Van Minh-Institute of Tropical Biology, Vietnam. A voucher specimen was deposited in the herbarium of Applied Biochemistry Laboratory, Department of Applied Chemistry, School of Biotechnology, International University, Vietnam National University Ho Chi Minh City, Vietnam with voucher No. HB-B10-08-09-14.

2.2. Chemicals

Methanol, KCl, EDTA–phosphate, and thiobarnituric acid were purchased from Chemsol Co., Ltd. (Viet Nam). CP, malondialdehyde, glutathione, ellman or 5,5′-dithiobis-(2-nitrobenzoic acid), and silymarin were supplied from Sigma Co., Ltd. (USA).

2.3. Animals

Adult (5–6 weeks) *Swiss albino* male mice (23 ± 2 g) were obtained from Pasteur Institute of Ho Chi Minh city and kept in animals' house of the university animal facility. All mice were acclimated for a week prior to commencement of experiments. They were provided standard pellet diet, water *ad libitum* and maintained a controlled 12 light–dark cycles at room temperature. All the ethical protocols and guideline for experimental animal handling and treatment were strictly followed the guidelines enunciated in the "Guideline for the Treatment of Animals in Behavioral Research and Teaching" and complied with ethical measures for animal research.

2.4. Preparation of mushroom extract

The fruiting body of *G. lucidium* was first rinsed thoroughly with tap water, and subjected to dry heat treatment before grinding it into powdery form using an electric grinder. The extraction was done by cold maceration of 2 kg in 96% of ethanol and kept aside for 3 days with frequent agitation. The filtrate obtained was concentrated using rotary evaporator. The residue subsequently referred to as the extract, was stored in a refrigerator until required for further use.

Table 1. Experimental treatment design

Treatment	Groups	Dosages (mg/kg body weight)	N	MDA and GSH (nM/g protein)
Cyclophosphamide-untreated mice CP (–)	Normal control	-	10	-
	Total extract	330	10	-
		230	10	-
		120	10	-
	Silymarin	100	10	-
Cyclophosphamide-treated mice CP (+)	Negative control	-	10	-
	Total extract	330	10	-
		230	10	-
		120	10	-
	Silymarin (positive control)	100	10	-

2.5. Cyclophosphamide-induced hepatotoxicity and treatment

The animal treatment was designed to study the hepatoprotective effect detailed in Table 1

After 8-day experimental period, the animals were sacrificed and liver tissues were immediately excised and rinsed in ice-cold saline then homogenized for the subsequent assays

2.6. Assessment of liver function

Livers were resected and then homogenized in KCl (1, 15%) in 1 min at 13,000 rpm. 2 mL of tissue homogenized was added to 1 mL of Tris (pH 7.4). The mixture was incubated at 37°C for 60 min, then stopped the reaction by 1 mL of 10% trichloroacetic acid. The next step was the centrifugation of the mixture at 5°C with 10,000 rpm/min for 10 min.

2.7. Estimation of malondialdehyde (MDA) level

Two milliliters of supernatant reacted with 1 mL of 0.8% thiobarbituric acid (TBA) at 100°C for 15 min and the absorbent were measured at $\lambda = 532$ nm. 2-Thiobarbituric acid reactive substance (TBARS) assay values are usually reported in MDA equivalents, a compound that results from the decomposition of polyunsaturated fatty acid lipid peroxide. The amount of MDA formed was quantified by reaction with TBA. It was formed according to the methods of Stroev and Makarova (1989) and Abraham and Sugumar (2008) with few modifications (Ngo & Nguyen, 2011; Nonanka et al., 2005).

2.8. Estimation of glutathione (GSH) level

About 1 mL of supernatant reacted with 0, 2 mL of Ellman 5,5′-dithiobis (2-nitrobenzoic acid) and 0, 8 mL of phosphate-EDTA. After 3 min, the absorbent was determined at $\lambda = 412$ nm. The method is based on the reduction of 5,5′-dithiobis (2-nitrobenzoic acid) (DTNB) with reduced GSH to produce a yellow compound. The reduced chromogen is directly proportional to GSH concentration and its absorbance can be measured at 412 nm. Noeman, Hamooda, and Baalash (2011) with some modifications (Ngo & Nguyen, 2011).

2.9. Histopathological examination

The livers were excised and soaked by optimal cutting temperature O.C.T compound. The samples were then freezed and fit into Tissue-Tek Cryo 3 (Sakura, USA). They were cut into slides by MX35 Premier and Microtome Blade (Thermo Scientific, Japan) before being melt by block heater (Stuart, United Kingdom). Slides were preserved in 4% paraformaldehyde, stained with hematoxylin and eosin. The samples were then observed under Leica microscope for histopathological study.

Table 2. The effect of *G. lucidium* total extract on MDA content in cyclophosphamide-untreated and -treated mice

Treatment	Groups	Dosages (mg/kg body weight)	N	MDA (nM/g protein)
Cyclophosphamide-untreated mice CP (−)	Normal control	–	10	58.82 ± 4.06
		330	10	46.00 ± 3.51#
	Total extract	230	10	46.82 ± 3.28#
		120	10	54.23 ± 4.24
	Silymarin	100	10	50.89 ± 4.37
Cyclophosphamide-treated mice CP (+)	Negative control	–	10	110.91 ± 4.46##
		330	10	61.17 ± 7.71*
	Total extract	230	10	59.46 ± 5.22*
		120	10	61.52 ± 5.31*
	Silymarin (positive control)	100	10	63.54 ± 5.09*

*$p < 0.05$ as compared to cyclophosphamide-treated negative control.

#$p < 0.05$ with respect to cyclophosphamide-untreated normal control.

##$p < 0.05$ with respect to cyclophosphamide-untreated normal control.

2.10. Statistical analysis

All data were presented as the mean ± Standard Error of Mean (SEM). Mean values were assessed for significance by Student's *t*-test at $p < 0.05$. One-way ANOVA and Dunnett's method were used for multiple comparisions of data. Statistical analysis was processed using the SigmaStat, version 3.5. Correlation between variables was evaluated using Pearson's correlation coefficient with level of significance $p < 0.05$.

3. Results

3.1. Liver parameters

The hepatoprotective effects of *G. lucidium* on CP-induced hepatic injury and normal mice are shown in Table 2.

As can be seen in Table 2, the MDA content in hepatic tissues of CP (+) mice in negative control group was found to be significantly ($p < 0.05$) increased (110.91 ± 4.46 nM/g protein) as compared with CP (−) normal control (58.82 ± 4.06 nM/g protein), reflecting the hepatic damage induced by CP. The MDA level in hepatic tissues in mice of CP (+) and *G. lucidium* co-treated group at the dose of 330, 230, and 120 mg/kg b.wt were figured out almost the same (61.17 ± 7.71, 59.46 ± 5.22, 61.52 ± 5.31 nM/g protein, respectively) and showed a statistically significant decrease of MDA content as compared with the CP (+) negative control, ($p < 0.05$). Meanwhile, silymarin served as positive control also produced a decrease of MDA in liver tissue significantly (63.54 ± 5.09 nM/g protein).

As can be observed from Table 3, GSH level in liver of total extract treatment were significantly recorded high ($p < 0.05$) with respect to CP (−) normal group, proving the effect of *G. lucidium* extract in enhancing the GSH content. However, GSH after a single dose of CP 150 mg/kg b.wt decreased significantly ($p < 0.05$) when compared with CP (−) mice normal control group. Statistically, 50% depletion of hepatic GSH in CP-induced mice was observed.

Data in Table 3 also showed a significant increase of GSH level in the mice with CY (+) and *G. lucidium* total extract co-treated group after 8-day treatment. However, mice treated with the dose of 230 mg/kg b.wt made the highest GSH level in liver (7,063.59 ± 402.09 nM/g protein) among the tested samples, followed by the dose of 120 mg/kg b.wt (6,114.42 ± 443.58 nM/g protein), and 330 mg/kg b.wt (5,803.09 ± 416.35 nM/g protein). Meanwhile, silymarin at the dose of 100 mg/kg b.wt used as a reference drug also produced a significant increase of GSH (5,935.06 ± 354.05 (nM/g protein).

Table 3. The effect of treatment with the *G.lucidium* total extract on GSH content in cyclophosphamide-treated and normal mice for 8 days

Treatment	Groups	Dosages (mg/kg body weight)	*N*	GSH (nM/g protein)
Cyclophosphamide-untreated mice CP (–)	Normal control	–	10	6,661.23 ± 390.33
	Total extract	330	10	7,192.96 ± 269.33
		230	10	7,316.76 ± 457.59
		120	10	7,427.36 ± 545.14
	Silymarin	100	10	7,311.02 ± 550.05
Cyclophosphamide-treated mice CP (+)	Negative control	–	10	4,279.55 ± 278.03#
	Total extract	330	10	5,803.09 ± 416.35*
		230	10	7,063.59 ± 402.09*
		120	10	6,114.42 ± 443.58*
	Silymarin (positive control)	100	10	5,935.06 ± 354.05*

*$p < 0.05$ with respect to cyclophosphamide-treated negative control.

#$p < 0.05$ with respect to the cyclophosphamide-untreated normal control.

3.2. Histopathological examiation

Histopathological observations of normal mice group showed normal hepatocyte with polygonal shape and no lobular inflammation. The central vein did not contain lymphocytes and none of necrosis of liver cells (Figure 1). Liver sections of CP (150 mg/kg b.wt) treated group expressed the portal tracts with moderate to marked inflammation. Hepatocellular necrosis appeared lobular inflammation with lyphocytic infiltration. Liver of mice in this group showed a severe active hepatitis (Figure 2(a) and (b)). Meanwhile, liver of mice from CP and *G. lucidium* total extract co-treated group

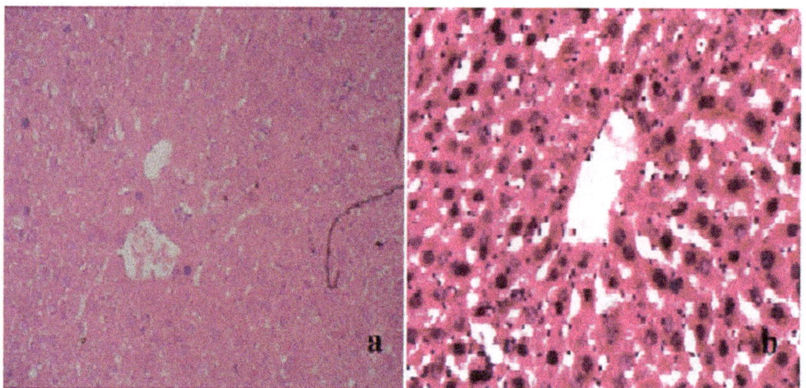

Figure 1. Normal group, section of mouse liver showing no lobular inflammation, none of necrosis and fibrosis. No abnormality of central vein and hepatocytes.

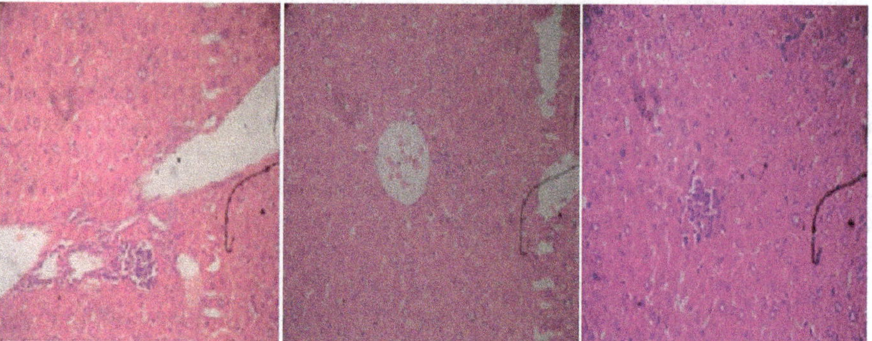

Figure 2a. Cyclophosphamide (150 mg/kg)-treated group, a section of mouse liver showing portal inflammation, hepatocellular necrosis, and lymphocytic inflammatory infiltrations (Hematoxylin and eosin-stained paraffin section; H&E 200).

(120 mg/kg b.wt) showed a mild active hepatitis. Portal space presented slight to moderate inflammation, no necrotic liver cells and small focal lobular inflammation (Figure 3(a) and (b)). Liver sections of silymarin treatment group were manifested in Figure 4. The central vein contained an adhesion of the lymphocytes on the endothelial cell, moderate portal, and lobular inflammation. No necrotic liver cell occurred, the liver were exposed to be a moderate active hepatitis.

4. Discussion

CP is one of the commonly used anticancer drugs for its therapeutic efficacy against various cancers. Earlier studies have proved that ameliorative dose of CP could cause liver toxicity (Mahsa & Shivanandappa, 2013). CP goes through a metabolic activation by hepatic microsomal cytochrome P450 mixed functional oxidase system to produce the two metabolites, phosphoramide mustard and acrolein. Phosphoramide mustard is believed to have an antineoplastic activity, while acrolein highly reactive metabolite with a short biological half-life, may be responsible for CP-induced liver injury (Ludeman, 1999; Selvakumar, Prahalathan, Mythili, & Varalakshmi, 2005). Experimental evidence recommends that oxidative stress is answerable for CP hepatotoxicity. It could generate reactive oxygen species (ROS) like superoxide anion, hydroxyl radical, and hydrogen peroxide (H_2O_2) during its oxidative metabolism and discourages the antioxidant defense mechanism in liver (Bhattacharya et al., 2003). A number of studies have shown that natural products with antioxidant activity protect against CP hepatotoxicity (Haque et al., 2003; Kumar & Kuttan, 2005; Sharma, Trikha, Athar, & Raisuddin, 2000).

G. lucidium has been of research interest because it is commonly use as traditional folk medicine to treat liver disorders. There have been various studies on the efficacy of hepatoprotective property including *in vitro* and animal assays using the extracts with varying solvents. A queous and alcoholic extracts of *G. lucidium* utilized the protective action against acute hepatitis. These achievements were attributed to the inhibitory activities of the *G. lucidium* extract on the membrane lipid peroxidation and the formation of free radical (Lakshmi, Ajith, Jose, & Janardhanan, 2006; Shieh et al., 2001; Wang, Liu, Che, & Lin, 2002). On the other hand, these components contained in the mushroom had been subjected to investigate the key role in liver protection. In a report by Soares et al. (2013), polysaccharides and triterpenoids might possess protective activity against liver injury induced by toxic

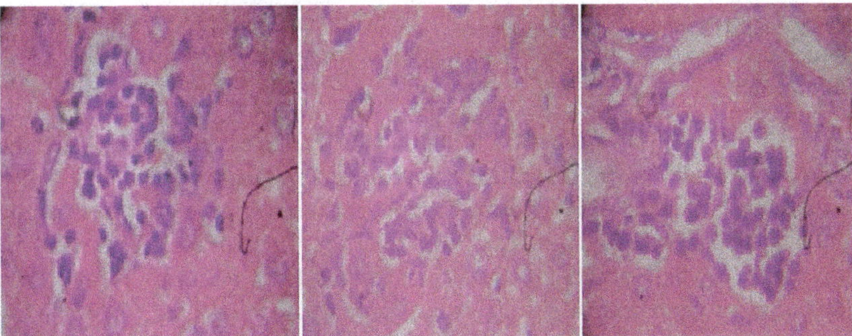

Figure 2b. Cyclophosphamide (150 mg/kg)-treated group, a section of mouse liver showing portal inflammation, hepatocellular necrosis, and lymphocytic inflammatory infiltrations (Hematoxylin and eosin-stained paraffin section; H&E 400).

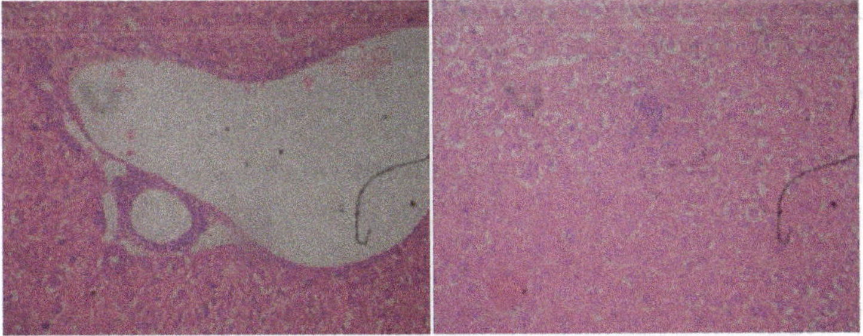

Figure 3a. Section of mouse liver treated with 120 mg/kg of *G. lucidium* total extract showing a mild active hepatitis with no inflammatory at central vein and slight portal inflammation. (Hematoxylin and eosin-stained paraffin section; H&E 200).

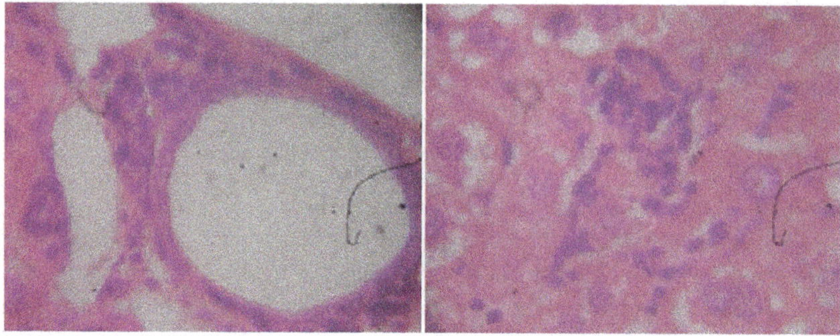

Figure 3b. Section of mouse liver treated with 120 mg/kg of *G. lucidium* total extract showing a mild active hepatitis with no inflammatory at central vein and slight portal inflammation. (Hematoxylin and eosin-stained paraffin section; H&E 400).

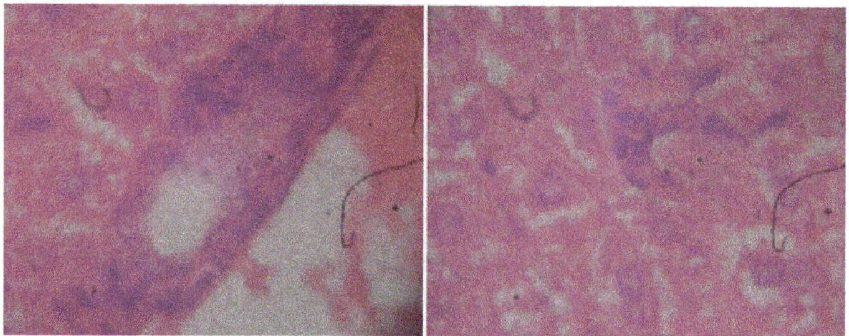

Figure 4. Section of mouse liver with silymarin treated group (100 mg/kg) showing a mild active hepatitis with small portal inflammation, slight lymphocytic infiltration, and no inflammatory of central vein. (Hematoxylin and eosin-stained paraffin section; H&E 400).

agents. Recent studies suggest CP generates ROS like superoxide anion, hydroxyl radical, and hydrogen peroxide (H_2O_2) which cause lipid peroxidation of cellular membrane. During their oxidative metabolism, they depress the antioxidant defense mechanisms in liver (Lakhanpal & Rana, 2005; Mishra & Singh, 2010).These achievements were attributed to the inhibitory activities of the *G. lucidium* extract on the membrane lipid peroxidation and the formation of free radical (Lakshmi et al., 2006; Shieh et al., 2001; Wang et al., 2002).

G. lucidium in Vietnam is naturally harvested in the forest and the fruiting body is developed on the dead ironwood tree. We evaluated the hepatoprotective activity of *G. lucidium* against the hepatic damage induced by CP. Administration of *G. lucidium* for eight consecutive days after CP injection showed a significant inhibition in the liver injury compared to CP-treated mice without *G. lucidium* supplement. At the dose of 230 mg/kg b.wt, the hepatoprotective activity was determined by the depletion of MDA and restored the decrease of endogenous hepatic GSH antioxidant content in the liver. When the polyunsaturated lipids were degenerated by ROS, an ascent of cellular MDA occurs. The production of this aldehyde, typically quantified as thiobarbituric acid reactive substances (TBARS). It is normally considered as a biomarker to evaluate lipid peroxidation which could be indicated to the free radical scavenging activity as well as suppressing oxidative stress in an organism (Devasagayam, Boloor, & Ramasarma, 2003). The results of this study could support the previous findings of Shi, Sun, He, Guo, and Zhang (2008), Nonanka et al. (2005) and Truong and Nguyen (2010) that oral supplement of *G. lucidium* could prevent the CP-induced lipid peroxidation. The current study also revealed a significant reduction of GSH following CP administration which might result from the direct conjugation of CP metabolites with glutathione as part of mechanism of hepatoprotective action.

5. Conclusion

The results of present study suggest that administration of *G. lucidium* developing on the dead ironwood tree harvested in the Central Highlands of Vietnam might have potential in protecting liver against cyclophosphamide-induced hepatotoxicity. Further study is necessary to elucidate target compounds from *G. lucidium* in Vietnam responsible for liver protection

Acknowledgments

I would like to thank MSc Huynh Ngoc Linh for thought-provoking comments and discussions. Special thanks to Anatomical Department, Pham Ngoc Thach University of Medicine, for providing necessary facilities to carry out the histological examination.

Funding

The authors received no direct funding for this research.

Competing Interests

The authors declare no competing interest

Author details

Hong Ngoc Pham[1]

E-mail: phamngocst@gmail.com

Le Son Hoang[1]

E-mail: hlson@hcmiu.edu.vn

Van Trung Phung[2]

E-mail: pvtrung@ict.vast.vn

[1] Department of Biochemistry, School of Biotechnology, International University - Vietnam National University, HCMC, Vietnam.

[2] Institute of Chemical Technology - Vietnam Academy of Science and Technology, HCMC, Vietnam.

References

Abraham, P., & Sugumar, E. (2008). Increased glutathione levels and activity of PON1 (phenyl acetate esterase) in the liver of rats after a single dose of cyclophosphamide: A defense mechanism? *Experimental and Toxicologic Pathology, 59*, 301–306. http://dx.doi.org/10.1016/j.etp.2007.06.006

Acharya, K., Chatterjee, S., & Ghosh, S. (2011). Comparative evaluation on the free radical scavenging activity of eleven Idian cultivated strains of *Pleurotus ostreatus*. *Pharmacologyonline, 1*, 440–450.

Acharya, K., Yonzone, P., Rai, M., & Acharya, R. (2005). Antioxidant and nitric oxide synthase activation properties of *Ganoderma applanatum*. *Indian Journal of Experimental Biology, 43*, 926–929.

Bhattacharya, A., Lawrence, R. A., Krishnan, A., Zaman, K., Sun, D., & Fernandes, G. (2003). Effect of dietary n-3 and n-6 oils with and without food restriction on activity of antioxidant enzymes and lipid peroxidation in livers of cyclophosphamide treated autoimmune-prone NZB/W female mice. *Journal of the American College of Nutrition, 22*, 388–399. http://dx.doi.org/10.1080/07315724.2003.10719322

Devasagayam, T. P. A., Boloor, K. K., & Ramasarma, T. (2003). Methods for estimating lipid peroxidation: An analysis of metrits and demertis. *Indian Journal of Biochemistry & Biophysics, 40*, 300–308.

Gao Y. H., Huang M., Lin Z. B., & Zhou, S. F. (2003). Hepatoprotective activity and the mechanisms of action of Ganoderma lucidium (Curt: Fr) P. Karst. (Ling Zhi, Reishi Mushroom) (Aphyllophormomycetidae). *International Journal of Medicinal Mushrooms, 5*, 111–131.

Haque, R., Bin-Hafeez, B., Parvez, S., Pandey, S., Sayeed, I., Raisuddin, S., & Raisuddin, S. (2003). Aqueous extract of walnut (*Juglans regia* L.) protects mice against cyclophosphamideinduced biochemical toxicity. *Human & Experimental Toxicology, 22*, 473–480. http://dx.doi.org/10.1191/0960327103ht388oa

Kumar, K. B. H., & Kuttan, R. (2005). Chemoprotective activity of an extract of *Phyllanthus amarus* against cyclophosphamide induced toxicity in mice. *Phytomedicine, 12*, 494–500. http://dx.doi.org/10.1016/j.phymed.2004.03.009

Lakhanpal, T. N., & Rana, M. (2005). Medicinal and nutraceutical genetic resources of mushrooms. *Plant Genetic Resources: Characterization and Utilization, 3*, 288–303. http://dx.doi.org/10.1079/PGR200581

Lakshmi, B., Ajith, T. A., Jose, N., & Janardhanan, K. K. (2006). Antimutagenic activity of methanolic extract of *Ganoderma lucidum* and its effect on hepatic damage caused by benzo[a]pyrene. *Journal of Ethnopharmacology, 107*, 297–303. http://dx.doi.org/10.1016/j.jep.2006.03.027

Lindequist, U., Niedermeyer, T. H. J., & Julich, W. D. (2005). The pharmacological potential of mushrooms. *eCAM, 2*, 285–299.

Ludeman, S. M. (1999). The chemistry of metabolites of cyclophosphamide. *Current Pharmaceutical Design, 5*, 627–643.

Mahsa, Z., & Shivanandappa, T. (2013). Amelioration of cyclophosphamide-induced hepatotoxicity by the root extract of Decalepis hamiltonii in mice. *Food and Chemical Toxicology, 57*, 179–184.

Mishra, S., & Singh, R. B. (2010). Effect of mushroom on the lipid profile, lipid peroxidation and liver functions of aging Swiss Albino rats. *The Open Nutraceuticals Journal, 3*, 248–253. http://dx.doi.org/10.2174/1876396001003010248

Ngo, Q. H., & Nguyen, T. T. H. (2011). Antioxidant effect of polysaccharide extracted from Ganoderma colossum on cyclophosphamide-induced hepatotoxicity. *Journal of HCMC Medicine, 15*, 50–55.

Noeman, S. A., Hamooda, H. E., & Baalash, A. A. (2011). Biochemical study of oxidative stress markers in the liver, kidney and heart of high fat diet induced obesity in rats. *Diabetology & Metabolic Syndrome, 3*, 17. http://dx.doi.org/10.1186/1758-5996-3-17

Nonanka, Y., Ishibashi, H., Nakai, M., Shibata, H., Kiso, Y., & Abe, S. (2005). Soothing effect of *Ganoderma lucidium* antlered form on cyclophosphamide-induced adverse reaction. *Gan To Kagaku Ryoho, 32*, 1586–1588.

Pal, J., Ganguly, S., Tahsin, K. S., & Acharya, K. (2000). In vitro free radical scavenging activity of wild edible mushroom, Pleurotus squarrosulus [Mont.] Singer. *Indian Journal of Experimental Biology, 47*, 1210–1218.

Selvakumar, E., Prahalathan, C., Mythili, Y., & Varalakshmi, P. (2005). Mitigation of oxidative stress in cyclophosphamide-challenged hepatic tissue by DL-α-lipoic acid. *Molecular and Cellular Biochemistry, 272*, 179–185. http://dx.doi.org/10.1007/s11010-005-7322-4

Sharma, N., Trikha, P., Athar, M., & Raisuddin, S. (2000). Inhibitory effect of Emblica officinalis on the *in vivo* clastogenicity of benzo[a]pyrene and cyclophosphamide in mice. *Human & Experimental Toxicology, 19*, 377–384. http://dx.doi.org/10.1191/096032700678815945

Shieh, Y. H., Liu, C. F., Huang, Y. K., Yang, J. Y., Wu, I. L., Lin, C. H., & Lin, S. C. (2001). Evaluation of the hepatic and renal-protective effects of *Ganoderma lucidum* in Mice. *The American Journal of Chinese Medicine, 29*, 501–507. http://dx.doi.org/10.1142/S0192415X01000526

Shi, Y., Sun, J., He, H., Guo, H., & Zhang, S. (2008). Hepatoprotective effects of *Ganoderma lucidum* peptides against D-galactosamine-induced liver injury in mice. *Journal of Ethnopharmacology, 117*, 415–419. http://dx.doi.org/10.1016/j.jep.2008.02.023

Soares, A. A., de Sá-Nakanishi, A. B., Bracht, A., da Costa, S. M. G., Koehnlein, E. A., de Souza, C. G. M., & Peralta, R. M. (2013). Hepatoprotective effects of mushrooms. *Molecules, 18*, 7609–7630. http://dx.doi.org/10.3390/molecules18077609

Stroev, E. A., & Makarova, V. G. (1989). Determination of lipid peroxidation rate in tissue homogenate laboratory. In *Laboratory manual in biochemistry* (pp. 243–256). Moscow: Mir.

Truong, T. M. C. & Nguyen, T. T. H. (2010). In vitro and *in vivo* study on antioxidant activity of Ganoderma lucidium (Lingzhi) and Trametes versicolor (Yunzhi). *Journal of HCMC Medicine, 14*, 135–141.

Wang, M. Y., Liu, Q., Che, Q. M., & Lin, Z. B. (2002). Effects of total triterpenoids extract from Ganoderma lucidium (Curt: Fr) P. Karst. (Reishi mushroom) on experimental liver injury models induced by carbon tetrachloride or D-galactosamine in mice. *International Journal of Medicinal Mushrooms, 4*, 337–342.

Wasser, S. P. (2005). Reishi or Ling Zhi (*Ganoderma lucidium*). *Encyclopedia of Dietary Supplements, 2005*, 603–623.

9

Pitch tracking of bird vocalizations and an automated process using YIN-bird[†]

Colm O'Reilly[1]* and Naomi Harte[1]

*Corresponding author: Colm O'Reilly, School of Engineering, Sigmedia, ADAPT Centre, Trinity College Dublin, Dublin Ireland
E-mail: oreilc16@tcd.ie

Reviewing editor: Hynek Burda, Universitat Duisburg-Essen, Germany
Additional information is available at the end of the article

Abstract: Pitch or fundamental frequency is an important feature of bird song, from which scientists can learn much about a population. To use pitch as a feature, researchers need confidence in their pitch extraction system. Pitch detection algorithms (PDAs) proven to work on human speech may not be suitable for all types of bird vocalizations. This paper discusses pitch estimation performance on a variety of common bird vocalizations. The presence of multiple partials or tones simultaneously, extended frequency sweeps through multiple octaves, and rapid pitch modulations are just some of the difficulties encountered when estimating the pitch of bird song. Carefully tuned parameters improve pitch tracking with YIN, but optimal parameters can change quickly even within one song. YIN is a PDA which estimates pitch of human speech very well. This paper presents YIN-bird, a modified version of YIN which exploits spectrogram properties to automatically set a minimum fundamental frequency parameter for YIN. Gross pitch errors on whistles and trills were reduced by up to 4% on a ground truth data-set of synthetic bird song with known pitch. This data-set was evaluated by expert listeners and described as "sounding like original & can hardly tell it is synthetic". A qualitative analysis showing YIN-bird not to be suitable for more complex bird vocalizations, such as nasals, is also presented.

ABOUT THE AUTHOR

Colm O'Reilly submitted his dissertation to Trinity College, the University of Dublin, Ireland for the degree of Doctor of Philosophy. He hopes to graduate mid-2017. Colm has been a research member of Sigmedia, a signal processing and media applications research group with the Department of Electronic and Electrical Engineering, since 2013. Colm graduated with a first class honors in Electronic Engineering from University College Dublin in 2011. He spent two years as a test development engineer with Analog Devices 2011–2013. His research interests include digital signal processing and machine learning. His PhD dissertation concentrated on digital signal processing techniques applied to bird song. His research involved collaborations with the school of Zoology at Trinity College. Work in this paper reports improvements in pitch tracking of bird vocalizations which is important for quantifying difference in calls or songs of different bird populations. The reported improved pitch tracker helps zoologists analyze acoustic evidence automatically for taxonomy review.

PUBLIC INTEREST STATEMENT

The analysis of bird song has increased in the speech processing community in the past five years. Much of the reported research has concentrated on the identification of bird species from their songs or calls. A lesser reported topic is the analysis of bird songs from subspecies of the same bird. A common way to quantify difference between bird populations is to analyze the pitch of bird song. Scientists sometimes extract pitch without knowledge of how well pitch trackers perform on bird song. This paper reports pitch tracking performance on different syllable types of bird vocalizations. This paper also presents YIN-bird an improved pitch tracker for birds. A well-known algorithm, YIN, accurately tracks the pitch of human speech, but is susceptible to octave errors when tracking bird song. YIN-bird optimizes YIN to improve pitch tracking performance of bird vocalizations. Syllables for which YIN-bird is not suitable for are also described.

Subjects: Zoology; Electrical & Electronic Engineering; Digital Signal Processing

Keywords: pitch tracking; bird vocalizations; bird song

1. Introduction

The ability to automatically analyze bird vocalizations would greatly benefit zoologists in their behavioral and ecological studies. The importance of birds' vocalizations cannot be overstated. Bird song is essential for communication, especially for mate attraction and territory defense (Catchpole & Slater, 2008). When visibility is limited, such as in rainforests with dense vegetation, acoustic communication may be the only means of species identification (Trifa, Kirschel, Taylor, & Vallejo, 2008). While the scientific study of bird song has made important contributions to the field of zoology, its intrigue has also sparked interest from speech and language researchers in an effort to improve the efficiency, accuracy, and repeatability of bird song analysis to monitor and assess bird communities (Connor, Li, & Li, 2012).

In the last few years, the speech processing community has researched many issues in bird vocalizations, notably species classification (Connor et al., 2012; Fagerlund & Laine, 2014; Graciarena, Delplanche, Shriberg, Stolcke, & Ferrer, 2010; Heller & Pinezich, 2008; Trifa et al., 2008), syllable or phrase classification (Anderson, Dave, & Margoliash, 1996; Chen & Maher, 2006; Tan, Alwan, Kossan, Cody, & Taylor, 2015; Kaewtip, Tan, Alwan & Taylor, 2013; Kogan & Margoliash, 1998; Ranjard & Ross, 2008; Tan, Kaewtip, Cody, Taylor, & Alwan, 2012), and song structure analysis (Lachlan et al., 2013; Sasahara, Cody, Cohen, & Taylor, 2012). The use of songs and calls to delimit species and monitor populations has several practical advantages, e.g. ease and economy of sound recording and analysis (Remsen, 2005).

Another topic in ornithology is determining how similar two populations of birds are based on their calls and songs. Catchpole and Slater (2008) mention the importance of vocalizations in mate choice and species recognition. This suggests acoustic signals may give early clues of species distinction (Lambert & Rasmussenm, 1998). Harte, Murphy, Kelly, and Marples (2013) investigated the issue of call similarity and concluded that classifier performance is related to similarity but not to a quantifiable indicator. Prosodic features like pitch have been used to quantify differences in bird populations. O'Reilly, Marples, Kelly, and Harte (2015) used pitch contour micro-structure to measure similarity of bird calls and songs inspired by dialect similarity measures used in Mehrabani, Boril, and Hansen (2010) and Mehrabani and Hansen (2015). McKay, Reynolds, Hayes, and Lee (2010) examined song in making a case for the Bahan subspecies of the Yellow-throated Warbler to be reclassified as a distinct species. Song divergence was important evidence in the reclassification process. Comparisons were on the basis of visual inspection of spectrograms. Sangster, King, Verbelen, and Trainor (2013) described a new species of owl, known as the Rinjani Scops Owl, based on analysis of vocalizations. In both McKay et al. (2010) and Sangster et al.'s studies (2013), various features were measured, like amplitudes at certain frequencies, number of syllables and phrases, pitch slope, and frequency. These are just some of the studies that would benefit from accurate and automatic measurement of pitch.

Quantitative measures of acoustic similarity were used to investigate patterns of shared vocal behavior in social species by Meliza, Keen, and Rubenstein (2013). Pitch- or fundamental frequency (f_0)-based methods performed best at separating distinct categories of superb starling calls. If two populations with a common origin are isolated, one can expect that the songs of each will accumulate modifications independently. Detecting those changes can help infer population histories and relationships (Ranjard & Ross, 2008). Understanding how vocalizations are shared among individuals of the same species requires quantitative methods for measuring how acoustic features vary across groups and individuals (Meliza et al., 2013). Tobias et al. (2010) developed a system of standardized criteria for species delimitation in birds using acoustic evidence of song structure like maximum frequency, minimum frequency, bandwidth, and peak frequency. The system Tobias et al. (2010) used was biometirc (e.g. size and shape), plumage (e.g. color and pattern) and voice (e.g. pitch and pace) as evidence. Sangster et al. (2013) also relied on frequency information to reclassify a species of owl. Lachlan et al. (2013) included pitch to evaluate chaffinch song. These examples highlight the growing interest in accurate acoustic analysis of bird song.

Thus, there is growing agreement among ornithologists that pitch analysis of bird vocalizations is useful, as many avian calls and songs are tonal (Meliza et al., 2013; Tchernichovski, Nottebohm, Ho, Pesaran, & Mitra, 2000). YIN (De Cheveigné & Kawahara, 2002) is a pitch detection algorithm (PDA) which was developed to estimate pitch of human speech or musical sounds. YIN, as discussed later in Section 3, has a strong potential for pitch tracking in bird song. However, it must be carefully tuned for each species and often even for different segments of a single song. This paper presents a modification to YIN to allow more fully automated pitch tracking. This offers advantages in large batch processing when outputs can't be checked in detail. The aim is to offer zoologists a tool for pitch tracking that requires less specialist knowledge and intervention. Improving automatic pitch estimation of bird vocalizations is also beneficial to engineers and scientists, allowing larger scale studies where results can achieve greater levels of statistical significance and precision. Knowledge of performance of current pitch tracking systems is also important. In many previous experiments, researchers may have used pitch tracking systems designed for human vocalizations and assumed the accuracy to be sufficient.

There were a number of objectives which influenced the work in this paper. The first objective was to develop an improved pitch tracking system for bird song, as YIN, which works well for speech, is prone to octave errors when estimating pitch of bird song due to its higher fundamental frequency. The improved system presented in this paper referred to subsequently as YIN-bird was developed. Second objective was to evaluate YIN-bird's use on different syllable types of bird song. This required generating ground truth pitch values for data. In speech, ground truth is calculated by measuring vibration rates of the larynx. That is not feasible for bird song. Instead, a data-set of synthesized bird song was generated and the pitch of synthesized data was used as ground truth. Once the data-set was generated, YIN-bird pitch estimation was performed and performance was evaluated on whistles, trills, and nasals. The final objective of this paper is to qualitatively evaluate performance of complex bird song for which ground truth could not be easily generated. This presents the reader with useful information on what types of vocalizations YIN-bird works well on and what types of vocalizations it does not.

YIN-bird, the improved system presented in this paper, is described in Section 4. For the first time, a ground truth database of synthesized bird vocalizations with known pitch was developed to allow a quantitative evaluation of YIN-bird and is discussed in Section 5.2. The performance of YIN compared to YIN-bird is thus evaluated using the standard error metrics in the signal processing field (described in Section 5.4). Common types of bird vocalizations are presented in Section 2. Using YIN-bird on a set of bird whistles improves gross pitch error from 1.67 to 0.58%. For trills, the figure reduces from 6.29 to 2.31%. Performance on other vocalization types is discussed in Section 6. Finally, a qualitative analysis of more complex bird vocalizations and pitch tracking performance on these sounds is discussed in Section 8.

2. Bird vocalizations

Like in humans, bird sounds are produced by the flow of air during expiration through a vocal system (Doupe & Kuhl, 1999). Some bird vocalizations, like the song of the Swamp Sparrow (*Melospiza georgiana*), are tonal. Others, such as the song of the Zebra Finch (*Taeniopygia guttata*), have a noisy spectrum quality with multiple frequency components, more closely resembling human speech (Doupe & Kuhl, 1999).

Bird vocalizations are produced by source filter mechanism that is similar to that of humans (Beckers, Suthers, & Ten Cate, 2003). Beckers (2011) has also shown that human and avian sound perception is comparable. Human voiced speech is made up of strong energy at f_0 with relative amplitudes at multiple harmonics due to the properties of humans' source filter system.

Bird vocalizations have higher pitch than humans which means the interval between harmonics is larger than for human speech. This implies that even gentle low pass filtering by a bird's vocal tract could potentially remove all harmonics leaving just a pure f_0 tone. In contrast, overtone or nasal sounds are produced when the bird uses a wider bandwidth filter which allows sounds at f_0 and multiple harmonics to be emitted during vocalization.

Birds produce a wide variety of vocalizations. These range from short, monosyllabic calls, to long complex song (Catchpole & Slater, 2008). Early researchers did not agree on a common set of units by which birds' song of various different species might be described (Thompson, LeDoux, & Moody, 1994). Thompson et al. (1994) presented a system for describing bird song units in the hope of greater standardization in the protocols by which researchers generate and name bird song units. A note or element refers to the smallest level of song (which can be analogous to phonetic units) and is defined by a sound represented by a continuous trace on the spectrogram. Notes can be grouped together to form syllables, which are units of sound separated by silent intervals (Doupe & Kuhl, 1999). Syllables and notes are themselves organized into third-order units known as phrases, and phrases are in turn clustered together into performances called song (Catchpole & Slater, 2008; Thompson et al., 1994). Labeling units of bird song still differs from scientist to scientist, and species to species, but an example of unit segmentation from Thompson et al. (1994) can be seen in Figure 1. Note sometimes syllables can be made up of just a single note rather than a group of notes, as seen in Figure 1.

Experiments here evaluate pitch tracking performance on different vocalization types at the syllable level. Syllables tend to fall into one of the following categories:

2.1. Whistles

Catchpole and Slater (2008) describe whistles as the most basic and common type of vocalization. A short whistle of constant pitch appears as a pure, unmodulated frequency trace (see (a) on the spectrogram in Figure 2). A sound which drops from a high to low frequency appears as a downward slope (see (b) in Figure 2). Whistles can be monotone, upslurred, downslurred, overslurred (where pitch rises then falls), or underslurred (where the opposite is true). Whistles often occur in repetition to form phrases. These phrases can contain a constant series of whistles with each whistle rising or falling in frequency. An accelerating or decelerating series of syllables is also possible (Pieplow & Spencer, 2013). Figure 3 shows an example of a decelerating downslurred series of upslurred whistles. Intervals between each whistle will augment over time (decelerating) with each whistle rising in frequency (upslurred whistle) and each whistle syllable will be a lower frequency than the previous syllable (downslurred series).

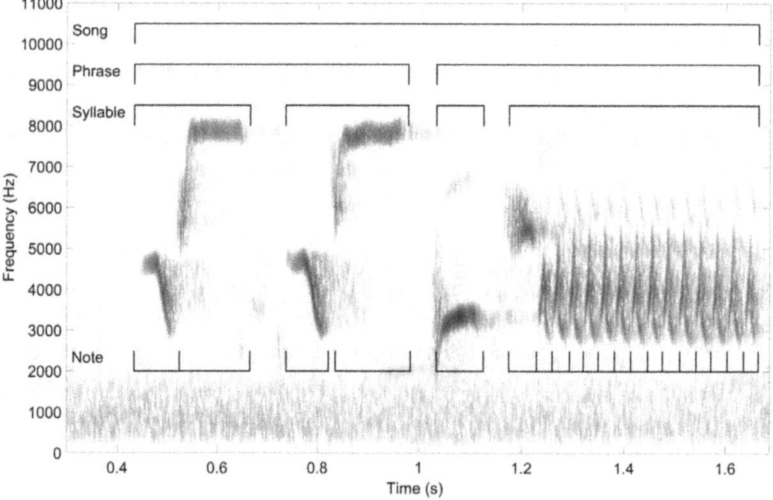

Figure 1. A Song Sparrow song showing note, syllable, phrase, and song boundaries.

Source: This plot is inspired by work in Thompson et al. (1994).

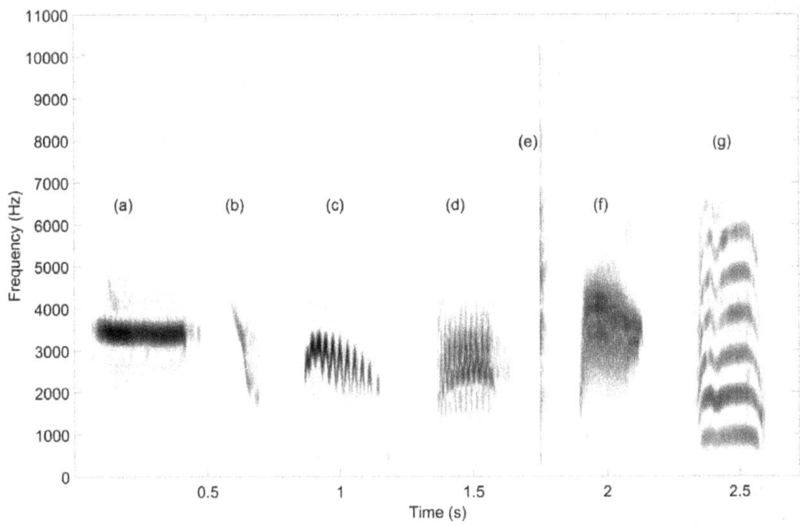

Figure 2. Spectrogram of different syllable types based on diagram from Catchpole and Slater (2008). (a) Whistle, (b) Downslurred whistle, (c) Trill, (d) Buzzy sound, (e) Noisy sound, (f) Noisy buzz, (g) Harmonic or Nasal sound.

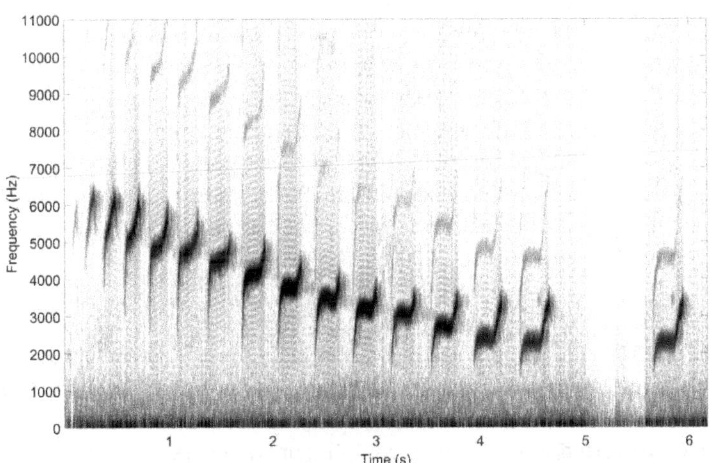

Figure 3. Decelerating downslurred series of upslurred whistles.

Source: Canyon Wren Song (XC, 2013).

2.2. *Hoots*

Hoots are just low-pitched whistles, typically less than 1 kHz (Pieplow & Spencer, 2013). These sounds are typical of the voices of doves and large owls.

2.3. *Trills*

Syllables that contain a series of elements or notes which rise and fall in frequency at a rate greater than 10 Hz will be perceived as a trill. Sounds with more rapid modulations are referred to as "buzzy" sounds. Buzzy sounds are less musical. An example of trilled vibrato and buzzy vibrato can be seen in Figure 2(c) and (d), respectively.

2.4. *Noise*

Not all bird sounds are tonal or periodic. Noisy sounds are constructed from short bursts of white noise and sound like a click. A noisy example is shown in Figure 2(e) and a noisy buzz sound is shown at (f). Noisy bird sounds are likely to be harsh on the ear (Pieplow & Spencer, 2013). As noisy sounds are unvoiced, they are excluded from pitch extraction experiments here.

2.5. Harmonics (or Nasals)

Many bird sounds are actually combinations of multiple simultaneous whistles (partials) of different frequencies that the human brain typically perceives as a single sound (because of the mathematical relationship between the frequencies of the different whistles). Harmonic sounds are represented on a spectrogram by a typical ladder pattern and have a noisy spectral quality (i.e. many simultaneous frequencies present). An example of a harmonic nasal sound is shown in Figure 2(g). The different whistles are called partials because they are only partial components of the sound (Pieplow & Spencer, 2013). If the energy at f_0 or the 2nd partial is prominent, the sound will be soft and melodic because the 2nd harmonic is an octave above f_0. These sounds blend well together. If higher partials have stronger energy, the sounds are more nasal, as the partials tend to clash perceptually (we will use the term "Nasals" to refer to these sounds). While whistles can be a pure tone or a combination of strong f_0 with lower amplitude harmonics, nasals refer to sounds with high energy at higher order partials and sound harsher than melodic whistles with harmonics.

These nasal sounds are very challenging for pitch tracking, as many nasals sounds have missing harmonics (including missing fundamental) or inharmonic partials (Marler & Slabbekoorn, 2004). These problems will be discussed in detail later in Sections 6 and 8.

2.6. Two-voiced sounds

Some birds have the ability to produce sounds with two f_0 values at once (Catchpole & Slater, 2008), which means there are two f_0. This results in vocalizations complete with two f_0, harmonics of both f_0 and heterodyne frequencies (Pieplow & Spencer, 2013). An example of this is shown in Figure 4 with a song of a Prothonotary Warbler. Note the labels showing two f_0 (A & B), harmonics (integer multiples of A & B), and heterodyne frequencies (sums and differences of f_0 and harmonics) (Pieplow & Spencer, 2013). Birds produce sound using their equivalent of the human voice box called the syrinx. Whereas the human larynx is situated at the top of the trachea, the syrinx is much lower down, at the junction of the two bronchi. This means that the syrinx has two potential sound sources (voices), one in each bronchus. The sounds are mixed when fed into the common trachea and buccal cavity (Catchpole & Slater, 2008). Complex two-voiced sounds contrast to many common bird songs that have one main frequency band (Sturdy & Mooney, 2000).

While there is scientific literature on the two-voiced phenomenon (Krakauer et al., 2009; Miller, 1977; Zollinger, Riede,& Suthers, 2008), its regularity is undocumented. Informally, zoologists suggest most birds use only one side of their syrinx, some switch between sides during song, and few birds use both sides simultaneously. The complexity of pitch tracking and the inaccuracy of ground truth pitch calculations exclude two-voiced sounds from quantitative analysis here, but a qualitative evaluation is given in Section 8.

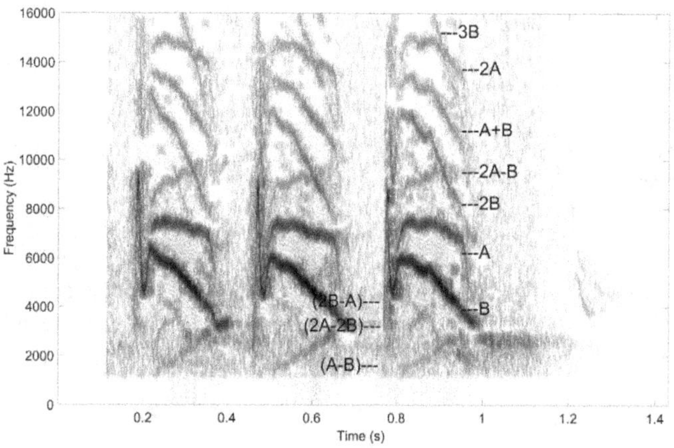

Figure 4. Song of Prothonotary Warbler. Two-voiced bird song with labels showing two f_0 (A & B), harmonics (integer multiples of A & B), and heterodyne frequencies (sums and differences of f_0 and harmonics)

Source: Pieplow and Spencer (2013).

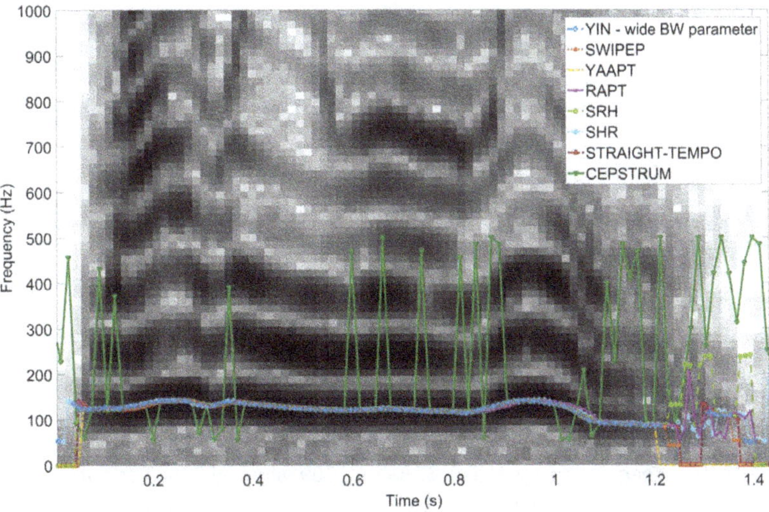

Figure 5. Pitch tracking on speech example from TIMIT database.

These syllable types are very broad categories. Some of them grade into one another, and some of them occur in combination, e.g. a note may be simultaneously buzzy, noisy, and harmonic. Nonetheless, this basic vocabulary is very useful when discussing the qualities of bird sounds (Pieplow & Spencer, 2013).

3. Pitch tracking

Pitch or f_0 estimation is a much debated topic in speech processing. In speech, the term fundamental frequency (f_0) describes the period of voiced speech, and is analogous to pitch. f_0 is the inverse of the smallest true period in the interval being analyzed (Talkin, 1995). Pitch is the perceived f_0 of a signal (Camacho, 2007). A sound which may not be periodic may still be perceived as having a pitch. However, period and pitch are considered equivalent over a wide range of possible values. Thus, f_0 estimation methods are often referred to as PDAs (De Cheveigné & Kawahara, 2002).

Pitch provides important information about a sound source. In speech, pitch can be used for a variety of tasks, like identifying gender, as males tend to speak with a lower f_0 than women (Wang & Lin, 2004). Speaker emotion can be inferred from pitch, e.g. low pitch can suggest the speaker is sad while high pitch suggests excitement. Pitch changes in a sentence influence how the sentence is interpreted, e.g. a rising pitch is generally observed when a question is asked (Murray & Arnott, 1993). In music, pitch estimation is used to name notes (Sethares, 2005) which can be used for automatic music transcription.

3.1. State-of-the-art pitch tracking

Throughout the last 30 years, PDAs have been a hot research topic. While there have been major developments in PDAs, debates still exist about which tool to use under certain conditions. Details relevant to the current work are presented here. However, for in-depth comparisons, the reader is directed to Talkin (1995), De Cheveigné and Kawahara (2002), Sethares (2005), Camacho and Harris (2008), and Luengo et al. (2007).

The YIN PDA is based on the well-known autocorrelation method with a number of modifications (De Cheveigné & Kawahara, 2002). Autocorrelation (AC)-based pitch estimators are preferred for the majority of cases as AC can deal with missing harmonics and inharmonic signals. AC can overcome the problem of giving high scores for subharmonics of the pitch. YIN uses average squared difference instead of AC and includes several modifications that combine to prevent errors. YIN looks for dips instead of peaks (which is why it's called YIN opposed to YANG) which makes it more immune to amplitude changes which affect AC (De Cheveigné & Kawahara, 2002). Other commonly used pitch algorithms include: Harmonic Product Spectrum (HPS) (Schroeder, 1968), Subharmonic summation (SHS) (Hermes, 1988), cepstrum (CEP) (Noll, 1967), and Subharmonic-to-harmonic ratio (SHR) (Sun, 2002).

Most pitch tracking software uses some form of the aforementioned techniques. PRAAT (Boersma, 1993) is a commonly used package that estimates pitch in two ways, autocorrelation and cross-correlation. RAPT (sometimes referred to as GET_F0) (Talkin, 1995) is a robust algorithm that uses a multi-rate approach and the normalized cross-correlation function (NCCF). eSRPD (Bagshaw, Hiller, & Jack, 1993) is a super resolution pitch determinator that uses NCCF and removes discontinuities during post-processing. TEMPO (Kawahara, Katayose, De Cheveigné, & Patterson, 1999) is a tool found in STRAIGHT (Kawahara, Estill, & Fujimura, 2001), a speech analysis and synthesis toolkit, and uses the instantaneous frequency of the outputs of a filterbank. It estimates pitch using both time interval and frequency cues, and is designed to minimize perceptual disturbance due to errors in source information extraction (Babacan, Drugman, d'Alessandro, Henrich, & Dutoit, 2013).

De Cheveign and Kawahara (2002) presented an evaluation of YIN against AC, CC, SHS, eSRPD, CEP, and TEMPO. YIN outperformed these (and other) pitch estimators when tested on Japanese, English, and French databases. Luengo et al. (2007) showed PRAAT, RAPT, and cepstrum to work better than SHR on clean speech and noisy speech. Work by Wei and Alwan (2009) showed PRAAT to outperform RAPT (GET_F0) and TEMPO under both white noise and babble noise conditions. Evaluation of pitch tracking performance by Camacho and Harris (2008) showed SWIPE' to perform the best followed by SHS, RAPT, TEMPO, and YIN with gross errors less than 2.10% when tested on clean speech. SHR, eSRPD, & CEP had a gross error greater than 3.5%. Pitch estimation was also trialed on musical instruments in Camacho and Harris (2008). The gross error rate for tests on musical instruments by octave for SWIPE' was 0.97% and YIN was 0.99%. SHR performed the worst at 36%. RAPT was excluded from musical instrument tests because the bandwidth of musical instruments is too large for the two-pass down-sampling method used by RAPT.

Babacan et al. (2013) evaluated pitch tracking on singing voice, with PRAAT and RAPT providing the best determination of voicing boundaries. RAPT reached the lowest number of gross pitch errors while YIN achieved the best accuracy on singing. Finally, YIN was shown to suffer the most on signing in reverberant conditions while STRAIGHT was the most robust.

An example of pitch tracking on speech using popular PDAs can be seen in Figure 5. All PDAs perform well except for cepstrum. This evaluation was not exhaustive and is included for descriptive purposes as opposed to evaluating pitch trackers on speech.

3.2. Pitch tracking for birds

Pitch extraction tools which have been proven to work for human speech and music may not work as well on birds. Bird vocalizations differ to speech in a number of ways. An important difference is the frequency range. Bird vocalizations tend to have a wider bandwidth and higher mean pitch than human vocalizations. In 2011, Tchnernichovski released software called Sound Analysis Pro (SAP) (Tchernichovski, Kashtelyan, Swigger, & Mitra, 2011) (also available as MATLAB toolkit SAT). SAP calculates a number of features, one being f_0 which is calculated using the YIN algorithm (De Cheveigné & Kawahara, 2002). Mandelblat-Cerf and Fee (2014) also used SAP for evaluating song imitation (also for zebra finches) where pitch again was a crucial feature. While zebra finch vocalizations may not be liable to pitch errors, YIN's performance on other types of bird vocalizations is undocumented. Babacan et al. (2013) discussed pitch tracking performance on singing sounds. While singing sounds are not identical to bird vocalizations, they are more comparable than speech and bird song. Work in (Babacan et al., 2013) showed YIN to have the lowest fine pitch error rate and second lowest gross pitch error and F0 frame error after RAPT.

Early work leading to this paper revealed that a number of pitch trackers do not match the spectrogram. An example of pitch tracking on bird song can be seen in Figure 6. The plot shows YIN and SWIPE' to accurately track the bird syllables while the rest of the PDAs do not. YIN performed slightly better voiced/unvoiced detection than SWIPE' here. Most parameters were the default settings except window size was set to 6.7 ms and frame rate to 1.7 ms. These values were chosen experimentally. Our choice of window size accommodates the trade-off between a large window which does not capture rapid pitch modulations and a small window which does not capture as many periods.

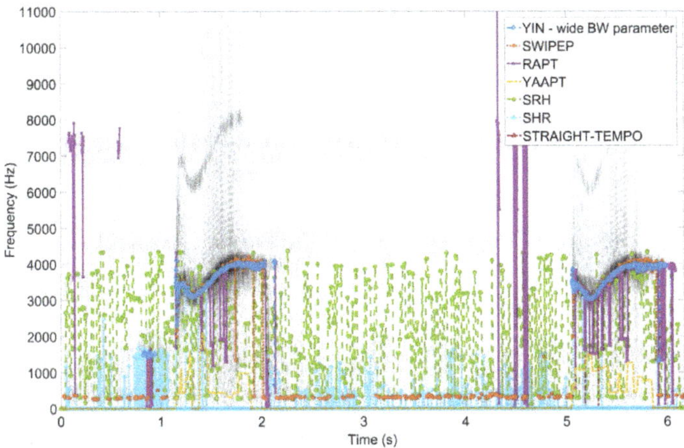

Figure 6. Pitch tracking on bird song of Eastern Wood Pewee.

Source: XC 1377 (2013).

When investigating other PDAs, we used default parameters provided by authors. In some instances, frequency range needed to be modified to allow pitch tracking of bird song which is a lot higher than humans. Further careful parameter selection for other PDAs may have brought their performance in line with YIN but Figure 6 shows YIN to be the optimal choice of pitch tracker for bird song with minimal initial tuning.

Based on these preliminary tests, the findings in Babacan et al. (2013), the use of YIN in SAP (Mandelblat-Cerf & Fee, 2014; Tchernichovski et al., 2000), and its reputation in the speech community as a good pitch estimator for speech and music, YIN was chosen as the baseline pitch estimator for extracting pitch of bird recordings in this paper. Although RAPT performed slightly better than YIN on singing sounds in Babacan et al. (2013), the pitch range of birds (> 1,666 Hz) is too large for RAPT (Camacho & Harris, 2008); hence, it was not chosen as the baseline here. Autocorrelation is very effective for pitch tracking, but some autocorrelation peaks suffer ambiguity, which leads to octave error or estimates too low in frequency (Lee & Ellis, 2012). Some bird vocalizations change frequency rapidly and over a wide range (1–5 kHz). Many syllables include extended frequency sweeps that sometimes exceed two octaves (Marler & Slabbekoorn, 2004), which makes bird vocalizations prone to these types of errors.

Other pitch packages have been used within the bird community. Lachlan et al. developed Luscinia available at Lachlan (2012). Luscinia is widely used in the community to analyze bird song (Lachlan et al., 2013). It provides pitch extraction but requires supervision. The Luscinia GUI allows the user to select elements that require pitch tracking, but even when elements are carefully selected the exported pitch tracks may not always match the spectrogram. Meliza et al. (2013) developed Chirp, a tool which allows the user draw a mask on the spectrogram to improve pitch estimation. Time-frequency reassignment spectrographic analysis, harmonic template matching, and Bayesian particle filtering are combined to produce pitch estimates (Meliza et al., 2013). While this method provides great pitch estimates when compared to the spectrogram, the masks need to be hand drawn on the spectrogram. Even with hand drawn masks, the pitch can still be error prone.

The focus of this paper is to present a pitch tracking algorithm suited to large-scale processing of tracks that require no manual intervention other than initial tuning.

4. Tuning YIN to bird song

YIN processes audio data and outputs a pitch estimate. Parameters can be specified for each file, with a more accurate pitch estimate when parameters are carefully selected to match the input characteristics. One of the more sensitive input parameters is minimum frequency threshold ($f_{0_{min}}$). As bird vocalizations have a wider bandwidth than human speech, a single $f_{0_{min}}$ for all segments of the input file may not be suitable. Our proposed system YIN-bird determines a suitable $f_{0_{min}}$ for each

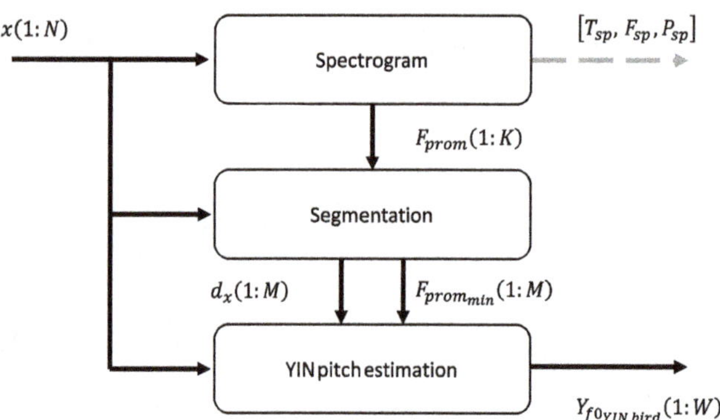

Figure 7. Block diagram of adaptive $f_{0_{min}}$ YIN (YIN-bird).

segment of a bird recording, through careful analysis of the input spectrogram. Using spectrogram information, each segment will be assigned a $f_{0_{min}}$ parameter which leads to a more accurate pitch estimate for each input file. Maximum frequency parameter ($f_{0_{max}}$) had little influence on our pitch estimates and was set to 0.4 $x\, f_s$ where f_s is the sampling frequency of the input file. A block diagram of the system is shown in Figure 7.

Step 1: involves calculating the spectrogram parameters $[T_{sp}, F_{sp}, P_{sp}]$, of audio recording $x(1:N)$, where x is the input signal, N is the number of samples of the input signal, T_{sp} is the spectrogram

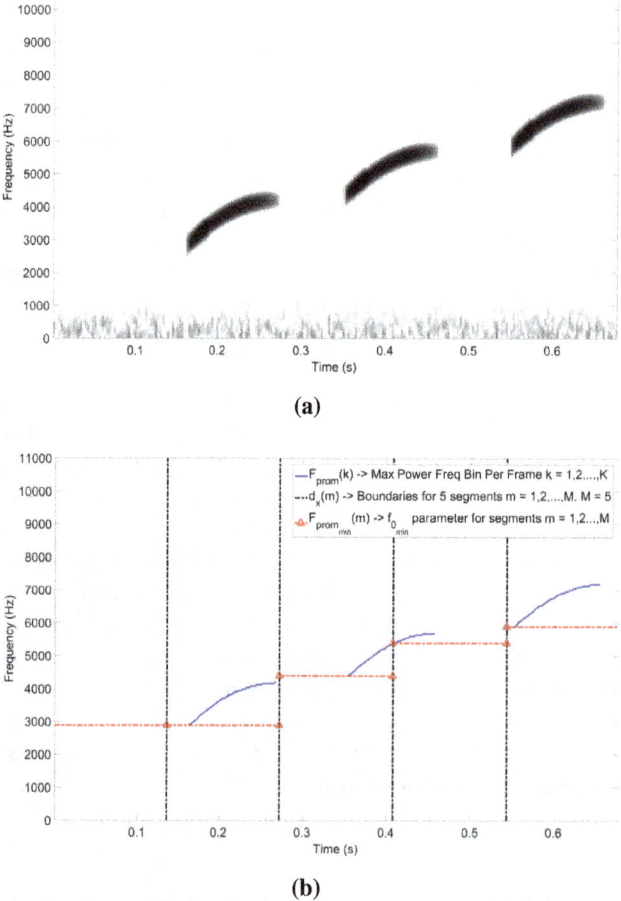

Figure 8. Elements of processing in YIN-bird. (a) Spectrogram of synthetic bird whistles input to YIN-bird and (b) Bird song prominent frequencies (continuous line (online version: blue)), segment boundaries (vertical broken line (online version: black)) and adaptive $f_{0_{min}}$ values (broken horizontal line with circular markers (online version: red)) used by YIN-bird.

frame time information, F_{sp} is the spectrogram frequency bins, and P_{sp} is a matrix containing the power of each frequency bin at each time frame. Figure 8(a) shows a spectrogram of bird syllables. Using the power (dB) and frequency (Hz) information, a prominent frequency (i.e. frequency bin with most power) for each frame is selected $F_{prom}(k)$ where $k = 1, \ldots, K$ and K is the number of frames in the spectrogram. Figure 8(b) has the prominent frequencies $F_{prom}(k)$ plotted with a continuous line (online version: blue). Information at frequencies of 200 Hz (value chosen by low pass filtering multiple bird recordings and checking for the presence of bird vocalizations) and below is assumed to be noise and is ignored. If the power of frame k's prominent frequency ($P_{F_{prom}}(k)$) is less than $\mathrm{mean}(P_{F_{prom}}(1{:}K))$ for a given recording, frame k's prominent frequency is ignored (in practice, assigned "not a number" (NaN) in MATLAB) as it is most likely an unvoiced frame or a frame without vocalization (See Equation (1)).

$$\begin{aligned} &\text{If: } P_{F_{prom}}(k) < \mathrm{mean}(P_{F_{prom}}(1{:}K)) \\ &\qquad \text{then } F_{prom}(k) = NaN \end{aligned} \tag{1}$$

where k is the spectrogram frame number, $P_{F_{prom}}(k)$ is the power at the prominent frequency of frame k, $\mathrm{mean}(P_{F_{prom}}(1{:}K))$ is the average of the power at each frame's most prominent frequency over a single recording, and $F_{prom}(k)$ is the prominent frequency at frame k.

Step 2: segments the audio file into chunks specified by the user. The segment size is selected based on the bird corpus being used (small segment size gives slower execution). In this paper, each segment contains 3,000 samples of input x (68 ms when fs is 44.1 kHz). 3,000 was chosen experimentally accounting for the trade-off between slow processing for short segments and less frequent updates of $f_{0_{min}}$ for larger segments. All files in our data-set were resampled to 44.1 kHz (original recordings varied from 22.5 to 44.1 kHz). Each segment is described as $d_x(m)$ where $m = 1, \ldots, M$ and M is the input number of samples (N) divided by 3,000. Segments are shown divided by broken vertical lines (online version: black) in Figure 8(b). Groups of prominent frequencies ($F_{prom}(1{:}K)$) are assigned to an appropriate $d_x(m)$. If K is 300 frames and M is 30 segments, then prominent frequency values $F_{prom}(1{:}10)$ will be grouped in $d_x(1)$. The minimum F_{prom} in each $d_x(m)$ is $F_{prom_{min}}(m)$. $F_{prom_{min}}(m)$ for each frame is plotted with a broken horizontal line and circular marker (online version: red) in Figure 8(b). In Figure 8(b), the first two segments have the same value for $f_{0_{min}}$ ($F_{prom_{min}}(1) = F_{prom_{min}}(2)$). This is because if there is no vocalization within a segment, F_{prom_m} will take its nearest neighbor's value (or nearest neighbor with a value). If there are two nearest neighbors, previous values take precedence over posterior neighbor's value.

Step 3: involves processing the whole audio file (x) with YIN multiple times. This is purely to make timing information of YIN-bird's output consistent with YIN. Each YIN estimation uses $f_{0_{min}}$ taken from $F_{prom_{min}}(m)$. $F_{prom_{min}}$ values are rounded to the nearest 100 Hz to reduce the number of times x is passed through YIN. If any two values in $F_{prom_{min}}$ are equal, this reduces the number of times YIN is called from M to $M - 1$. Once all the pitch estimates have been collected, each segment $d_x(m)$ is assigned its pitch estimate from YIN's output when $f_{0_{min}}$ equals $F_{prom_{min}}(m)$. Finally, an output pitch vector from YIN-bird is concatenated, $Y_{f0_{YINbird}}(1{:}W)$ (where W is number of pitch values, reliant on YIN's hop size (or frame rate, equal to 1.7 ms here) parameter).

5. Experimental setup

These experiments have two main goals: to evaluate the accuracy of pitch tracking on different types of bird vocalizations and to evaluate the benefit of using an adaptive $f_{0_{min}}$ parameter (YIN-bird). To evaluate the accuracy of pitch tracking, a synthesized bird song data-set had to be generated with known ground truth pitch.

5.1. Data

Examples of birds that produce sounds discussed in Section 2 are given at earbirding.com (Pieplow & Spencer, 2013). Recordings of these birds were downloaded from xeno-canto.org, a popular website

Table 1. Bird vocalization data

Category	No. of examples	Length (min:sec)
Whistles & hoots	107	40:09
Trills	65	13:02
Nasals	63	12:32

Table 2. Species which make up Whistles and Hoots data-set

Whistles & Hoots
American Robin—Turdus migratorius
American Robin—Turdus migratorius caurinus
Black-capped Chickadee—Poecile atricapillus
Black-capped Chickadee—Poecile atricapillus occidentalis
Black-chinned Sparrow—Spizella atrogularis
Black-chinned Sparrow—Spizella atrogularis cana
Canyon Wren—Catherpes mexicanus
Canyon Wren—Catherpes mexicanus mexicanus
Cedar Waxwing—Bombycilla cedrorum
Common Ground Dove—Columbina passerina
Common Ground Dove—Columbina passerina albivitta
Common Ground Dove—Columbina passerina griseola
Common Ground Dove—Columbina passerina pallescens
Common Ground Dove—Columbina passerina passerina
Dusky-capped Flycatcher—Myiarchus tuberculifer
Dusky-capped Flycatcher—Myiarchus tuberculifer nigriceps
Eastern Wood Pewee—Contopus virens
Field Sparrow—Spizella pusilla
Great Horned Owl—Bubo virginianus
Lesser Goldfinch—Spinus psaltria
Lesser Goldfinch—Spinus psaltria colombiana
Lesser Nighthawk—Chordeiles acutipennis
Lesser Nighthawk—Chordeiles acutipennis aequatorialis
Mountain Chickadee—Poecile gambeli
Mountain Chickadee—Poecile gambeli gambeli
Mourning Dove—Zenaida macroura
Mourning Dove—Zenaida macroura marginella
Northern Cardinal—Cardinalis cardinalis
Northern Cardinal—Cardinalis cardinalis superbus
Northern Saw-whet Owl—Aegolius acadicus
Northern Saw-whet Owl—Aegolius acadicus brooksi
Phainopepla—Phainopepla nitens
Phainopepla—Phainopepla nitens lepida
Spotted Sandpiper—Actitis macularius
Tufted Titmouse—Baeolophus bicolor

dedicated to sharing bird sounds from around the world (XC, 2013). Recordings were preprocessed manually using Adobe Audition to remove silence and unwanted birds where regions which did not contain birds of interest were highlighted and deleted. No other preprocessing or noise reduction was used. Thus, the recordings varied in quality and background noise levels, as is typical of bird recordings taken in the wild. The data were grouped into "Whistles & hoots", "Trills" and "Nasals". The data are summarized in Table 1. The species each data-set is made up of are given in Tables 2–4.

5.2. Synthesized bird sounds

Bird vocalizations contain voiced and unvoiced parts. One way to view pitch tracker accuracy is to superimpose the pitch tracker output on the spectrogram. A good estimate will show the pitch tracking the lowest spectral peak of the voiced parts. This is subjective. To overcome this, we wanted a ground truth pitch to compare pitch tracker outputs too. Ground truth pitch refers to the true fundamental frequency of the periodic parts of the data-set. Unfortunately, no data-set with ground truth pitch for birds exists. The data in Table 1 inspired our creation of a synthesized bird sounds data-set complete with ground truth pitch. The synthesis system used here was taken from "Spectral Modeling

Table 3. Species which make up Thrills data-set

Thrills
Ash-throated Flycatcher—Myiarchus cinerascens
Ash-throated Flycatcher—Myiarchus cinerascens cinerascens
Carolina Wren—Thryothorus ludovicianus
Carolina Wren—Thryothorus ludovicianus ludovicianus
Common Nighthawk—Chordeiles minor
Common Nighthawk—Chordeiles minor henryi
Dark-eyed Junco—Junco hyemalis
Dark-eyed Junco—Junco hyemalis aikeni
Dark-eyed Junco—Junco hyemalis dorsalis
Dark-eyed Junco—Junco hyemalis shufeldti—[Oregon]
Dark-eyed Junco—Junco hyemalis thurberi
Eastern Kingbird—Tyrannus tyrannus
Eastern Screech Owl—Megascops asio
Marsh Wren—Cistothorus palustris
Marsh Wren—Cistothorus palustris tolucensis
Mountain Pygmy Owl—Glaucidium gnoma
Scarlet Tanager—Piranga olivacea
Western Tanager—Piranga ludoviciana

Table 4. Species which make up Nasals data-set

Nasals
Black-billed Magpie—Pica hudsonia
California Quail—Callipepla californica
California Quail—Callipepla californica achrustera
Greater Pewee—Contopus pertinax
Killdeer—Charadrius vociferus
Killdeer—Charadrius vociferus ternominatus
Killdeer—Charadrius vociferus vociferus
Mountain Pygmy Owl—Glaucidium gnoma
Pinyon Jay—Gymnorhinus cyanocephalus
Red-breasted Nuthatch—Sitta canadensis
Sinaloa Crow—Corvus sinaloae
Sora—Porzana carolina

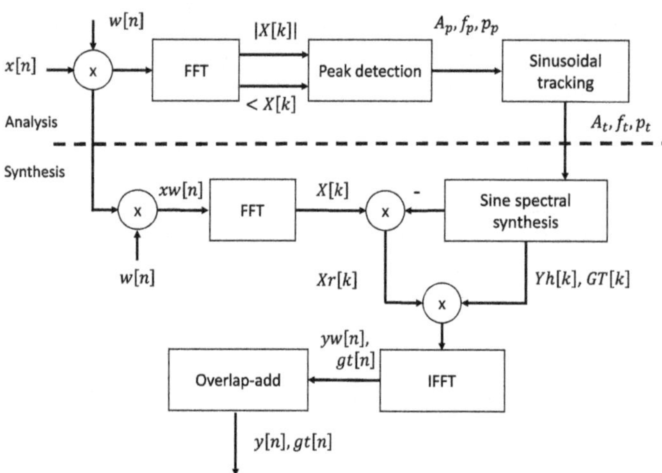

Figure 9. Block diagram of SpR system used to synthesis the bird data with ground truth pitch.

Source: Serra (1989).

Synthesis (SMS) Tools" a python implementation for analysis, transformation, and synthesis of musical sounds based on various spectral modeling approaches by Serra (1989). SMS contains a synthesis method called "sine plus residual (SpR)", a method which uses peak detection.

The aim is to synthesize the periodic parts of the vocalizations (whistles, trills, and nasal data) and save a ground truth frequency. The non-periodic residual is then added to the synthesized periodic part to give a more natural sound. SpR requires parameters, like number of sines (we set this to 8), which is the number of sine waves used in the synth phase, and the power threshold, which sets the amplitude threshold required for a given peak to be highlighted as periodic. Power threshold was usually set to −60 dB but this varied for some recordings as manual supervision was required to ensure optimal rejection of non-periodic peaks and accurate ground truth prediction. Every frame will have peaks, but only the high amplitude peaks (amplitude at frequency bin per frame) will be highlighted as peaks to be synthesized as sine waves. In theory, high amplitude unvoiced peaks can be highlighted as periodic frames and low amplitude voiced frames can be missed as periodic frames for synthesis. In practice, only voiced parts will have a high concentrated amplitude at a frequency bin, whereas unvoiced sounds will have energy distributed across a wider bandwidth. Carefully choosing the threshold for different files prevents unvoiced parts being synthesized and being included in the ground truth pitch signal. Inevitably some gaps in the ground truth will appear due to amplitude fluctuations of bird recordings. −60 dB rejected most of the non-periodic peaks but depending on the amplitude this level was supervised so minimal non-periodic parts were incorrectly labeled as periodic ground truth. Varying the amplitude threshold does not bias evaluation of YIN-bird, it just ensures a more accurate ground truth.

In summary, the output synth sound will contain synthesized sine waves and the original residual (non-periodic) audio added together in the frequency domain. The lowest sine track frequencies are saved as the ground truth. In some cases, the ground truth needs manual correcting, e.g. when the fundamental is missing, the 2nd harmonic would be incorrectly picked as the fundamental frequency. This method is proposed as the optimal way to generate a ground truth without resorting to fully manual labeling of pitch.

The SpR code can be found at Sinusoid plus residual python code (n.d.). All files in our data-set required resampling to 44.1 kHz to work with the SpR code. Each raw wave file ($x[n]; f_s = 44.1$ kHz) was passed through Serra's SpR system. A block diagram of the system can be seen in Figure 9. Input $x[n]$ was windowed using a Hamming window ($w[n]$). Taking the FFT of each window (STFT) resulted in magnitude $|X[k]|$ and phase $< X[k]$. $|X[k]|$ and $< X[k]$ were passed to a peak detector. The amplitude, frequency, and phase (A_p, f_p, P_p) of the peaks were then passed to the sinusoidal tracking block where sine tracks were identified. These periodic sine tracks were used by the sine spectral synthesis block to synthesize the periodic part of the bird vocalization. The non-periodic residual was found by subtracting the modeled periodic parts from the spectrum of the windowed input. The residual was added to the sine model to give a more realistic synthesized bird sound. The final synthesized sound ($y[n]$) was constructed by adding the residual spectrum ($X_r[k]$) to the periodic sine spectral synthesis output ($Yh[k]$) and calculating its IFFT to give $y_w[n]$. This windowed signal passed through an overlap and add block to get the synthesized output signal ($y[n]$). The ground truth pitch ($g[n]$) was identified as the lowest frequency peak of the sine model over time. For unvoiced regions, $g[n]$ was assigned a value of "NaN". The parameters used for window and hop size were the same as YIN-bird so the ground truth signal would be the same length as the YIN-bird output.

5.3. Listener tests

Listener tests were performed to evaluate how well the synthetic sounds match the original recordings from Table 1. Listeners were asked to listen to original recordings of bird vocalizations followed by the synthesized version and compare the pair of audio clips on a scale of $\{-3, \ldots, 3\}$ (see Table 5 for scale description). The scale was influenced by work in Sakamoto and Saito (2002), where listeners were asked to evaluate the speaker recognizability of synthetic speech using a similar scale. Three training examples were included with a recommended score revealed after listeners gave their answer. Participants were asked "how are you listening?". Thirty-six percent used over-ear headphones, 22% in-ear headphones, 21% laptop speakers, 7% HQ external speakers, and 14% regular external speakers. Of 23 respondents, 13 described themselves as "Expert" listeners, 15 as "Intermediate" listeners, and 1 as "beginner" listener with regard to their understanding of bird song. Survey results didn't show any difference between listener's experience with bird song. There were 26 synthetic examples tested. The survey was designed on surveygizmo.com and remains available at Bird Synthesis Listening Survey (n.d). On a worst to best scale of $\{-3, \ldots, 3\}$, the average score of the 23 listeners was 2.17 which describes the synthetic sound as "Sounds very much like original, could be fooled into thinking it is a real bird". This test was used to clarify that the synthetic sounds are similar enough to the original recordings, that they can be used in our pitch estimation experiments.

5.4. Error metrics

Performance of the two pitch tracking systems was assessed using four standard error metrics (Babacan et al., 2013; Wei & Alwan, 2009).

- **Gross Pitch Error (GPE)** is the percentage of frames for which the absolute pitch error is higher than a certain threshold. For speech, this threshold is usually 20%. As bird vocalizations tend to have higher pitch, the threshold was reduced to 10%. Only frames considered voiced by both the pitch tracker and ground truth were included in this calculation.
- **Fine Pitch Error (FPE)** is the standard deviation of the absolute error in Hz. Frames that have gross pitch errors were excluded. Only frames with ground truth and YIN estimates being voiced were used to calculate FPE.
- **Voicing Decision Error (VDE)** is the percentage of frames for which an incorrect voiced/unvoiced decision is made.
- **F0 Frame Error (FFE)** is the percentage of frames where either a GPE or VDE is observed.

Table 5. Listener testing evaluation scale

Evaluation	Rate	Description
Very different	−3	Doesn't sound like original, clear it's a synth version
Fairly different	−2	Sounds slightly different than original, most likely a synth
Little different	−1	Sounds like original, most likely a synth
Fair	0	Sounds like original, might be a synth
Little similar	1	Sounds like original, unsure if a synth
Fairly similar	2	Sounds very much like original, could be fooled into thinking it is a real bird
Very similar	3	Sounds identical to original, confident it's a real bird recording

5.5. Experiment parameters

The commonly used YIN system was compared with YIN-bird. For YIN, parameters wide enough to accommodate all bird vocalizations were used. $f_{0_{min}}$ was 500 Hz, window size was 6.7 ms, hop size was 1.7 ms (approximately 75% overlap), and quality was "good" which means estimates with aperiodic value of less than 2×0.2 (1×0.2 for "best", convention used in YIN code) were considered voiced. For the trills, the window size was reduced to 2 ms for increased time resolution, as pitch changes more rapidly for these types of sounds.

YIN-bird used the same window sizes as used with YIN above. No $f_{0_{min}}$ needed to be specified. The buffer size was set to 3,000 samples, meaning that for every 3,000 samples (68 ms) of the input audio file there would be a new value for $f_{0_{min}}$ parameter.

6. Quantitative results

Pitch estimates using YIN, with parameters mentioned in Sections 5 and 5.5, were compared to ground truth pitch ($g[n]$) in Hz from the synthesized values. Pitch estimates using YIN-bird were also compared to the same ground truth. The results are shown in Table 6.

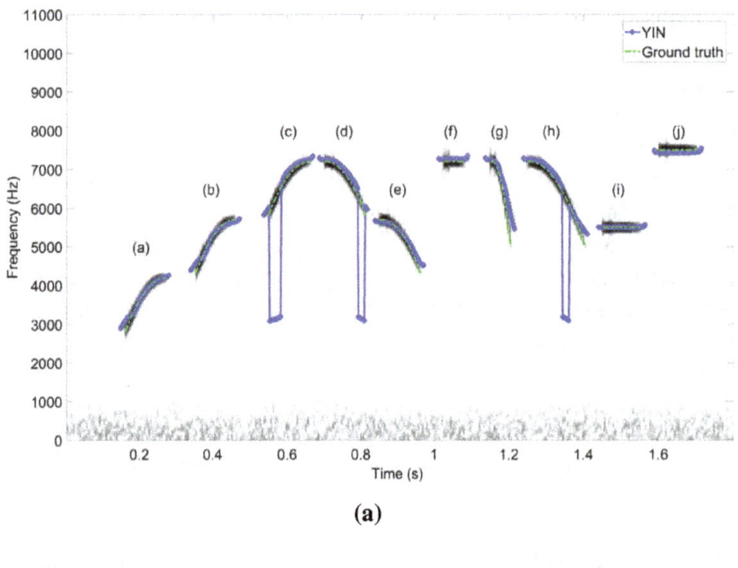

(a)

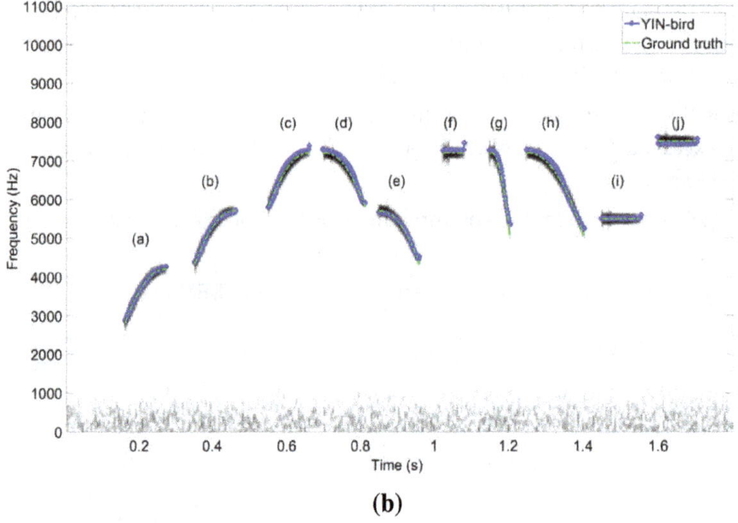

(b)

Figure 10. Performance comparison between (a) YIN and (b) YIN-bird on bird whistles.

Notes: Pitch is plotted with circular markers (online version: blue) and ground truth with broken line (online version: green).

Table 6. Error rates using YIN and YIN-bird

	GPE (%)	FPE (Hz)	VDE (%)	FFE (%)
Whistles				
YIN	1.67	40.97	25.72	26.37
YIN-bird	0.58	39.41	23.68	24.09
Trills				
YIN	6.29	88.89	37.93	41.12
YIN-bird	2.31	63.75	35.61	36.78
Nasals				
YIN	31.00	42.60	33.28	48.94
YIN-bird	6.21	58.67	32.69	35.60

When using YIN-bird for whistles, the GPE score shows an improvement of 1.09%. For trills, the improvement is 3.98%. Typical YIN and YIN-bird performance on whistle sounds is shown in Figure 10. This shows how YIN performs on synthetic bird syllables created in MATLAB. Note syllables (c), (d), and (h) experience octave errors or errors too low in Figure 10(a) ($g[n]$) is plotted with a broken line (online version: green) and the YIN pitch estimate is plotted with circular markers (online version: blue). The same errors are observed using SAP (Tchernichovski et al., 2011). Similar errors are produced by real data. These errors are corrected in Figure 10(b), where pitch values obtained YIN-bird are plotted.

Fine pitch error and voice detection error are included in Table 6 to show that the addition of an adaptive $f_{0_{min}}$ in YIN-bird does not diminish FPE and VDE. YIN-bird reduces "pitch being too low" errors exclusively so VDE will not improve directly with YIN-bird. As FFE combines GPE and VDE, it can be used as an overall measure of pitch estimation performance (Babacan et al., 2013; Wei & Alwan, 2009). For whistles and trills, the FFE improvement is 2.28 and 4.34%, respectively. An example of pitch tracking improvement for trills can be seen in Figure 11(a) and (b).

YIN-bird has reduced GPE and FPE for the ground truth data-set of whistles and trills. Not all bird sounds are a single tone. Nasal sounds contain many harmonics. Pitch tracking on nasal sounds with multiple partials is a challenge, especially when f_0 is missing, as is possible. Although the GPE results can be presented as an improvement for nasals, Figure 12(a) and (b) show how the pitch estimations jump between bands for both YIN and YIN-bird for nasal sounds. YIN-bird tends to identify the pitch as the strongest partial instead of f_0. If f_0 is weak or missing, YIN-bird will set $f_{0_{min}}$ to the prominent partial, thus estimating f_0 to be the prominent harmonic rather than the absent f_0. YIN sometimes identifies a weak f_0 but other times estimates a higher partial. The correct ground truth for nasals is also difficult to establish. Our synthesis system is prone to suggesting the strongest partial to be the ground truth as opposed to the true f_0 and manual ground truth corrections were required for some nasal examples. This is why nasal results are presented with caution. Although FPE is worse for YIN-bird, that in itself is not an indicator YIN-bird performs poorly, Figure 12(b) however presents strong evidence against trusting YIN-bird for nasals.

It is worth noting that the validity of the comparison to ground truth heavily relies on the accuracy of the ground truth used. Section 5.2 discussed the establishment of the ground truth pitch for the synthesized songs and admitted that on some occasions errors may be present (e.g. the highlighted areas of Figure 13). The only alternative would be hand labeling combined with expert listening. In the absence of such a data-set, we feel the synthesized data-set represents the best possible trade-off.

Where energy in partials is higher than that at the fundamental, timbre of the sound will change; how this affects the birds perception is not known. Perhaps to some birds, quality is more important

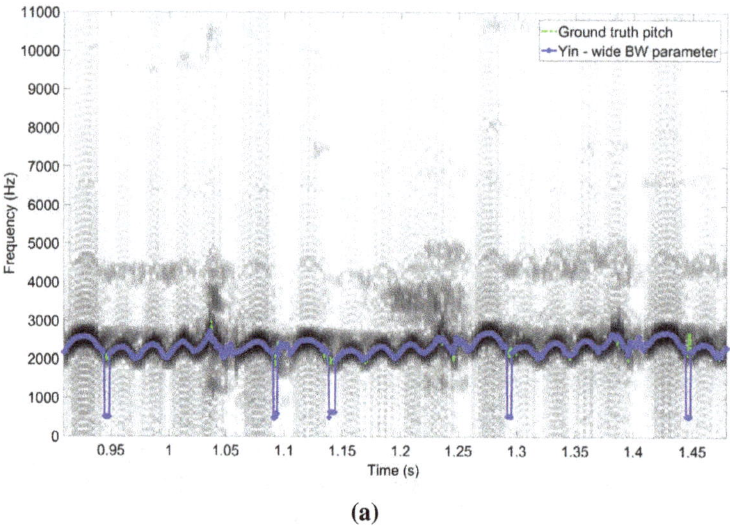

(a)

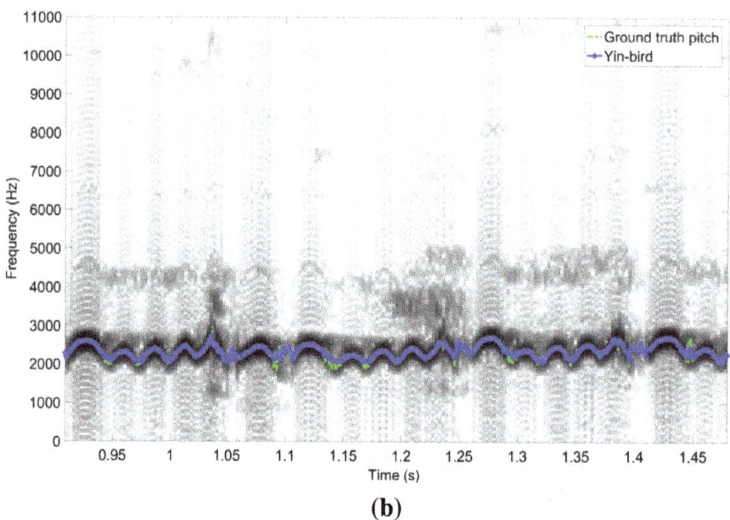

(b)

Figure 11. Performance comparison between (a) YIN and (b) YIN-bird on bird trill example from a Ash throated Flycatcher XC286838 (2013).

Notes: Pitch is plotted with circular markers (online version: blue) and ground truth with broken line (online version: green).

than pitch. If f_0 is missing, the interval between harmonics could be used to calculate f_0, but if harmonics are mistuned or missing, then this method will fail also Kent (2004).

7. Automatic pitch extraction from populations

Results in Section 6 show YIN-bird outperforms YIN on a range of whistles and trills taken from a wide sample of birds. A major stated motivation in developing YIN-bird was to develop a tool to enable larger-scale automatic pitch extraction for the study of a single species. To show how YIN-Bird changes the pitch values obtained on calls from a large number of samples of a single species, pitch extraction for two populations of bird, the Dot-winged Antwren (DWA), scientific name *Microrhopias quixensis*, from Trifa et al. (2008), and the Wangi-Wangi Olive-backed Sunbird (WWOBSB), scientific name *Cinnyris jugularis infrenatus*, from O'Reilly et al. (2015), using YIN and YIN-bird was compared. The DWA data-set contained 100 recordings from 21 individuals amounting to 232 s. The WWOBSB data-set contained 261 recordings from 10 individuals amounting to 132 s. The probability density function (PDF) of the difference of consecutive pitch estimates (Δf_0) is displayed in Figure 14. Octave or halving errors will lead to higher values of Δf_0 and hence it is a useful indicator of the smoothness

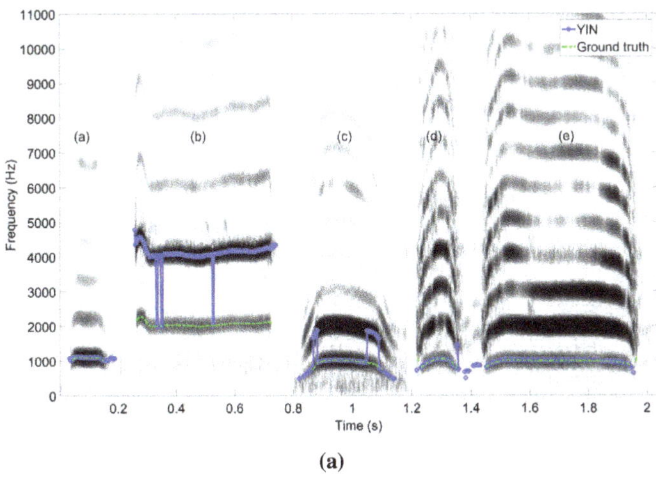

(a)

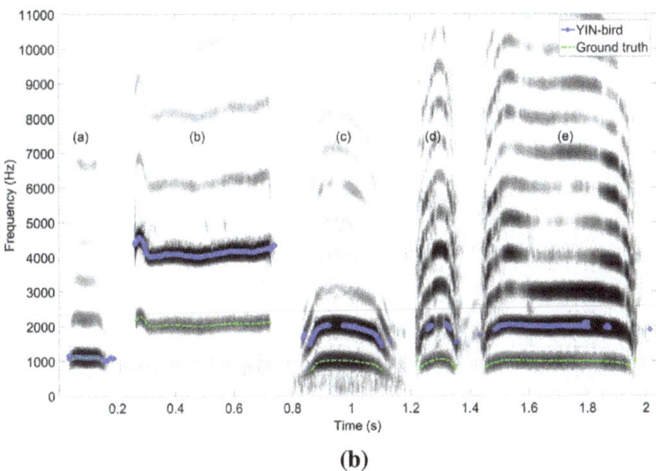

(b)

Figure 12. (a) YIN and (b) YIN-bird pitch estimation on nasal sounds.

Notes: Pitch is plotted with circular markers (online version: blue) and ground truth with a broken line (online version: green).

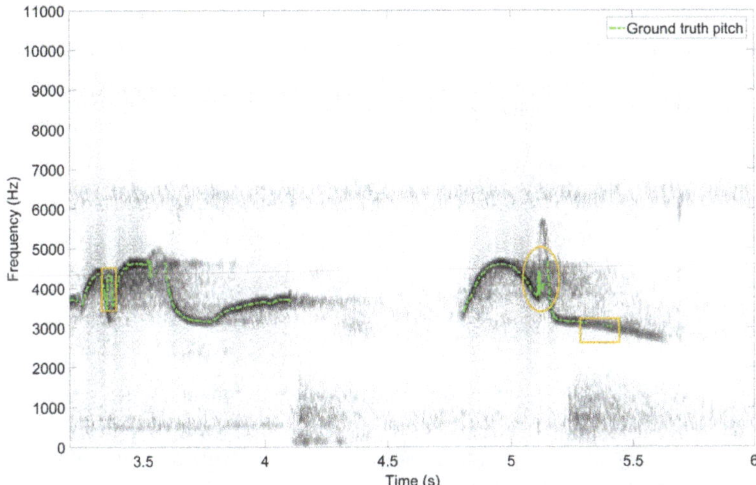

Figure 13. Example of ground truth pitch errors due to reverb and amplitude variation of Eastern Wood Pewee synth example (XC 7704, 2013).

Note: The GT errors the rectangles and circle.

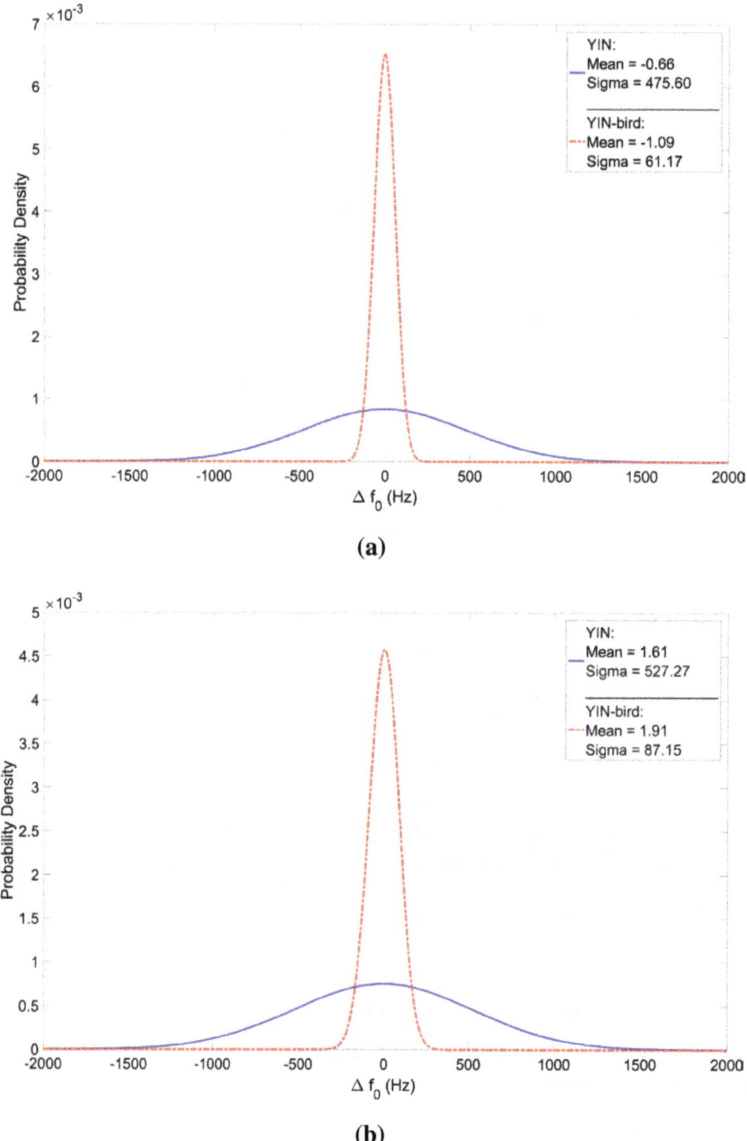

Figure 14. PDF of Δf_0 values using YIN (continuous line; online version: blue) and YIN-bird (broken line; online version: red). (a) Dot-winged Antwren (DWA) and (b) Wangi-Wangi Olive-backed Sunbird (WWOBSB).

of a pitch contour. If pitch estimates of a syllable are accurate, consecutive values of Δf_0 are typically not greater than 100 Hz. For example, if a monotone whistle at 2 kHz suffers an octave error in the middle of the vocalization, Δf_0 would be approximately 1 kHz. Lower sigma thus suggests that there are less octave errors. Of course some bird populations with rapid pitch modulations can have a wide sigma but sigma of accurate pitch values will be lower than sigma for Δf_0 values with a large number of octave errors for a given population of birds.

In Figure 14(a), sigma for Δ YIN pitch values is 475.60 Hz which is greater than 61.17 Hz, sigma for Δ YIN-bird pitch values. The same is observed for pitch estimates of WWOBSB, where Δ for YIN-bird has the tighter distribution, as shown in Figure 14(b). Lower probability of high Δ values suggests smoother pitch tracks which suggests less halving or octave errors.

8. Qualitative analysis

Syllable types that could be synthesized with accurate ground truth pitch were used to quantitatively evaluate YIN-bird performance. Pitch performance on syllable types that posed problems

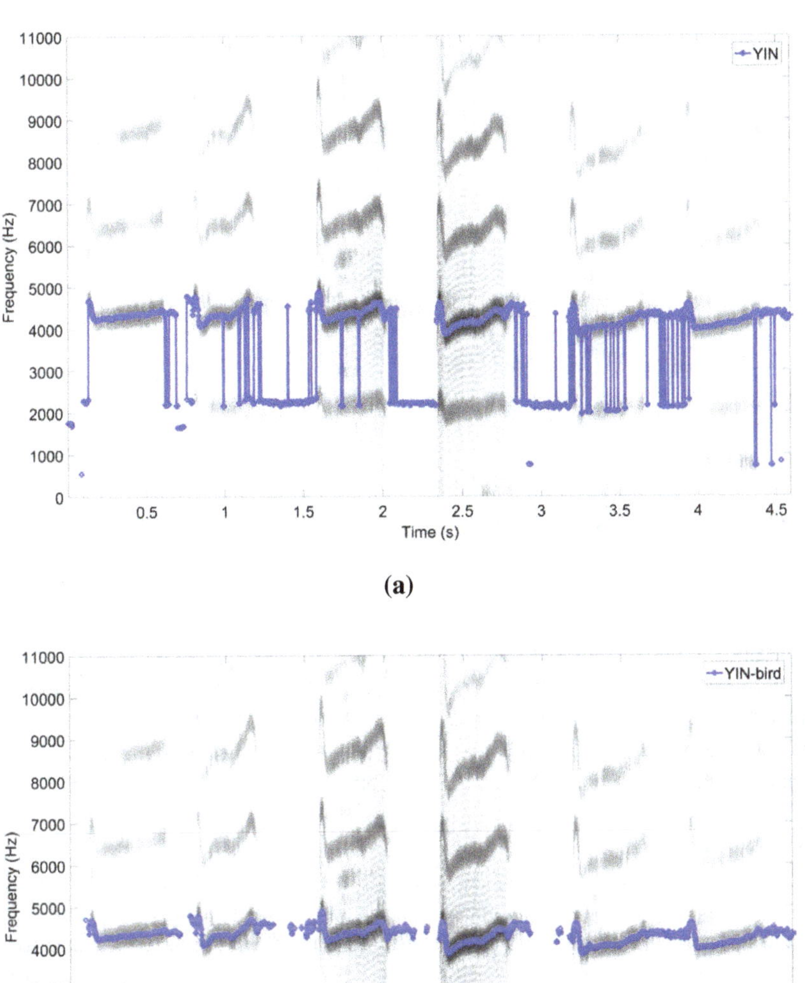

Figure 15. Pitch tracking on Killdeer song with missing f_0. (a) YIN pitch and (b) YIN-bird pitch.

Source: XC 173884 (2013).

when calculating ground truth pitch is discussed qualitatively here. These sounds can still be successfully synthesized even though determining a ground truth pitch was not possible.

8.1. Harmonic (or Nasal) pitch tracking

As mentioned in Section 6, not all bird sounds are a single tone. Harmonic (or Nasal) sounds with strong energy at f_0 are straightforward for PDAs, just like harmonics in speech do not diminish PDAs performance for humans. When the 2nd harmonic ($2 \times f_0$, sometimes referred to as the 1st overtone or 2nd partial) is stronger than f_0, the PDA jumps between f_0 and the prominent partial as shown in Figure 15(a) at syllables 2, 3, and 5 using YIN. YIN-bird will set $f_{0_{min}}$ estimate to the 2nd harmonic and thus will track a single partial but incorrectly return the 2nd harmonic as f_0 as shown in Figure 15(b) at syllables 2, 3, and 5. An example of missing f_0 can be seen in Figure 15 at syllable 1. In Figure 15, f_0 is missing from the first syllable and appears faintly for the other syllables. The 2nd harmonic is the prominent partial throughout the phrase in Figure 15. YIN pitch estimates jumps between f_0 and 2nd harmonic in Figure 15(a), while YIN-bird tracks the 2nd harmonic in Figure 15(b).

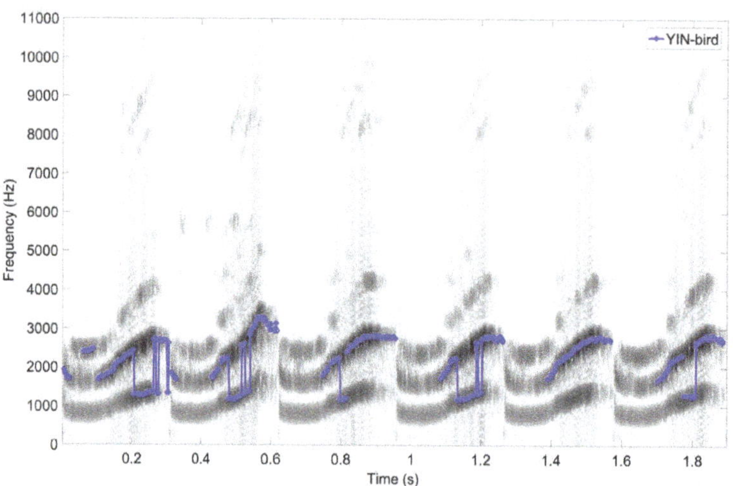

Figure 16. YIN-bird performance on song of Sora where the prominent frequency changes between f_0 and 2nd harmonic.

Source: XC 292447 (2013).

The problem of missing fundamental, missing harmonic, and stronger higher order harmonics needs to be addressed as neither YIN or YIN-bird offer solutions to these problems. The advantage of YIN-bird is it will at least identify the prominent frequency partial and is less likely to jump between bands. Although YIN-bird is less likely to jump between partials while pitch tracking, when the prominent frequency changes throughout a vocalization, the pitch estimation will jump as seen in Figure 16.

8.2. Two-voiced pitch tracking

Two-voiced sounds were introduced and two-voiced production explained in Section 2. YIN-bird pitch tracking performance on a two-voiced example can be seen in Figure 17. PDAs get confused between the two tracks and output estimates that try and find an estimate average or common denominator between the two pitch tracks. This ultimately does not give you the pitch of the signal.

Sometimes, two-voiced sounds can resemble harmonic sounds. Nowicki and Capranica (1986) showed that a Chickadee's sound which resembled a harmonic sound was actually produced by heterodyne frequencies resulting in cross-modulation between the two syringeal sides. Pitch tracking of this phenomenon can be seen in Figure 18. Note the pitch estimate is at f_0 sometimes even though it is missing from the vocalization. Pitch estimation using YIN-bird on this example tracks the partial at 3 kHz as the pitch (i.e. the prominent partial). In a true harmonic series, f_0 is usually the

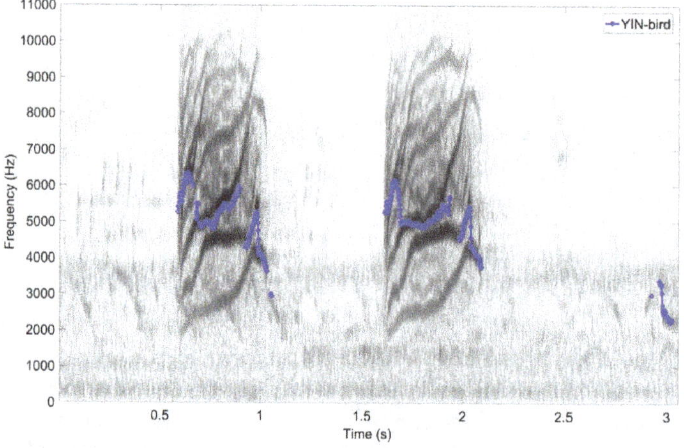

Figure 17. Typical performance of YIN-bird with two-voiced song shown on Pine Siskin song example.

Source: XC 97358 (2013).

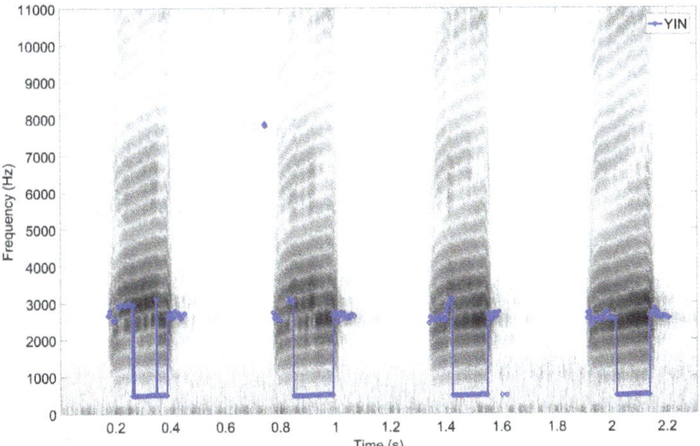

Figure 18. Red-breasted Nuthatch song. Harmonic-like sound produced by syringeal coupling.

Source: Pieplow and Spencer, (2013); Nowicki and Capranica, (1986), XC 76418 (2013).

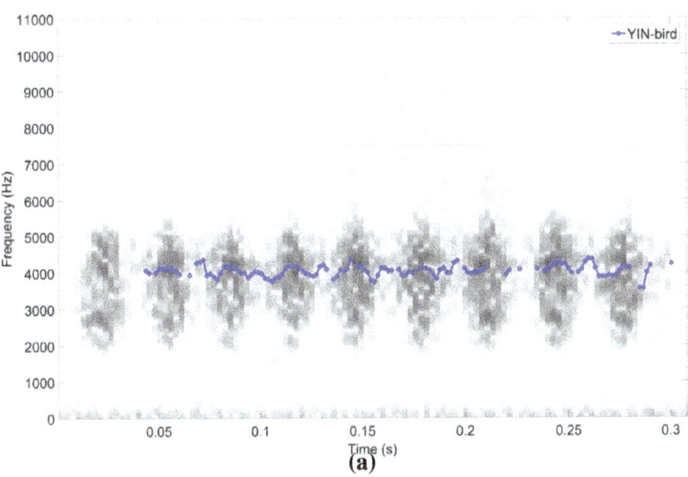

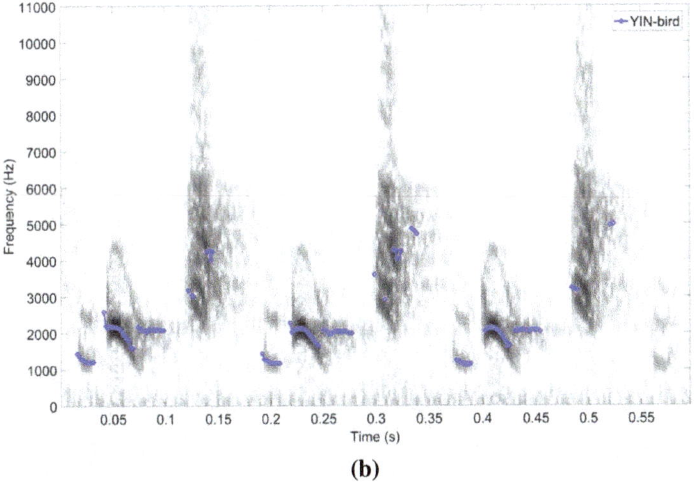

(b)

Figure 19. Complex vocalizations of Nightingale.

Source: XC (2013).

lowest and loudest partial but here a Red-breasted Nuthatch uses both sides of the syrinx simultaneously but not independently; instead, vibrating sources are coupled together to create what looks like a harmonic series but technically isn't because f_0 is missing and the prominent frequencies are around 3 kHz (Pieplow & Spencer, 2013).

8.3. Birds with complex vocalizations

Nightingales, Sedge Warblers, Sky larks, and Wood Warblers are just some of the birds that produce complex vocalizations. Figure 19 shows four complex phrases produced by Nightingales. Figure 19(a) shows a series of rapid harsh sounds which sound like amplitude-modulated narrow band noise. Figure 19(b) shows three phrases repeated. The first syllable is a fast whistle with a 2nd harmonic. The second syllable is made up of multiple partials but not a harmonic series. This is more than likely syringeal coupling or two-voiced biphonation and sounds harsh.

Examples of Sedge Warbler and Sky Lark vocalizations can be seen in Figure 20. The syllables in Figure 20(a) show what appears to be syringeal coupling. Greater amplitude at the higher partials makes them harsh sounding. Figure 20(b) shows rapid warble sounds. The time resolution had to be refined in order to display these syllables.

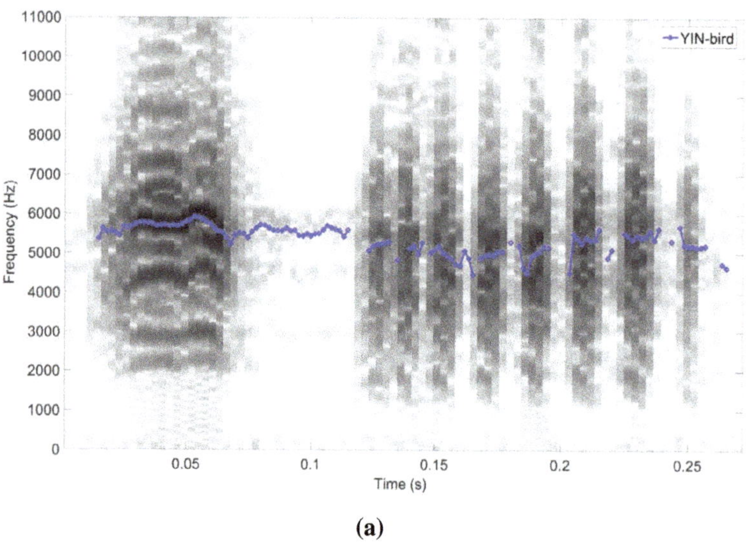

(a)

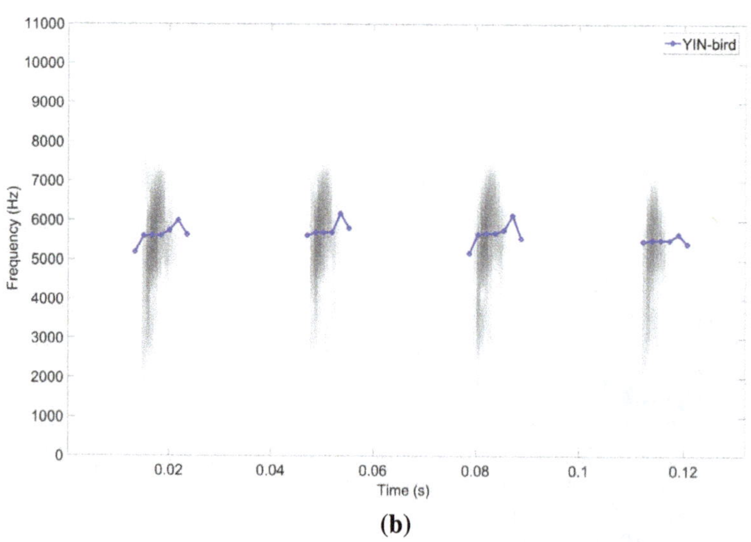

(b)

Figure 20. Complex vocalizations from (a) Segde Warbler and (b) Sky Lark.

Source: XC (2013).

These examples show the complexity of nightingale, Sedge Warbler, and Sky Lark's song. Each syllable in isolation may not be complicated but the speed at which they vocalize and how rapidly the syllable type changes from whistles to trills or two-voiced leads to their overall complexity.

Throughout this section, an in-depth description of the challenges in tracking pitch of bird song is presented. Much work remains to be done before pitch tracking of complex song, like the song of Nightingale, Sedge Warbler, and Sky Lark, is performed automatically and accurately.

9. Conclusion

Bird vocalizations may sound no more complex than human speech, but recordings are usually subject to adverse conditions such as contaminant vocalizations and non-homogeneous noise backgrounds (Kogan & Margoliash, 1998). This and the larger bandwidth make pitch tracking of bird song more complicated than simply applying human PDAs to bird recordings. Pitch is not only important for analysis and synthesis, but is used in measuring bird population similarity. This relies on accurate pitch estimation. Bird frequency range varies dramatically from species to species, and even within syllables in a song repertoire from a single bird. Hence, static YIN parameters are not useful in bird recordings. Results presented here have demonstrated that automatically determining the $f_{0_{min}}$ parameter on a segment-by-segment basis for YIN (performed by YIN-bird) improves pitch estimation. This improvement can in turn improve accuracy on bird species and phrase comparisons, allowing fully automatic batch processing of large numbers of recordings from different species. The synthesized bird calls and ground truth pitch have been shared for research at Bird synthesis database (n.d.). This will allow other researchers to compare the performance of PDAs on bird song in a quantitative manner for the first time.

A qualitative description of a range of vocalizations that pose problems for PDAs was also presented in detail and examples shown. Knowledge of when to use YIN-bird, depending on what type of vocalizations are being analyzed, is useful. Pitch estimates of whistles, harmonics with strong f_0, and trills should be treated with confidence while pitch tracking on two-voiced and nasal sounds remains problematic.

The value of YIN-bird lies not only in this demonstrated performance improvement. The possibility of fully automatic processing of bird song to extract pitch will allow researchers to process larger amounts of data, improving certainty in tasks such as species comparison based on pitch measurements. We encourage the use of the YIN-bird algorithm in popular bioacoustics software. Accurate extraction of pitch is also beneficial to statistical analysis of bird song as minimum, maximum, and peak frequency can be extracted from pitch contours generated by YIN-bird. Another publication under review presents YIN-bird extraction of these frequencies to automate a bird species delimitation system presented in Tobias et al. (2010).

Acknowledgements
Thank you to David J. Kelly and Nicola M. Marples from the Department of Zoology, TCD, for help with the listener tests and bird-related queries.

Funding
The ADAPT Centre for Digital Content Technology is funded under the SFI Research Centres Programme [grant number 13/RC/2106] and is co-funded under the European Regional Development Fund.

Authors' Contributions
Colm O'Reilly performed analysis experiments on all samples, interpreted data, wrote the manuscript, and is the corresponding author. Naomi Harte supervised the development of work, helped in data interpretation, and manuscript evaluation and revision.

Competing Interests
The authors declare no competing interest.

Author details
Colm O'Reilly[1]
E-mail: oreilc16@tcd.ie
Naomi Harte[1]
E-mail: nharte@tcd.ie
[1] School of Engineering, Sigmedia, ADAPT Centre, Trinity College Dublin, Dublin, Ireland.
[†] Portion of work presented at INTERSPEECH 2016

References

Anderson, S. E., Dave, A. S., & Margoliash, D. (1996). Template-based automatic recognition of birdsong syllables from continuous recordings. *The Journal of the Acoustical Society of America, 100*, 1209–1219.

Babacan, O., Drugman, T., d'Alessandro, N., Henrich, N., & Dutoit, T. (2013). A comparative study of pitch extraction algorithms on a large variety of singing sounds. In *2013 IEEE International Conference on Acoustics, Speech and Signal Processing (ICASSP)* (pp. 7815–7819). IEEE.

Bagshaw, P. C., Hiller, S. M., & Jack, M. A. (1993). Enhanced pitch tracking and the processing of f0 contours for computer aided intonation teaching. *In Proceedings of the 3rd European Conference on Speech Communication and Technology* (pp. 1003–1006).

Beckers, G. J. (2011). Bird speech perception and vocal production: A comparison with humans. *Human Biology, 83*, 191–212.

Beckers, G. J., Suthers, R. A., & Ten Cate, C. (2003). Pure-tone birdsong by resonance filtering of harmonic overtones. *Proceedings of the National Academy of Sciences, 100*, 7372–7376.

Boersma, P. (1993). Accurate short-term analysis of the fundamental frequency and the harmonics-to-noise ratio of a sampled sound. In *Proceedings of the Institute of Phonetic Sciences* (Vol. 17, pp. 97–110). Amsterdam.

Bird synthesis database. (n.d.). Retrieved 29 September, 2016, from http://www.mee.tcd.ie/sigmedia/Resources/SynthBirdsData

Bird Synthesis Listening Survey. (n.d). Retrieved 29 September, 2016, from http://www.surveygizmo.com/s3/2570269/Bird-synth-comparison

Camacho, A. (2007). *SWIPE: A sawtooth waveform inspired pitch estimator for speech and music* (Ph.d. dissertation). Gainesville: University of Florida.

Camacho, A., & Harris, J. G. (2008). A sawtooth waveform inspired pitch estimator for speech and music. *The Journal of the Acoustical Society of America, 124*, 1638–1652.

Catchpole, C. K., & Slater, P. J. (2008). *Bird song: Biological themes and variations* (2nd ed.), Cambridge University Press. ISBN 9780521872423.

Chen, Z., & Maher, R. C. (2006). Semi-automatic classification of bird vocalizations using spectral peak tracks. *The Journal of the Acoustical Society of America, 120*, 2974–2984.

Connor, E. F., Li, S., & Li, S. (2012). Automating identification of avian vocalizations using time-frequency information extracted from the gabor transform. *The Journal of the Acoustical Society of America, 132*, 507–517.

De Cheveigné, A., & Kawahara, H. (2002). Yin, a fundamental frequency estimator for speech and music. *The Journal of the Acoustical Society of America, 111*, 1917–1930.

Doupe, A. J., & Kuhl, P. K. (1999). Birdsong and human speech: Common themes and mechanisms. *Annual Review of Neuroscience, 22*, 567–631.

Fagerlund, S., & Laine, U. K. (2014). New parametric representations of bird sounds for automatic classification. In *2014 IEEE International Conference on Acoustics, Speech and Signal Processing (ICASSP)* (pp. 8247–8251). IEEE.

Graciarena, M., Delplanche, M., Shriberg, E., Stolcke, A., & Ferrer, L. (2010). Acoustic front-end optimization for bird species recognition. In *2010 IEEE International Conference on Acoustics Speech and Signal Processing (ICASSP)* (pp. 293–296). IEEE.

Harte, N., Murphy, S., Kelly, D. J., & Marples, N. M. (2013, August 25–29). Identifying new bird species from differences in birdsong. In *INTERSPEECH 2013–14th Annual Conference of the International Speech Communication Association* (pp. 2900–2904). Lyon.

Heller, J. R., & Pinezich, J. D. (2008). Automatic recognition of harmonic bird sounds using a frequency track extraction algorithm. *The Journal of the Acoustical Society of America, 124*, 1830–1837.

Hermes, D. J. (1988). Measurement of pitch by subharmonic summation. *The Journal of the Acoustical Society of America, 83*, 257–264.

Kaewtip, K., Tan, L. N., Alwan, A., & Taylor, C. E. (2013). A robust automatic bird phrase classifier using dynamic time-warping with prominent region identification. In *2013 IEEE International Conference on Acoustics, Speech and Signal Processing (ICASSP)* (pp. 768–772). IEEE.

Kawahara, H., Estill, J., & Fujimura, O. (2001). Aperiodicity extraction and control using mixed mode excitation and group delay manipulation for a high quality speech analysis, modification and synthesis system straight. *MAVEBA* (pp. 59–64). Firenze

Kawahara, H., Katayose, H., De Cheveigné, A., & Patterson, R. D. (1999). Fixed point analysis of frequency to instantaneous frequency mapping for accurate estimation of f0 and periodicity. *In EuroSpeech, 99*, 2781–2784.

Kent, R. D. (2004). *The MIT encyclopedia of communication disorders*. Cambridge: MIT Press.

Kogan, J. A., & Margoliash, D. (1998). Automated recognition of bird song elements from continuous recordings using dynamic time warping and hidden markov models: A comparative study. *The Journal of the Acoustical Society of America, 103*, 2185–2196.

Krakauer, A. H., Tyrrell, M., Lehmann, K., Losin, N., Goller, F., & Patricelli, G. L. (2009). Vocal and anatomical evidence for two-voiced sound production in the greater sage-grouse centrocercus urophasianus. *Journal of Experimental Biology, 212*, 3719–3727.

Lachlan, R. F. (2012) . *Luscinia*. Retrieved 29 September, 2016, from http://luscinia.sourceforge.net/

Lachlan, R. F., Verzijden, M. N., Bernard, C. S., Jonker, P.-P., Koese, B., Jaarsma, S., ... Ten Cate, C. (2013). The progressive loss of syntactical structure in bird song along an island colonization chain. *Current Biology, 23*, 1896–1901.

Lambert, F., & Rasmussen, P. (1998). A new scops owl from Sangihe Island, Indonesia. *Bulletin of the British Ornithologists' Club, 118*, 204–216.

Lee, B. S., & Ellis, D. P. W. (2012, September 9–13). Noise robust pitch tracking by subband autocorrelation classification. In *INTERSPEECH 2012–13th Annual Conference of the International Speech Communication Association* (pp. 707–710). Portland, OR.

Luengo, I., Saratxaga, I., Navas, E., Hernaez, I., Sanchez, J., & Sainz, I. (2007). Evaluation of pitch detection algorithms under real conditions. In *Acoustics, Speech and Signal Processing, 2007. ICASSP 2007. IEEE International Conference on*, 4, IV-1057–IV-1060.

Mandelblat-Cerf, Y., & Fee, M. S. (2014). An automated procedure for evaluating song imitation. *PloS one, 9*(5), e96484.

Marler, P.R., & Slabbekoorn, H. (2004). *Nature's music: The science of birdsong*. San Diego, CA: Academic Press.

McKay, B. D., Reynolds, M. B. J., Hayes, W. K., & Lee, D. S. (2010). Evidence for the species status of the bahama yellow-throated warbler (dendroica "dominica" flavescens). *The Auk, 127*, 932–939.

Mehrabani, M., Boril, H., & Hansen, J. H. (2010). Dialect distance assessment method based on comparison of pitch pattern statistical models. In *2010 IEEE International Conference on Acoustics Speech and Signal Processing (ICASSP)* (pp. 5158–5161). IEEE.

Mehrabani, M., & Hansen, J. H. (2015). Automatic analysis of dialect/language sets. *International Journal of Speech Technology*, 1–10.

Meliza, C. D., Keen, S. C., & Rubenstein, D. R. (2013). Pitch-and spectral-based dynamic time warping methods for comparing field recordings of harmonic avian vocalizations. *The Journal of the Acoustical Society of America, 134*, 1407–1415.

Miller, D. B. (1977). Two-voice phenomenon in birds: Further evidence. *The Auk, 94*, 567–572.

Murray, I. R., & Arnott, J. L. (1993). Toward the simulation of emotion in synthetic speech: A review of the literature on human vocal emotion. *The Journal of the Acoustical Society of America, 93*, 1097–1108.

Noll, A. M. (1967). Cepstrum pitch determination. *The Journal of the Acoustical Society of America, 41*, 293–309.

Nowicki, S., & Capranica, R. R. (1986). Bilateral syringeal coupling during phonation of a songbird. *The Journal of Neuroscience, 6*, 3595–3610.

O'Reilly, C., Marples, N. M., Kelly, D. J., & Harte, N. (2015, September 6–10). Quantifying difference in vocalizations of bird populations. In *INTERSPEECH 2015–16th Annual Conference of the International Speech Communication Association* (pp. 3417–3421). Dresden.

Pieplow, N., & Spencer, A. (2013). *Earbirding.com - the seven basic tone qualities and how to read spectrograms?* Retrieved from http://earbirding.com/blog/archives/4621, http://earbirding.com/blog/specs

Ranjard, L., & Ross, H. A. (2008). Unsupervised bird song syllable classification using evolving neural networks. *The Journal of the Acoustical Society of America, 123*, 4358–4368.

Remsen, Jr, J. (2005). Pattern, process, and rigor meet classification. *The Auk, 122*, 403–413.

Sakamoto, M., & Saito, T. (2002, September 16–20). Speaker recognizability evaluation of a voicefont-based text-to-speech system. In *INTERSPEECH 2002–3rd Annual Conference of the International Speech Communication Association*. Denver, CO.

Sangster, G., King, B. F., Verbelen, P., & Trainor, C. R. (2013). A new owl species of the genus otus (aves: Strigidae) from lombok, indonesia. *PloS one, 8*(2), e53712.

Sasahara, K., Cody, M. L., Cohen, D., & Taylor, C. E. (2012). Structural design principles of complex bird songs: a network-based approach. *PLoS ONE, 7*, e44436.

Schroeder, M. R. (1968). Period histogram and product spectrum: New methods for fundamental-frequency measurement. *The Journal of the Acoustical Society of America, 43*, 829–834.

Serra, X. (1989). *A system for sound analysis/transformation/ synthesis based on a deterministic plus stochastic decomposition* (Ph. d. dissertation). Stanford, CA: Stanford University. Retrieved from http://mtg.upf.edu/ technologies/sms

Sethares, W. A. (2005). *Tuning, timbre, spectrum, scale*. London: Springer Science & Business Media.

Sturdy, C. B., & Mooney, R. (2000). Bird communication: Two voices are better than one. *Current Biology, 10*, R634–R636.

Sinusoid plus residual python code. (n.d.). Retrieved 29 September, 2016, from https://github.com/MTG/sms-tools/blob/master/ software/models_interface/sprModel_function.py

Sun, X. (2002). Pitch determination and voice quality analysis using subharmonic-to-harmonic ratio. In *2002 IEEE International Conference on Acoustics, Speech, and Signal Processing (ICASSP)* (Vol. 1, pp. I-333–I-336). IEEE.

Talkin, D. (1995). A robust algorithm for pitch tracking (rapt). *Speech Coding and Synthesis Journal, 495*, 518.

Tan, L. N., Alwan, A., Kossan, G., Cody, M. L., & Taylor, C. E. (2015). Dynamic time warping and sparse representation classification for birdsong phrase classification using limited training data a. *The Journal of the Acoustical Society of America, 137*, 1069–1080.

Tan, L. N., Kaewtip, K., Cody, M. L., Taylor, C. E., & Alwan, A. (2012, September 9–13). Evaluation of a sparse representation-based classifier for bird phrase classification under limited data conditions. In *INTERSPEECH 2012–13th Annual Conference of the International Speech Communication Association* (pp. 2522–2525). Portland, OR.

Tchernichovski, O., Kashtelyan, E., Swigger, D., & Mitra, P. P. (2011). *Sound analysis pro (sap) software download.* Retrieved from http://soundanalysispro.com/

Tchernichovski, O., Nottebohm, F., Ho, C. E., Pesaran, B., & Mitra, P. P. (2000). A procedure for an automated measurement of song similarity. *Animal Behaviour, 59*, 1167–1176.

Thompson, N. S., LeDoux, K., & Moody, K. (1994). A system for describing bird song units. *Bioacoustics: The International Journal of Animal Sound and its Recording, 5*, 267–279.

Tobias, J. A., Seddon, N., Spottiswoode, C. N., Pilgrim, J. D., Fishpool, L. D., & Collar, N. J. (2010). Quantitative criteria for species delimitation. *Ibis, 152*, 724–746.

Trifa, V. M., Kirschel, A. N., Taylor, C. E., & Vallejo, E. E. (2008). Automated species recognition of antbirds in a mexican rainforest using hidden markov models. *The Journal of the Acoustical Society of America, 123*, 2424–2431.

Wang, M., & Lin, M. (2004). An analysis of pitch in chinese spontaneous speech. In *International Symposium on Tonal Aspects of Languages: With Emphasis on Tone Languages*.

Wei, C., & Alwan, A. (2009). Reducing f0 frame error of f0 tracking algorithms under noisy conditions with an unvoiced/voiced classification frontend. In *IEEE International Conference on Acoustics, Speech and Signal Processing, 2009. ICASSP 2009* (pp. 3969–3972). IEEE.

XC. (2013). *Xeno-canto.org bird library*. Retrieved from http:// www.xeno-canto.org/

Zollinger, S. A., Riede, T., & Suthers, R. A. (2008). Two-voice complexity from a single side of the syrinx in northern mockingbird mimus polyglottos vocalizations. *Journal of Experimental Biology, 211*, 1978–1991.

10

An evaluation of glutathione transferase *associated with Dichlorvos degradation in* African palm weevil (*Rynchophorus phoenicis*) larva

Olufemi Samuel Bamidele[1]*, Joshua Oluwafemi Ajele[1] and Folasade Mayowa Olajuyigbe[1]

*Corresponding author: Olufemi Samuel Bamidele, Enzymology Research Unit, Department of Biochemistry, The Federal University of Technology, P. M. B 704, Akure, Nigeria
E-mail: osbamidele@futa.edu.ng

Reviewing editor: Yasser Gaber, Beni-Suef University, Egypt
Additional information is available at the end of the article

Abstract: This study was conducted to investigate the metabolic defensive mechanism in the larvae of African palm weevil (*Rynchophorus phoenicis*) administered with dichlorvos (2,2-dichlorovinyl dimethylphosphate) solution. Bioassay experiment with dichlorvos was conducted on the larva and glutathione-utilizing enzyme activities were determined in the major organs: fat body, gut, and head of *R. phoenicis* larva 48 h after treatment with 0–0.060 μg g^{-1} body weight dichlorvos solution. Glutathione transferase was purified from the gut of larvae by ion-exchange chromatography on diethylaminoethyl-Sephadex A50 and affinity chromatography on glutathione-Sepharose 4B columns. The purified enzyme was homogenous as revealed by sodium dodecylsulfate polyacrylamide gel electrophoresis. Initial velocity studies were carried out on the purified enzyme using standard procedures. Bioassay experiment indicated alterations of glutathione peroxidase, glutathione reductase, and glutathione transferase activities in the major organs of larva caused by dichlorvos. Glutathione transferase activity in the gut of larva was three times higher than that of glutathione peroxidase and glutathione reductase activities, an indication of possible detoxification role of glutathione

ABOUT THE AUTHORS

The Enzymology Research Group, under the supervision of Professor Joshua O. Ajele, focuses on the understanding of the metabolic functions of various enzymes and functional proteins that drive life processes. This broad research area involves the isolation, purification, and characterization of these biomolecules from plants, micro-organisms, insects, and higher animals including humans. The group is currently focusing on the identification and heterologous expression of enzymes implicated in the antioxidant defense system of insect pest towards the development of effective and safe insecticides in the face of increasing resistance to insect pest control and other environmental health challenges.

PUBLIC INTEREST STATEMENT

Insect pest constitutes a large group of plant enemies. Palm (*Elaeis guineensis*), an economically important perennial crop is a host of African palm weevil (*Rynchophorus phoenicis*). The larva of this pest cause more harm to palm than the adult. Palm weevil is controlled by insecticides. Insecticide sometimes failed due to the defense system inherent in the insect. In order to gain insight into the biochemistry of GST in the larva, palm weevil larva was treated with a known insecticide—dichlorvos and an assessment of some important enzymes that are linked to molecular defense in the larva were carried out. Glutathione-transferase, glutathione reductase, and glutathione peroxidase were altered by dichlorvos in the larva. Glutathione transferase associated with the breakdown of dichlorvos in the gut of larva was isolated and purified. The purified enzyme from the gut of larva showed a mechanism in which two substrates interact with the enzyme to form the products in any order. This study confirmed the presence of glutathione transferase in the larva and the enzyme may enhance removal of insect poison from the larva.

transferase in the organ. A 49.7 kDa homodimeric glutathione transferase was identified from the gut of larva and was tagged *rpl*GSTc. Mechanism of action of *rpl*GSTc with 1-chloro-2,4-dinitrobenzene, and glutathione as substrates conformed to the random sequential mechanism. These results confirmed the presence of GST associated with the degradation of dichlorvos in the gut of *R. phoenicis* larva.

Subjects: Biochemistry; Enzymology; Entomology

Keywords: glutathione peroxidase (GPX); glutathione reductase (GR); glutathione transferase (GST); insecticide; purification; *R. phoenicis* larva

1. Introduction

Palm (*Elaeis guineensis*) is an economically important plant cultivated in Nigeria. The primary products of palm are palm oil, palm kernel oil, and palm kernel cake (Soyebo, Farinde, & Dionco-Adetayo, 2005). The full yield potentials of palm products are affected by insect pests. African palm weevil, *Rynchophorus phoenicis (Coleoptera: Curculionidae)* is identified as one of the major insect pests of palm (Al-Ayied, Alswailem, Shair, & Al Jabr, 2006). Similar hosts of this insect are date palm (*Phoenix dactylifera* L.), raffia palm (*Raphia* spp.), and coconut palm (*Cocos nucifera* L.) (Bong, Er, Yiu, & Rajan, 2008; Gries et al., 1994). The life cycle of *R. phoenicis* is similar to other *Rynchophorus* species (Giblin-Davis et al., 1996). Damage to the host is caused by the grubs (larvae). These larvae make tunnels in the trunk and feed on the tissues of the palm. Decaying of the tissues results in the production of a foul smell and if unchecked, leads to death of palm (Mariau, Chenon, Julia, & Philippe, 1981). Huge loss of palm products and revenue may be caused by infestation of palm by *R. phoenicis* specie (Faleiro, 2010). However, insecticides are employed for the control of *R. pheonicis* and other related insect pests of palm (Barranco, de la Peña, Martín, & Cabello, 1998; Cabello, de la Pena, Barranco, Belda, & de la Pena, 1997).

Generally, insects can metabolize and thereby degrade toxic or otherwise detrimental chemicals for surviving in a chemically unfriendly environment. While all insects probably possess capacity to detoxify toxic chemicals, the amount can be expected to vary among species, with developmental stage, and with the nature of insect's recent environment (Sívori, Casabé, Zerba, & Wood, 1997). Versatility in the adaptation of insects to environment is provided by the phenomenon of induction of detoxification systems in insects (Liu, Zhu, Xu, Pridgeon, & Gao, 2006; Sívori et al., 1997). It is obvious from reports on resistance to insecticides in various species of insects that the most important factor in the defensive system of insects is an increased capacity to detoxify insecticides, most likely as a result of the production of additional enzymes of detoxification (Syvanen, Zhou, Wharton, Goldsbury, & Clark, 1996).

Glutathione (GSH)-related enzymes play pivotal role in the protection of biological cell against damage by toxic compound. The main protective roles of glutathione against oxidative stress consist of glutathione acting as a cofactor of several detoxifying enzymes such as glutathione peroxidase (GPX) and glutathione transferase (GST), participation in amino acid transport through the plasma membrane; scavenging hydroxyl radical and singlet oxygen directly, detoxifying hydrogen peroxide, and lipid peroxides by the catalytic action of glutathione peroxidase; regeneration of the most important antioxidants, Vitamins C and E, back to their active forms and reducing the tocopherol radical of Vitamin E directly, or indirectly, via reduction of semidehydroascorbate to ascorbate (Masella, Di Benedetto, Varì, Filesi, & Giovannini, 2005). The capacity of glutathione to regenerate the most important antioxidants is linked with the redox state of the glutathione disulfide-glutathione couple (GSSG/2GSH) (Pastore, Federici, Bertini, & Piemonte, 2003).

GSTs (EC 2.5.1.18) catalyze the conjugation of glutathione to electrophilic centers of non-polar compounds, making them more water soluble and eliminated from the cells (Hayes, Flanagan, & Jowsey, 2005; Salinas & Wong, 1999). They are involved in the detoxification of various endogenous and xenobiotic compounds, such as drugs, insecticides, organic pollutants, secondary metabolites, and other toxins (Blanchette, Feng, & Singh, 2007; Hayes et al., 2005). They are also involved in the biosynthesis and intracellular transport of hormones and protection against oxidative stress (Cnubben, Rietjens, Wortelboer, van Zanden, & van Bladeren, 2001; Hayes et al., 2005). GSTs are known to exist in dimers (homo- or heterodimers) (Vanhaelen, Francis, & Haubruge, 2004). Each GST subunit consists of two different domains, i.e. one-third of the N-terminal protein which provides the binding site for GSH (domain I), and two-third of the C-terminus which determines the substrate specificity (Armstrong, 1997). To date, at least nine classes of GSTs have been identified in mammals, eight cytosolic, and one microsomal class. In insects, first two distinct classes were described, class I and class II GSTs with 40–90% similarity between members of the same class (Chelvanayagam, Parker, & Board, 2001; Hemingway, 2000). Class I insect GSTs has also been referred to as the Delta class and class II as Sigma (Chelvanayagam et al., 2001). Another insect GST class was established (class III), which is also named Epsilon class GST (Sawicki, Singh, Mondal, Beneš, & Zimniak, 2003). GSTs have been described to play a major role as a detoxification mechanism for insecticides, thus contributing to insecticide resistance in economically important pest species in diverse agronomic cropping systems (Huang et al., 1998; Vontas, Small, & Hemingway, 2001; Vontas, Small, Nikou, Ranson, & Hemingway, 2002; Yu, 2002). GSTs have been shown to be involved in the detoxification of several classes of insecticides, i.e. organophosphates, pyrethroids, carbamates, and chlorinated hydrocarbons such as DDT (Ranson, Prapanthadara, & Hemingway, 1997). GST-based resistance to insecticides was described to be facilitated by the increase in the level of expression of one or more GSTs (Hemingway, 2000).

Some previous reports have shown bio-pesticide as an alternative to synthetic insecticide for the control of similar insect pest (Abuhussein, 2008; Faleiro, 2006; Ghoneim, Beam, Tanani, & Nassar, 2001; Nassar & Abdllahi, 2001). Although success had been recorded with the adult, but the *R. phoenicis* larvae inhabiting the core of the host calls for concern. The metabolic adaptive feature of this larva to its host may be connected to the activity of detoxifying enzyme present in the larva. Evolvement of adult *R. phoenicis* from its larva having resistant trait may be a reason why an alternative control method is sought. Bamidele, Ajele, Kolawole, and Akinkuolere (2013) previously reported the alteration of endogenous antioxidant enzymes and non-antioxidant enzymes activities in the major organs of *R. phoenicis* larvae caused by dichlorvos (DDVP). A further step is to gain deeper insight into the biochemistry of GST from *R. phoenicis* larva by elucidating the mechanism of defense against insecticide inherent in the larva. This study is designed to investigate organ distribution of some GSH-utilizing enzymes as well as to establish the involvement of glutathione transferase in the degradation of dichlorvos in the gut of larvae.

2. Materials and methods

2.1. Materials

Reduced glutathione (GSH), oxidized glutathione (GSSG), reduced nicotinamide adenine dinucleotide phosphate (NADPH), 1-chloro-2,4-dinitrobenzene (CDNB), 1,2-dichloro-4-nitrobenzene (DCNB), bovine serum albumin (BSA), GSTrap 4B (1×1 cm) column, and molecular weight marker were purchased from Sigma Chemicals Company, St. Louis, USA. Diethylaminoethyl (DEAE)-Sephadex A50 was from Pharmacia Fine Chemicals, Uppsala, Sweden. Insecticide: "Sniper 1000 EC" containing 2,2-dichlorovinyl dimethylphosphate (DDVP, 1,000 g L^{-1}) was manufactured by Hubei Sanonda Co. Ltd, Shashi Hubei, China. Other chemicals were of the highest purity commercially available.

2.2. Collection of insect larvae

African palm weevil (*R. phoenicis*) larvae (of 10.08 g mean body weight) were collected from Igbokoda palm plantation, Ondo State (7.2°N, 5.1°E) in June, 2014. Larvae were transported to the Enzymology Research Laboratory, The Federal University of Technology, Akure (FUTA) in an insect box constructed with iron-wire mesh ($25 \times 40 \times 45$ cm) and acclimatized to an air-conditioned laboratory room at 25°C, relative humidity (75%), and exposed to 12 h L: D cycle for one week prior to the experiment. The larvae were reared on degraded palm fiber collected from infested palm tree. Identification of larva was carried out in the Entomology Research Laboratory, Department of Biology, FUTA.

2.3. Larvae grouping

Individual weight of larva was taken and recorded prior to the experiment. Random selection method was used (based on recorded weights) to place active and stress-free larvae in groups. The larvae were divided into six groups consisting of 25 larvae per group. Group one was used as the control. The larvae were of an average weight of 10.08 g and the body length range of 7.0–8.0 mm in each group.

2.4. Administration of larvae with DDVP solution

Administration of larvae with DDVP solution was carried out according to Kostaropoulos, Papadopoulos, Metaxakis, Boukouvala, and Papadopoulou-Mourkidou (2001). Stock DDVP solution was prepared by dissolving "Sniper" solution in acetone and was diluted with normal saline (0.9% NaCl) to the desired concentrations of DDVP before administration. *R. phoenicis* larvae were administered 2 µL with different concentrations of DDVP (0, 0.20, 0.30, 0.40, 0.50, and 0.60 µg g^{-1} body weight) at two to three abdominal segments using a microsyringe. Care was taken to avoid puncturing the alimentary tract. Control larvae received 2 µL of normal saline containing 40% acetone (saline/acetone). Each DDVP concentration was administered into 25 individuals. Knocking down effect was recorded 48 h after treatment. The remaining live larvae were used for the analysis of GPX, GR, and GST. After the insects had been separated, they were immediately stored at −4°C. All tests were conducted at room temperature.

2.5. Dissection and preparation of larvae cytosolic fraction

Surviving larvae ($n = 5$) 48 h after exposure were demobilized by freezing, quickly dissected, and separated into three fractions; fat body (FB), gut (GT), and head (H) using dissecting kit. Fractions were stored below −20°C until use. The tissues were thereafter homogenized 1:3 (w/v) in ice-cold buffer: (25 mM potassium phosphate buffer, pH 7.2 containing 1 mM EDTA, and 1 mM 2-mercaptoethanol). Crude cytosolic enzyme was prepared subsequently by differential centrifugation. Homogenate was centrifuged at 10,000 g for 30 min at 4°C using Eppendorf Table Top Centrifuge model 5418R (Germany); floating lipid was carefully removed from supernatant through a funnel plugged with glass wool. The supernatant obtained after filtration was stored in aliquots at below −4°C and subsequently used as crude enzyme. Protein concentration was determined by the method of Bradford, (1976) using bovine serum albumin (BSA) as the standard.

2.6. Glutathione peroxidase (GPX) activity assay

The GPX (EC 1.11.1.9) activity was measured with H_2O_2 as substrate according to Paglia and Valentine (1987). This reaction was monitored indirectly as the oxidation rate of NADPH at 340 nm for 3 min using Jenway 6280 UV/Visible spectrophotometer (USA). Enzyme activity was expressed as micromoles of NADPH consumed per minute per milligram of protein, using an extinction coefficient of 6.220 M^{-1} cm^{-1}. A blank without homogenate was used as a control for the non-enzymatic oxidation of NADPH upon addition of hydrogen peroxide in 100 mM Tris–HCl buffer, pH 8.0.

2.7. Glutathione reductase (GR) activity assay

Glutathione reductase (GR) was assayed according to the method described by Saydam, Kirb, and Demir (1997). The assay mixture (3.0 mL) consisted in final concentration of 100 mM potassium phosphate buffer (pH 7.4), 1 mM EDTA, 1 mM GSSG, 0.16 mM NADPH, and 30 µL of the crude enzyme. NADPH oxidation was monitored at 340 nm for 3 min at 25°C using Jenway 6280 UV/Visible spectrophotometer (USA) and the enzyme activity was expressed as µmol min^{-1} mg^{-1} protein.

2.8. Glutathione transferase (GST) activity assay

Measurement of GST activity was conducted according to the procedure described by Habig, Pabst, and Jakoby (1974). The GSH conjugation reaction was initiated by the addition of enzyme solution to a reaction medium consisting of CDNB (1 mM) and GSH (1 mM) as substrates in 100 mM phosphate buffer pH 6.5. The change in absorbance of product for 3 min was measured at 340 nm and 25°C using Jenway 6280 UV/Visible spectrophotometer (USA). The amount of conjugated product formed was calculated using the extinction coefficient 9.6 mM^{-1} cm^{-1}. Enzyme activity was defined as µmol of product formed per min per mg protein.

2.9. Purification of GST

Palm weevil (*R. phoenicis*) larvae were administered 2 µL sub—lethal DDVP solution (0.35 µg g^{-1}) at two to three abdominal segments using a microsyringe. Control larvae received 2 µL of normal saline containing 40% acetone (saline/acetone). Larvae were collected at 2 h intervals for a period of 12 h from the control and treated larvae. They were immediately stored at −4°C and were later used for GST assay and SDS-PAGE analysis. Treatment of larvae was conducted at room temperature. Crude cytosolic enzyme was prepared subsequently by differential centrifugation as earlier described. Crude enzyme solution (6 mL; 30.45 mg mL^{-1}) was applied to an ion-exchange column of DEAE-Sephadex A50 (1.5 × 25 cm) previously equilibrated with 20 mM phosphate buffer pH 7.0 containing 1 mM EDTA and 1 mM 2-mercaptoethanol. Flow-through fractions were collected at a flow rate of 1 mL min^{-1}. After washing the column, the bound protein was eluted with a gradient of (0–1.0 M) sodium chloride in the elution buffer. Fractions with GST activity were pooled, desalted by dialysis, and concentrated by ultrafiltration on Amicon PM membrane. Concentrated protein was further purified on pre-packed GSH-Sepharose 4B affinity (1 × 1 cm) column (GSTrap 4B) previously equilibrated with phosphate buffer saline (PBS), pH 7.4 (140 mM NaCl, 2.7 mM KCl, 10 mM Na_2HPO_4, 1.8 mM KH_2PO4) according to the manufacturer instruction. Protein was eluted with the same buffer at a flow rate of 1 mL min^{-1} until no protein was detected in the flow through-fractions. Bound protein was eluted with 10 mM GSH in 50 mM Tris–HCl buffer, pH 8.0 (Konishi et al., 2005). The fractions with GST activity were pooled and subsequently used for electrophoretic and inhibition studies. All purification steps were conducted at 4°C.

2.10. Gel profile of GST

Purified GST was analyzed by SDS-PAGE using omniPAGE Vertical Electrophoresis system (England) according to the method of Laemmli (1970). Purified GST was mixed with protein sample buffer (0.0625 M Tris–HCl, pH 6.8, 2% SDS, 10% glycerol, 5% β-mercaptoethanol and 0.125% Bromophenol blue) and boiled for 2 min. A total of 21.50 µg of the mixture was subjected to 10% SDS-PAGE. After electrophoresis, gel was stained with 0.1% Coomassie Brilliant Blue R-250. Gels were thereafter destained in a destaining solution overnight.

2.11. Initial velocity of GST

Initial velocity rate measurements were performed in 100 mM potassium phosphate buffer pH 6.5 containing EDTA and 2-mercaptoethanol at 25°C according to Bastien, Fournier, Baudras, and Baudras (1999). The reaction mixture (3 mL) consisted of buffer, varied concentrations of GSH (0.25–1 mM), and different concentrations of CDNB (0.25–1 mM) which were kept constant. Reaction rate measurements were also carried out with varied concentrations of CDNB (0.25–1 mM) at constant varied concentrations of GSH (0.25–1 mM) which were kept constant. Substrates were freshly prepared during the experiment and ethanol concentration was kept at 4% in the reaction medium. Enzyme solution (30 µL) was added last to initiate the reaction. The appearance of product was monitored at 340 nm for 3 min. Reaction mixture without enzyme was used as a reference to correct non-enzymatic reaction. Each reaction rate was measured at least three times and then averaged. Data were fitted to the rectangular hyperbola equation and analyzed using Graphpad prism 6 (Graphpad Software, San Diego, CA, USA). The kinetic constants (V_{max} and K_m) were also determined.

2.12. Statistical analysis

Data obtained from the experiments were presented as mean ± SD of the results from three independent experiments. Kinetic data were analyzed using Graph pad prism 6 (Graphpad Software, San Diego, CA, USA). Data were statistically analyzed using one-way analysis (ANOVA) followed by Duncan's New Multiple Test where appropriate.

3. Results

3.1. GPX activity

Effect of DDVP on glutathione peroxidase activities in the FB, GT, and H tissues of *R. phoenicis* larva is shown in Figure 1. In the FB of DDVP-treated *R. phoenicis* larvae, GPX activities increased significantly ($p < 0.05$) with increased DDVP concentration when compared with the control. GPX activity of 24.94 µmol min^{-1} mg^{-1} protein in the FB of larva was produced by 0.33 µg g^{-1} DDVP. In the GT of larva, initial DDVP concentrations (0.22 and 0.33 µg g^{-1}) caused increased GPX activity while higher concentrations caused a decline in GPX activity when compared with the control. GPX activity in the H tissue of larva decreased significantly ($p < 0.05$) with increased DDVP concentrations when compared with the control. Highest GPX activity was recorded in the GT of larva.

3.2. GR activity

Effect of DDVP on glutathione reductase activity in the FB, GT, and H of *R. phoenicis* larva is shown in Figure 2. GR activity was DDVP-dose dependent in the three organs of *R. phoenicis* larva. GR activity was significantly increased in the gut of larva but declined at DDVP concentration greater than 0.44 µg g^{-1} when compared with the control. Similar result was observed in the head of *R. phoenicis* larva.

3.3. GST activity

Figure 3 shows the effect of DDVP on glutathione transferase activity in the FB, GT, and H of *R. phoenicis* larva. GST activity in the GT and H increased significantly ($p < 0.05$) with increased DDVP concentration when compared with the control. Due to significant increase in GST activity in the gut of larva, further investigation of the effect of sub-lethal DDVP concentration (0.35 µg g^{-1}) on GST activity in the GT for duration of eight hours was done. Bioassay results showed an increase in GST activity in the gut tissue of *R. phoenicis* larva as a function of time of DDVP treatment (Figure 4). The initial GST activity was similar for DDVP-treated and control larvae. The specific activity of crude GST from the gut of larvae treated with DDVP after 2, 4, 6, and 8 h were 20.49, 21.43, 25.34 and 22.52 µmol min^{-1} mg^{-1} protein, respectively. GST activity in the treated larva was four times higher than in the control. GST was detected by SDS-PAGE (Figure 5).

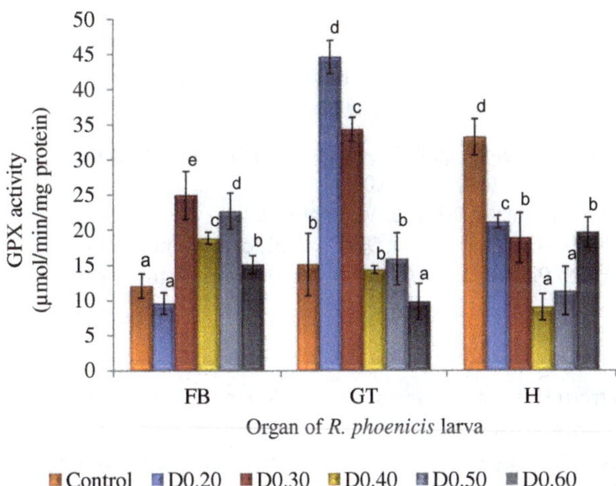

Figure 1. Glutathione peroxidase activity of fat body (FB), gut (GT), and head (H) tissues of *Rynchophorous phoenicis* larva treated with varied concentration of DDVP solution.

Notes: Each value is the mean of five repetitions ± SD. Significant difference in the treated groups from their corresponding control was indicated by alphabets (a-f) at $p < 0.05$.
Key: Control-larva treated with acetone in normal saline; D0.20-larva treated with 0.20 µg g^{-1} (w/v) DDVP; D0.30-larva treated with 0.30 µg g^{-1} (w/v); DDVP D0.40-larva treated with 0.40 µg g^{-1} (w/v); DDVP D0.50-larva treated with 0.50 µg g^{-1} (w/v); DDVP D0.60-larva treated with 0.60 µg g^{-1} (w/v) DDVP.

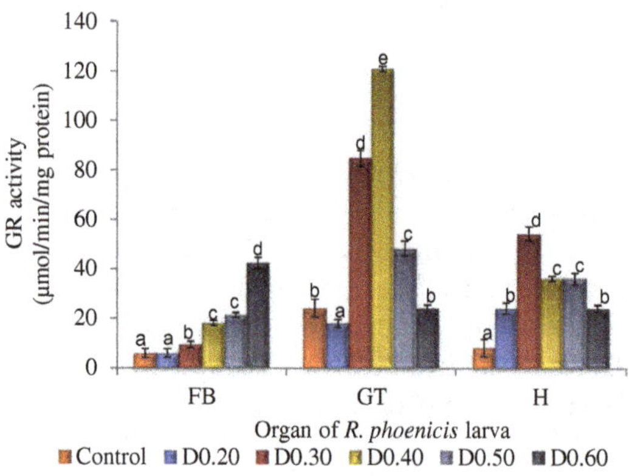

Figure 2. Glutathione reductase activity of fat body (FB), gut (GT), and head (H) tissues of *Rynchophorous phoenicis* larva treated with varied concentration of DDVP solution.

Notes: Each value is the mean of five repetitions ± SD. Significant difference in the treated groups from their corresponding control was indicated by alphabets (a-f) at $p < 0.05$.
Key: Control-larva treated with acetone in normal saline; D0.20-larva treated with 0.20 µg g^{-1} (w/v) DDVP; D0.30-larva treated with 0.30 µg g^{-1} (w/v); DDVP D0.40-larva treated with 0.40 µg g^{-1} (w/v); DDVP D0.50-larva treated with 0.50 µg g^{-1} (w/v); DDVP D0.60-larva treated with 0.60 µg g^{-1} (w/v) DDVP.

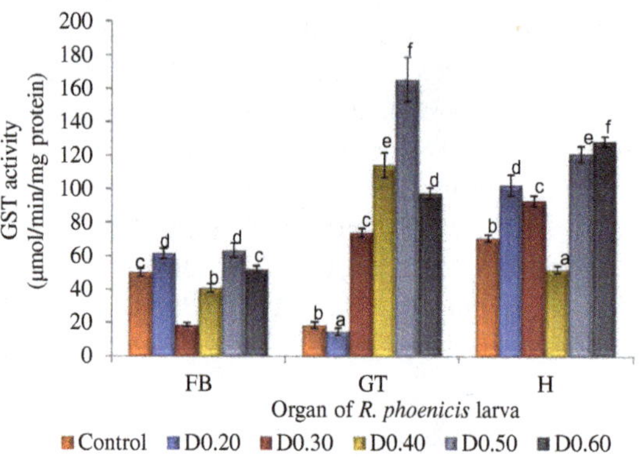

Figure 3. Glutathione transferase activity of fat body (FB), gut (GT), and head (H) tissues of *Rynchophorous phoenicis* larva treated with varied concentration of DDVP solution.

Notes: Each value is the mean of five repetitions ± SD. Significant difference in the treated groups from their corresponding control was indicated by alphabets (a-f) at $p < 0.05$.
Key: Control-larva treated with acetone in normal saline; D0.20-larva treated with 0.20 µg g^{-1} (w/v) DDVP; D0.30-larva treated with 0.30 µg g^{-1} (w/v); DDVP D0.40-larva treated with 0.40 µg g^{-1} (w/v); DDVP D0.50-larva treated with 0.50 µg g^{-1} (w/v); DDVP D0.60-larva treated with 0.60 µg g^{-1} (w/v) DDVP.

3.4. Purification of DDVP-induced GST

The elution profile of the crude enzymes from the gut of control and DDVP-treated larvae on DEAE-sephadexA50 are shown in Figure 6(A) and (B), respectively. Glutathione transferase activity was detected in the flow-through fractions as indicated by the two major activity peaks "a" and "b" in Figure 6(A). Minor GST activity peak "c" indicated bound GST eluted with a linear gradient of NaCl in buffer solution. Elution profile of GST from gut of DDVP-treated larva (Figure 6(B)) was similar to the control (Figure 6(A)) but an observable difference existed with the bound protein. After protein was eluted with a (0–1.0 M) gradient of NaCl in buffer solution, GST activity detected in the fractions obtained from bound protein was indicated as the GST activity peaks "c" and "d" in Figure 6(B). GST activity peak "c" was considered one of the GST isoenzymes associated with DDVP degradation in the gut of *R. phoenicis* larva. The specific activity of this ion-exchange purified GST was 11.65 ± 0.72 µmol min^{-1} mg^{-1} protein. Further purification by GSH-Sepharose 4B affinity gel chromatography revealed a major GST activity peak eluted with 10 mM GSH in buffer (Figure 7). The yield of purification of GST was 8.37 ± 1.61% of the total GST activity obtained from GSH-affinity

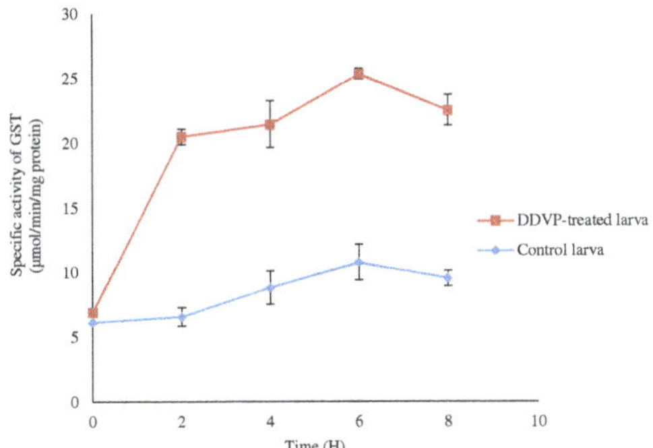

Figure 4. Time-course glutathione transferase activity (GST) in the gut tissue of *Rynchophorous phoenicis* larva.

Notes: Each value is the mean of three repetitions ± SD. Experiment was conducted at room temperature.

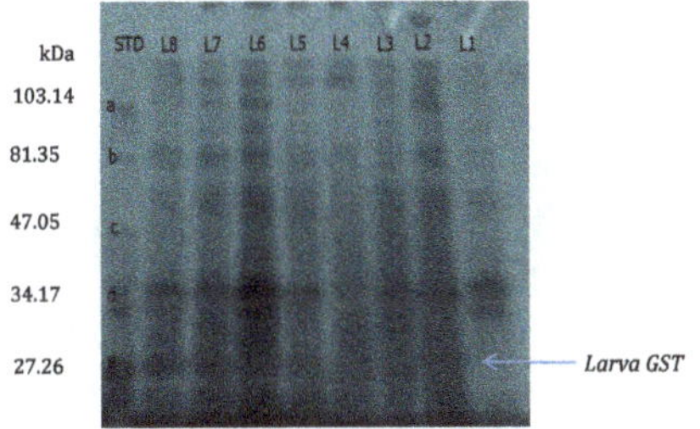

Figure 5. Gel profile of crude extracts of gut of DDVP-treated *Rynchophorus phoenicis* larva.

Notes: STD: Standard molecular mass, L1: Crude extract from control larva at 0 h and crude extract from larva L2–2 h, L3–4 h, L4–6 h, L5–8 h, L6–10 h, L7–12 h after treatment with DDVP.

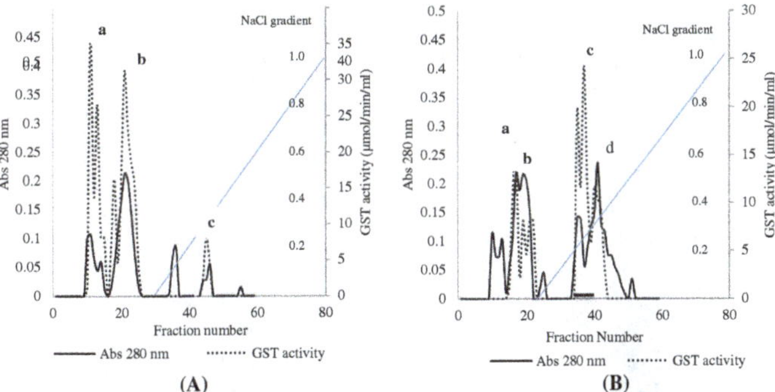

Figure 6. Elution profiles of glutathione transferase from the gut of (A) control *R. phoenicis* larva and (B) DDVP-treated *R. phoenicis* larva using ion exchange chromatography.

Notes: Crude enzyme was applied to a DEAE-Sephadex A50 (1.5 × 25 cm) column previously equilibrated with 25 mM potassium phosphate buffer, pH 7.2 containing 1 mM EDTA and 1 mM 2-mercaptoethanol. The protein was eluted at a flow rate of 1 mL min^{-1} and fractions of 5 mL each were collected. Fractions pooled: 33–37 (6B).

chromatography (Table 1). The specific activity and purification fold of purified GST were 52.16 ± 5.28 µmol min^{-1} mg^{-1} protein and 20.37 ± 2.43, respectively (Table 1). The affinity bound protein was tagged *rpl*GSTc and used for the SDS-PAGE. *rpl*GSTc showed a single band on a 10% gel and the estimated molecular weight was approximately 26.2 kDa (Figure 8(A)). Native gel electrophoresis revealed an estimated molecular weight of 49.7 kDa (Figure 8(B)).

3.5. Initial velocity of rplGSTc

Double reciprocal plot of *rpl*GSTc-catalyzed reaction with respect to GSH is shown in Figure 9(A). The plot of $1/v_0$ vs. 1/[CDNB] at constant varied concentrations of GSH were linear and the point of intersection or convergence of lines is on the 1/[CDNB] axis. The slopes and intercepts of double reciprocal

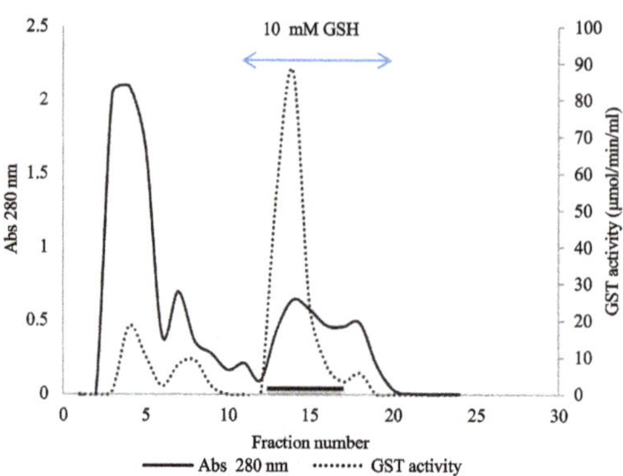

Figure 7. Elution profile of glutathione transferase on affinity chromatography-GSTrap 4B. Ion-exchange-purified enzyme was applied to a GSH-Sepharose 4B (1 × 1 cm) column previously equilibrated with PBS, pH 7.4.

Notes: The protein was eluted at a flow rate of 1 mL min^{-1} and fractions of 1 mL each were collected. Bound protein was eluted with 10 mM GSH in 50 mM Tris-HCl buffer, pH 8.0. Fractions pooled: 13–18.

Table 1. Summary of purification of *rpl*GSTc

Step	Total protein (mg)	Total activity (μmol min^{-1})	Specific activity (μmol min^{-1} mg^{-1})	Yield (%)	Fold
Crude GST	3,110.01 ± 194.2	8,011.22 ± 367.12	2.57 ± 0.16	100.00 ± 0.00	1.00 ± 0.00
Ion exchange chromatography	185.62 ± 11.5	2,164.25 ± 99.17	11.65 ± 0.72	27.10 ± 1.69	4.54 ± 0.28
GSTrap 4B	12.75 ± 2.34	665.05 ± 18.15	52.16 ± 5.28	8.32 ± 1.61	20.37 ± 2.43

Notes: Purification was conducted at 4°C. Values are mean ± SD of three repetitions.

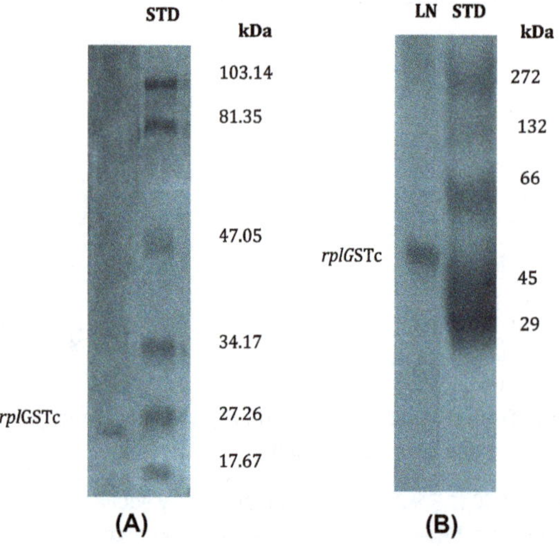

Figure 8. Electrophorectograms of purified *rpl*GSTc on (A) 10% polyacylamide gel slab at room temperature and (B) Native gel of 10% polyacrylamide.

Notes: Standard proteins ranging from molecular weight (MW) 17.27–103.14 kDa. Standard proteins ranging from molecular weight (MW) 29–272 kDa.

plot in Figure 9(A) vs. 1/[GSH] were also linear (Figure 9(a)). Similarly, the plot of $1/v_o$ vs. 1/[GSH] at constant varied concentrations of CDNB is shown in Figure 9(B). Double reciprocal plot of *rpl*GSTc-catalyzed reaction with respect to CDNB were linear and the point of convergence of lines is on the 1/[GSH] axis. Secondary replots of intercepts and slopes in Figure 9(B) vs. 1/[CDNB] were linear (Figure 9(b)). The estimated K_m^{CDNB} and K_m^{GSH} from the secondary replots were 0.191 ± 0.03 and 0.136 ± 0.02 mM, respectively, while maximum velocity (V_{max}) was 46.21 ± 7.23 μmol min^{-1} mL^{-1}.

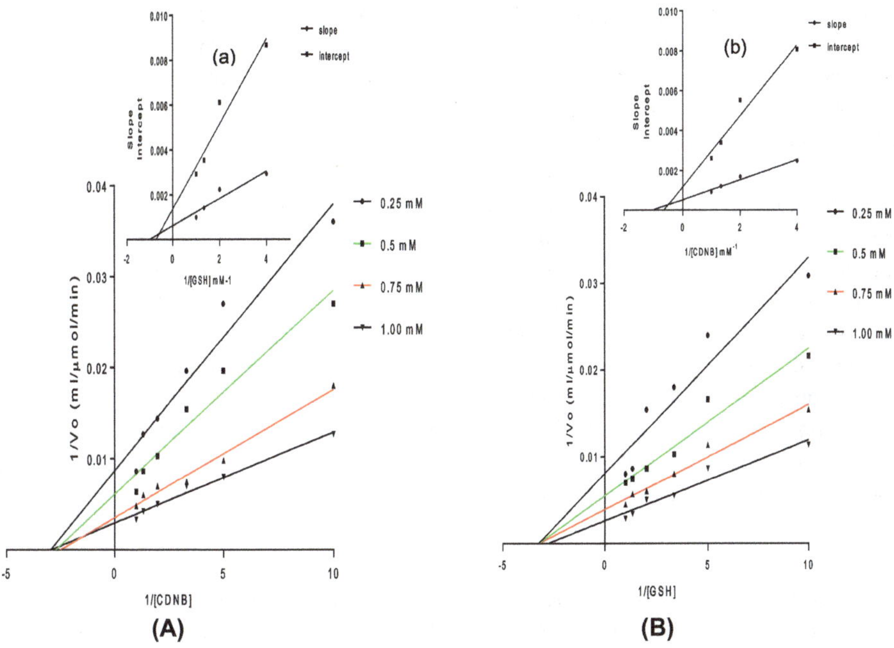

Figure 9. Double reciprocal plot of *rpl*GSTc-catalyzed reaction. (A) Plots of $1/v_o$ vs. 1/[CDNB] at constant varied concentrations of GSH. (a) Secondary plot: plots of intercept or slopes in (A) vs. 1/[GSH]. (B) Plots of $1/v_o$ vs. 1/[GSH] at constant varied concentrations of CDNB. (b) Secondary plot: plots of intercepts or slopes in (B) vs. 1/[CDNB].

4. Discussion

In this report, alteration of activities of some GSH-utilizing enzymes was observed in the fat body, gut, and head of DDVP-treated larva. This suggestively indicated the presence of GPX, GR and GST in the organs of the larva. Dose-dependent responses of GSH-utilizing enzymes against the concentration of DDVP were recorded. However, linearity was not observed with the increasing concentration of DDVP and the corresponding levels of GPX, GR, and GST activities in the organs. This observation might be on the account of sensitivity of individual larva to DDVP, genetic makeup, and different rate of elimination of the insecticide. Similar finding have been well documented by some authors (Bamidele et al., 2013; Vanhaelen et al., 2004; Zhang, Rashid, Ji, Rashid, & Wang, 2013). Due to the significantly increased activities of GPX, GR, and GST in the gut of larva, GSH-utilizing enzymes seem to play significant role in the gut of larva by probably degrading and detoxifying insecticide and its metabolites leading to removal of toxic compounds from the larval system. Also the various levels of glutathione-utilizing enzymes in the three organs showed that the individual organ handled the chemical assault differently. This is partly responsible for the degree of free radical scavenging and detoxifying activities in the larva.

Analysis of crude enzyme preparation from gut of treated larva by SDS-PAGE revealed increased intensity and broadness of GST bands at different time intervals after exposure. This observation conforms to the increased specific activity of GST measured over the period of exposure to the insecticide. This also suggested the capacity of GST in the gut of larva to degrade the chemical insecticide. Earlier reports by some authors suggested that increased level of GSTs is one of the major reasons for the development of resistance by metabolic detoxification of insecticides (Zhang et al., 2013). Development of insecticide resistance is caused by increased levels of GSTs in diamondback moth and *H. arnigera* (Furlong & Wright, 1994; Mohan & Gujar, 2003). Enhanced activities of GSTs that confer insecticide resistance from both qualitative and quantitative alterations in gene expression are provided. Evidence for over-expression of one or more GST isoenzymes in resistant insects was observed in *M. domestica* (Fournier, Bride, Poire, Berge, & Plapp, 1992; Zhang et al., 2013).

Purification processes revealed the possible GST associated with DDVP degradation in the gut of treated larva. It is evident therefore that GST among other GSH-utilizing enzymes is linked with the detoxification of insecticides and removal of possible metabolites from the gut of larva. The yield of the purified *rpl*GSTc was significantly lower than that obtained in *Myallopa florae* (77%), *Syrphus ribesii* (65%) (Vanhaelen et al., 2004; *L. paeta* (41%) (Wu, Dou, Wu, & Wang, 2009). This might be due to loss of GST activity and unrecovered enzyme bound to resin during the purification steps. But, the yield of the purified *rpl*GSTc was higher than *S. inneta* (3.3%) (Valles, Perera, & Strong, 2003). The purification step in this report were previously used by Konishi et al. (2005) and Shukor, Wajidi, Avicor, and Jaal (2014). The estimated molecular weights of *rpl*GSTc from native gel and SDS-PAGE revealed that the soluble cytosolic GST might be homodimeric enzyme and the result is similar to the size of GST of many other insect species such as *S. invicta* (Valles et al., 2003); fall webworm *H. cunea* (Yamamoto, Miake, & Aso, 2007) and some Psocids (Dou, Wu, Hassan, & Wang, 2009; Wu et al., 2009). The mechanism of *rpl*GSTc appears to conform to the random sequential model. This is indicated by the pattern of convergence of lines of double reciprocal plots in both cases of GSH and CDNB as variable substrates. This pattern would permit greater flexibility in the conjugation of various xenobiotics to water soluble, less toxic products (Adewale & Afolayan, 2005).

Conclusively, Dichlorvos (DDVP) caused the alteration of some glutathione-utilizing enzymes in the major organs of *R. phoenicis* larva. The elevated levels of these enzymes suggested possible protective handling of the toxic tissue-damaging insect poisons in the gut of the larva. A novel 49.2 kDa homodimeric GST isoenzyme (*rpl*GSTc) associated with DDVP degradation was purified and the enzyme revealed a random sequential mechanism toward its substrates. The present results indicate that GST might be involved in the degradation of dichlorvos in the gut of *R. phoenicis* larvae.

Funding

This research was supported by the Department of Biology, The Federal University of Technology, Akure.

Competing Interests

The authors declare no competing interest.

Author details

Olufemi Samuel Bamidele[1]
E-mail: osbamidele@futa.edu.ng
Joshua Oluwafemi Ajele[1]
E-mail: joajele@futa.edu.ng
Folasade Mayowa Olajuyigbe[1]
E-mail: folajuyigbe@futa.edu.ng
[1] Enzymology Research Unit, Department of Biochemistry, The Federal University of Technology, P. M. B 704, Akure, Nigeria.

Cover image

Source: Authors.

References

Abuhussein, M. O. (2008). The effect of neem seed extract (Azadiractin) on red palm weevil, *Rhynchophorus ferrugineus* (Olivier) (Rhchophoridae: Coleoptera). *International Journal of Science and Research, 17*, 89–94.

Adewale, I. O., & Afolayan, A. (2005). Studies on glutathione transferase from grasshopper (*Zonocerus variegatus*). *Pesticide Biochemistry and Physiology, 85*, 52–59.

Al-Ayied, H. Y., Alswailem, A. M., Shair, O., & Al Jabr, A. M. (2006). Evaluation of phylogenetic relationship between three phenotypically different forms of Red date palm weevil *Rhynchophorus ferrugineus* Oliv. using PCR-based RAPD technique. *Archives of Phytopathology and Plant Protection, 39*, 303–309. http://dx.doi.org/10.1080/03235400500222461

Armstrong, R. N. (1997). Structure, catalytic mechanism, and evolution of the glutathione transferases. *Chemical Research in Toxicology, 10*, 2–18. http://dx.doi.org/10.1021/tx960072x

Bamidele, O., Ajele, J., Kolawole, A., & Akinkuolere, O. (2013). Changes in the tissue antioxidant enzyme activities of palm weevil (*Rynchophorous phoenicis*) larva by the action of 2,2-dichlorovinyl dimethyl phosphate. *African Journal of Biochemistry Research, 7*, 128–137.

Barranco, P., de la Peña, J., Martín, M. M., & Cabello, T. (1998). Eficacia del control químico de la nueva plaga de las palmeras *Rhynchophorus ferrugineus* (Olivier, 1790) (Coleopterae.: Curculionidae). *Boletín de Sanidad Vegetal, Plagas, 24*, 23–40.

Bastien, N., Fournier, D., Baudras, A., & Baudras, B. (1999). Mechanism of an insect glutathione S-transferase: Kinetic analysis supporting a rapid equilibrium random sequential mechanism with housefly I1 isoform. *Insect Biochemistry and Molecular Biology, 29*, 71–79.

Blanchette, B., Feng, X., & Singh, B. R. (2007). Marine glutathione S-transferases. *Marine Biotechnology, 9*, 513–542. http://dx.doi.org/10.1007/s10126-007-9034-0

Bong, C. J., Er, C., Yiu, P., & Rajan, A. (2008). Growth performance of the red-stripe weevil *Rhynchophorus schach* Oliv. (Insecta: Coleoptera: Curculionidae) on meridic diets. *American Journal of Agricultural and Biological Sciences, 3*, 403–409.

Bradford, M. M. (1976). A rapid and sensitive method for the quantitation of microgram quantities of protein utilizing the principle of protein-dye binding. *Analytical Biochemistry, 72*, 248–254.

Cabello, T., de la Pena, J., Barranco, P., Belda, J., & de la Pena, J. (1997). Laboratory evaluation of imidacloprid and oxamyl against *Rhynchophorus ferrugineus*. *Tests Agrochemical Cultiva, 18*, 6–7.

Chelvanayagam, G., Parker, M. W., & Board, P. G. (2001). Fly fishing for GSTs: A unified nomenclature for mammalian and insect glutathione transferases. *Chemico-Biological Interactions, 133*, 256–260.

Cnubben, N. H. P., Rietjens, I. M. C. M., Wortelboer, H., van Zanden, J., & van Bladeren, P. J. (2001). The interplay of glutathione related processes in antioxidant defense. *Environmental Toxicology and Pharmacology, 10*, 141–152. http://dx.doi.org/10.1016/S1382-6689(01)00077-1

Dou, W., Wu, S., Hassan, M. W., & Wang, J. J. (2009). Purification and biochemical characterization of glutathione S-transferases from three strains of *Liposcelis bostrychophila Badonnel* (Psocoptera: Liposcelididae): Implication of insecticide resistance. *Pesticide Biochemistry and Physiology, 94*, 10–14. http://dx.doi.org/10.1016/j.pestbp.2009.02.005

Faleiro, J. R. (2006). A review of the issues and management of red palm weevil, *Rhynchophorus ferrugineus* (Curculionidae: Coleoptera) in coconut and date palm during the last one hundred years. *International Journal of Tropical Insect Science, 26*, 135–154.

Faleiro, J. R. (2010). *Consultation on strengthening of national capacities for the management of red palm weevil in North Africa (Morocco, Lybia and Tunisia)* (p. 43). Rome: IPM Mission, FAQ.

Fournier, D., Bride, J. M., Poire, M., Berge, J. B., & Plapp, F. W. (1992). Insect glutathione S-transferases: Biochemical characteristics of the major forms from houseflies susceptible and resistant to insecticides. *The Journal of Biological Chemistry, 267*, 1840–1845.

Furlong, M. J., & Wright, D. J. (1994). Examination of stability of resistance and cross-resistance patterns to acylurea insect growth regulators in field populations of the diamondback moth, Plutella xylostella, from Malaysia. *Pesticide Science, 42*, 315–326. http://dx.doi.org/10.1002/ps.v42:4

Ghoneim, R. S., Beam, A. S., Tanani, M. A., & Nassar, M. M. (2001). Respiratory Metabolic responsiveness during pupal stage of the red palm weevil *Rhynchophorus ferrugineus* (Curculionidae: Coleoptera) to certain plant extracts. *Medical Faculty, Ladbouwwo University, Gent, 66*, 492–502.

Giblin-Davis, R. M., Oehlschlager, A. C., Perez, A., Gries, G., Gries, R., Weissling, T. J., ... Gonzalez, L. M. (1996). Chemical and behavioral ecology of palm weevils (Curculionidae: Rhynchophorinae). *The Florida Entomologist, 79*, 153–167. http://dx.doi.org/10.2307/3495812

Gries, G., Gries, R., Perez, A. L., Gonzales, L. M., Pierce, H. D., Cameron Oehlschlager, A., ... Kouame, B. (1994). Ethyl propionate: Synergistic kairomone for African palm weevil, *Rhynchophorus phoenicis* L. (Coleoptera: Curculionidae). *Journal of Chemical Ecology, 20*, 889–897. http://dx.doi.org/10.1007/BF02059585

Habig, W. H., Pabst, M. J., & Jakoby, W. B. (1974). Glutathione S-transferases: The first enzymatic step in mercapturic acid formation. *The Journal of Biological Chemistry, 240*, 7130–7139.

Hayes, J. D., Flanagan, J. U., & Jowsey, I. R. (2005). Glutathione transferases. *Annual Review of Pharmacology and Toxicology, 45*, 51–88. http://dx.doi.org/10.1146/annurev.pharmtox.45.120403.095857

Hemingway, J. (2000). The molecular basis of two contrasting metabolic mechanisms of insecticide resistance. *Insect Biochemistry and Molecular Biology, 30*, 1009–1015. http://dx.doi.org/10.1016/S0965-1748(00)00079-5

Huang, H. S., Hu, N. T., Yao, Y. E., Wu, C. Y., Chiang, S. W., & Sun, C. N. (1998). Molecular cloning and heterologous expression of a glutathione S-transferase involved in insecticide resistance from the diamondback moth, *Plutella xylostella*. *Insect Biochemistry and Molecular Biology, 28*, 651–658. http://dx.doi.org/10.1016/S0965-1748(98)00049-6

Konishi, T., Kato, K., Araki, T., Shiraki, K., Takagi, M., & Tamaru, Y. (2005). A new class of glutathione S-transferase from the hepatopancreas of the red sea bream *Pagrus major*. *Biochemical Journal, 388*, 299–307. http://dx.doi.org/10.1042/BJ20041578

Kostaropoulos, I., Papadopoulos, A. I., Metaxakis, A., Boukouvala, E., & Papadopoulou-Mourkidou, E. (2001). Glutathione S-transferase in the defence against pyrethroids in insects. *Insect Biochemistry and Molecular Biology, 31*, 313–319. http://dx.doi.org/10.1016/S0965-1748(00)00123-5

Laemmli, U. K. (1970). Cleavage of structural proteins during the assembly of the head of bacteriophage T4. *Nature, 227*, 680–685. http://dx.doi.org/10.1038/227680a0

Liu, N. N., Zhu, F., Xu, Q., Pridgeon, J. W., & Gao, X. W. (2006). Behavioral change, physiological modification, and metabolic detoxification: Mechanisms of insecticide resistance. *Acta Entomologica Sinica, 49*, 671–679.

Mariau, D., Chenon, R., Julia, J. F., & Philippe, R. (1981). Oil palm and coconut pests in West Africa. *Oleagineux, 36*, 169–227.

Masella, R., Di Benedetto, R., Varì, R., Filesi, C., & Giovannini, C. (2005). Novel mechanisms of natural antioxidant compounds in biological systems: Involvement of glutathione and glutathione-related enzymes. *The Journal of Nutritional Biochemistry, 16*, 577–586. http://dx.doi.org/10.1016/j.jnutbio.2005.05.013

Mohan, M., & Gujar, G. T. (2003). Local variation in susceptibility of the diamondback moth, *Plutella xylostella* (Linnaeus) to insecticides and role of detoxification enzymes. *Crop Protection, 22*, 495–504. http://dx.doi.org/10.1016/S0261-2194(02)00201-6

Nassar, M., & Abdllahi, M. A. (2001). Evaluation of Azadiractin for the control of red palm weevil, *Rhynchophorus ferrugineus* (Oliever) (Curculionidae: Coleoptera). *Journal of the Egyptian German Society of Zoology, 6*, 163–173.

Paglia, D. E., & Valentine, W. N. (1987). Studies on the quantitative and qualitative characterization of erythrocytes glutathione peroxides. *Journal of Laboratory and Clinical Medicine, 70*, 158–169.

Pastore, A., Federici, G., Bertini, E., & Piemonte, F. (2003). Analysis of glutathione: Implication in redox and detoxification. *Clinica Chimica Acta, 333*, 19–39.

Ranson, H., Prapanthadara, L., & Hemingway, J. (1997). Cloning and characterization of two glutathione S-transferases from a DDT-resistant strain of *Anopheles gambiae*. *Biochemical Journal, 324*, 97–102. http://dx.doi.org/10.1042/bj3240097

Salinas, A. E., & Wong, M. G. (1999). Glutathione S-transferases—A review. *Current Medicinal Chemistry, 6*, 279–309.

Sawicki, R., Singh, S. P., Mondal, A. K., Beneš, H., & Zimniak, P. (2003). Cloning, expression and biochemical characterization of one Epsilon-class (GST-3) and ten Delta class (GST-1) glutathione S-transferases from *Drosophila melanogaster*, and identification of additional nine members of the Epsilon class. *Biochemical Journal, 370*, 661–669. http://dx.doi.org/10.1042/bj20021287

Saydam, N., Kirb, A., & Demir, Ö. (1997). Determination of glutathione, glutathione reductase, glutathione peroxidase and glutathione S-transferase levels in human lung cancer tissues. *Cancer Letters, 119*, 13–19. http://dx.doi.org/10.1016/S0304-3835(97)00245-0

Shukor, S. A. F. F., Wajidi, Mustafa S., Avicor, W., & Jaal, Z. (2014). Purification and expression of glutathione S-transferase from a Malaysian population of *Aedes albopictus* (Diptera: Culicidae). *Biologia, 69*, 530–535.

Sívori, J. L., Casabé, N., Zerba, E. N., & Wood, E. J. (1997). Induction of Glutathione S-transferase Activity in *Triatoma infestans. Memórias do Instituto Oswaldo Cruz, 92*, 797–802. http://dx.doi.org/10.1590/S0074-02761997000600013

Soyebo, K. O., Farinde, A. J., & Dionco-Adetayo, E. D. (2005). Constraints of oil production in Ife central local government area of Osun State, Nigeria. *Journal of Social Sciences, 10*, 55–59.

Syvanen, M., Zhou, Z., Wharton, J., Goldsbury, C., & Clark, A. (1996). Heterogeneity of the glutathione transferase genes encoding enzymes responsible for insecticide degradation in the housefly. *Journal of Molecular Evolution, 43*, 236–240. http://dx.doi.org/10.1007/BF02338831

Valles, S. M., Perera, O. P., & Strong, C. A. (2003). Purification, biochemical characterization, and cDNA cloning of a glutathione S-transferase from the red imported fire ant, *Solenopsis invicta. Insect Biochemistry and Molecular Biology, 33*, 981–988. http://dx.doi.org/10.1016/S0965-1748(03)00104-8

Vanhaelen, N., Francis, F., & Haubruge, E. (2004). Purification and characterization of glutathione S-transferases from two syrphid flies (*Syrphus ribesii* and *Myathropa florae*). *Comparative Biochemistry and Physiology Part B: Biochemistry and Molecular Biology, 137*, 95–100. http://dx.doi.org/10.1016/j.cbpc.2003.10.006

Vontas, J. G., Small, G. J., & Hemingway, J. (2001). Glutathione S-transferases as antioxidant defence agents confer pyrethroid resistance in *Nilaparvata lugens. Biochemical Journal, 357*, 65–72. http://dx.doi.org/10.1042/bj3570065

Vontas, J. G., Small, G. J., Nikou, D. C., Ranson, H., & Hemingway, J. (2002). Purification, molecular cloning and heterologous expression of a glutathione S-transferase involved in insecticide resistance from the rice brown planthopper, *Nilaparvata lugens. Biochemical Journal, 362*, 329–337. http://dx.doi.org/10.1042/bj3620329

Wu, S., Dou, W., Wu, J. J., & Wang, J. J. (2009). Purification and partial characterization of glutathione S-transferase from insecticide resistant field populations of Liposcelis paeta Pearman (*Psocoptera: Liposcelididae*). *Archives of Insect Biochemistry and Physiology, 70*, 136–150. http://dx.doi.org/10.1002/arch.v70:2

Yamamoto, K., Miake, F., & Aso, Y. (2007). Purification and characterization of a novel sigma-class glutathione S-transferase of the fall webworm, *Hyphantria cunea. Journal of Applied Entomology, 131*, 466–471. http://dx.doi.org/10.1111/jen.2007.131.issue-7

Yu, S. J. (2002). Biochemical characteristics of microsomal and cytosolic glutathione S-transferases in larvae of the fall armyworm, *Spodoptera frugiperda* (J. E. Smith). *Pesticide Biochemistry and Physiology, 72*, 100–110. http://dx.doi.org/10.1006/pest.2001.2585

Zhang, Y. L., Rashid, A. K., Ji, Y. L., Rashid, M., & Wang, D. (2013). Cantharidin impedes activity of glutathione S-transferase in the mid-gut of *Helicoverpa armigera* Hübner. *International Journal of Molecular Sciences, 14*, 5482–5500.

11

Catechins from green tea modulate neurotransmitter transporter activity in Xenopus oocytes

Y.X. Wang[1,2,3], T. Engelmann[1,2,3], Y.F. Xu[1,2] and W. Schwarz[1,2,3*]

*Corresponding author: W. Schwarz, Shanghai Research Center for Acupuncture and Meridians, Shanghai, China; Shanghai Key Laboratory of Acupuncture Mechanism and Acupoint Function, Fudan University, Shanghai, China; Institute for Biophysics, Goethe-University Frankfurt, Frankfurt am Main, Germany

E-mail: schwarz@biophysik.org

Reviewing editor: Tsai-Ching Hsu, Chung Shan Medical University, Taiwan

Additional information is available at the end of the article

Abstract: The GABAergic and glutamatergic systems play key roles in controlling activity of the central nervous system. Important membrane proteins in the mammalian central nervous system transporting extracellular GABA and glutamate are the GABA transporter GAT1 and the glutamate transporter EAAC1. We investigated the effect of catechins of green tea (*Camellia sinensis*) on the activity of GAT1 and EAAC1 by detecting the respective electrogenic transporter-mediated current under voltage clamp. Epigallocatechin-3-gallate inhibited GAT1-mediated current to 50% at about 100 μM. The EAAC1-mediated current could be stimulated up to 80% by (-)-epicatechin; 50% of maximum stimulation was achieved by about 5 μM. Inhibition of GAT1 and stimulation of EAAC1 will counteract hyperexcitability.

Subjects: Cell Biology; Neurobiology; Biophysics

Keywords: green tea; catechins; neurotransmitter transporter; voltage clamp

1. Introduction

The dominating inhibitory neurotransmitter in adult mammalian brain is γ-aminobutyric acid (GABA), and the dominating excitatory neurotransmitter is glutamate (Kandel, Schwartz, & Jessell, 2000). The GABAergic and glutamatergic systems play key roles in the occurrence of pathological conditions like epilepsy. To control synaptic transmission and concentration of extracellular neurotransmitter highly potent Na^+ gradient-driven transporters mediate the removal of respective extracellular neurotransmitter by uptake into presynaptic neurons or surrounding glial cells (Zhou & Danbolt, 2013). Transporters controlling extracellular concentrations of GABA and glutamate in the mammalian central nervous system are the GABA transporter 1 (GAT1) and the glutamate transporter EAAC1 (also named as EAAT3) (Bjørn-Yoshimoto & Underhill, 2016; Borden et al., 1994; Jensen, Chiu, Sokolova, Lester, & Mody, 2003; Lane et al., 2014). In addition to transporting glutamate, EAAC1

ABOUT THE AUTHORS

The focus of our research activities is the investigation of cellular events that might be involved in the effects of treatments by traditional Chinese medicine. This includes mechanisms of acupuncture in the periphery (acupuncture points) as well as in the central nervous system. In addition to acupuncture, effects of tradition Chinese herbal chemicals are tested. We investigate the modulation of activity of ion channels and carrier proteins by electrophysiological methods.

PUBLIC INTEREST STATEMENT

Chemicals extracted from plants used in traditional Chinese medicine have become an important source for effective drugs used in Western medicine. An example is the anti-malaria drug artemisinin extracted from *Artemisia annua* (woodworm), which was honoured by Nobel Prize to Youyou Tu in 2015 and formed the basis of modern anti-malaria drugs. In our work, we screened a number of chemicals extracted from herbs that are used in Chinese medicine to treat neuronal diseases such as epilepsy. We found that catechines from green tea have some potency to modulate neurotransmitter transporters that are essential in controlling brain activity.

mediates also neuronal uptake of cysteine and exerts critical neuroprotective function by interference with neuronal glutathione metabolism (Aoyama & Nakaki, 2013).

GAT1 utilizes the inward movement of 2 Na^+ ions and cotransport of 1 Cl^- ion to transport 1 GABA molecule into the cell. Therefore, the activity can be monitored by recording the electrogenic currents. Also, EAAC1 is an electrogenic transporter by coupling the uptake of 1 glutamate molecule to the inward movement of 3 Na^+ ions and 1 H^+ and the counter-transport of 1 K^+ ion.

For treatment of pathological conditions drugs have been designed to specifically modulate the function of a neurotransmitter transporter. For example, for treatment of epilepsy, drugs that increase the inhibitory synaptic activity by elevation of the concentration or dwell time of GABA in the synaptic cleft (see Löscher, 1998) became potent anti-epileptica. Tiagabine (TGB), a highly selective inhibitor of GAT1 that can inhibit the re-uptake of GABA, is one of these drugs (Borden et al., 1994; Schachter, 2001).

In analogy to inhibition of GAT1 and the associated elevation of the extracellular inhibitory GABA, stimulation of EAAC1 results in a reduction of the synaptically excitatory glutamate (see, e.g. Schwarz & Gu, 2003). Hence, drugs that inhibit GAT1 and stimulate EAAC1 will counteract hyperexcitability. It has been demonstrated that, e.g. stimulation of δ-opioid receptor inhibits GAT1-mediated current (Pu et al., 2012) and stimulates EAAC1-mediated current (Xia, Pei, & Schwarz, 2006), a mechanism that might contribute to pain suppression (Pu, Xu, & Schwarz, 2015; Schwarz & Gu, 2003; Yang et al., 2008).

Chinese medicine has become a promising source for the development of new drugs. This strategy has recently been recognized by Nobel Prize to Youyou Tu in 2015 (see Tu, 2011) for her discovery of artemisinin from *Artemisia annua* as an effective anti-malarial drug. In Asia, green tea consumption has a long history not only due to its flavour, it is also assumed to have potential health care benefits. Extracts of green tea (*Camellia sinensis*) have been used in Chinese medicine to prevent or treat various diseases (Cabrera, Artacho, & Giménez, 2006; Chen et al., 2011; Mereles & Hunstein, 2011), such as prostate cancer (Bettuzzi et al., 2006) and epilepsy (D'avila, Esteves Lopez, Patriarcha, & Araujo Restini, 2011; Noor, Mohammed, Khadrawy, Aboul Ezz, & Radwan, 2015; Xie et al., 2012); the effects had been attributed to the antioxidant effects of catechins. The most effective and most abundant component in the tea leaves is (-)-epigallocatechin-3-gallate (EGCG) (see Table 1). Although the effectiveness is controversially discussed (see e.g. Mereles & Hunstein, 2011; Mähler et al., 2013), interference of catechins from green tea with the GABAergic (Adachi, Tomonaga, Tachibana, Denbow, & Furuse, 2006) and glutamatergic system (Chou, Huang, Tien, & Wang, 2007)

Table 1. Relative effect of 40 or 50 μM (-)-epicatechin and EGCG on GAT1- and EAAC1-mediated current, respectively, and of 1 μM TGB on GAT1-mediated current at −100 mV with respect to the current in the absence of drug. Data represent averages ± SEM of *N* experiments

Current mediated by	(-)-Epicatechin	EGCG	Tiagabine
GAT1	1.00 ± 0.05	0.66 ± 0.08	0.34 ± 0.08
	(*N* = 3)	(*N* = 5)	(*N* = 3)
EAAC1	1.79 ± 0.10	1.06 ± 0.05	Ineffective
	(*N* = 5)	(*N* = 10)	
Structure	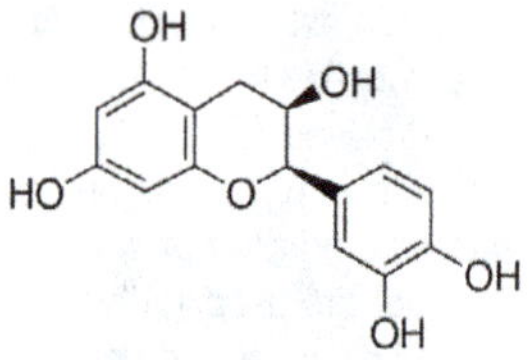	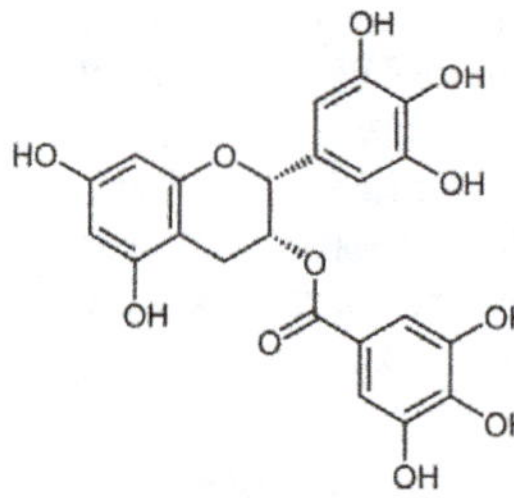	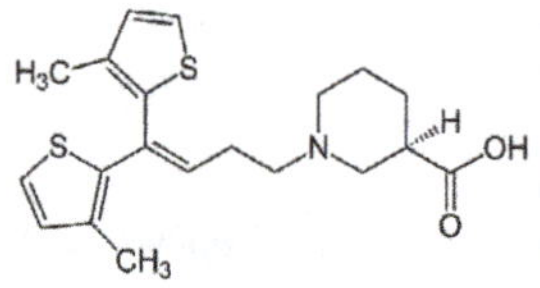

could be demonstrated. Direct modulation of GABAA receptor as well as modulation of signalling pathways (e.g. (Hossain, Hamamoto, Aoshima, & Hara, 2002; Mandel, Weinreb, Amit, & Youdim, 2004) were described. In our investigation, we evaluated whether catechins might interfere with the activity of the neurotransmitter transporters, and hence could be involved in the neuroprotective effects of green tea extracts.

2. Materials and methods

2.1. *Xenopus oocytes preparation and microinjection*

Xenopus oocytes were used as expression system for the GABA transporter GAT1 and the glutamate transporter EAAC1 and as a model system to test the effects of drugs extracted from green tea. The procedure of oocyte preparation and microinjection of the respective cRNA were as described previously (Gu et al., 2010; Pu et al., 2012). For expression of GAT1 or EAAC1 14 ng of the respective cRNA was microinjected per oocyte. After incubation for 24–48 h, the oocytes were ready for the voltage-clamp experiments. Un-injected oocytes served as controls. In all our experiments we never could detect any GABA- or glutamate-induced responses in control oocytes. We like to point out that such tests are needed because occasionally batches of oocytes may show endogenous contributions (see, e.g. Steffgen et al., 1991).

2.2. *Electrophysiological recording*

To investigate the function of the neurotransmitter transporters, membrane currents were measured by conventional TEVC using Turbo TEC-03 with CellWorks software (NPI electronic, Tamm, Germany). Glass microelectrodes were filled with 3 M KCl, and balanced in ORi solution (see Solutions) for at least 30 min before recording. Steady-state current–voltage dependencies were determined by averaging membrane currents during the last 20 ms of 200 ms rectangular voltage pulses from −150 to +10 mV in 10 mV increments that were applied from a holding potential of −60 mV. Figure 1 shows original current traces in response to voltage pulses in the absence and presence of GABA (Figure 1(A)) and glutamate (Figure 1(B)), and the corresponding steady-state current–voltage dependencies. Transporter-dependent current was determined as the difference of total membrane current (see the lower traces in Figure 1) in the presence (middle traces) and absence (upper traces) of the respective neurotransmitter. The data were collected after analogue filtering at 100 or 300 kHz and analysed by Origin software (OriginLab Corp., USA). All experiments were performed at room temperature (about 25°C).

Figure 2 illustrates the protocol of a typical experiment for an oocyte with expressed GAT1. The holding current at −60 mV (Figure 2(A)) shows that perfusion of the oocyte chamber with solution containing different concentrations of GABA led to concentration-dependent increases of inward-directed currents. The difference of steady-state current in the presence and absence of GABA represents the GAT1-mediated current. To correct for possible drifts of current with time, the current in the absence of GABA was determined as the average of current in pure ORi before and after the application of GABA. The respective current–voltage dependencies of this experiment are shown in Figure 2(B). Corresponding protocols were also applied to measuring EAAC1-mediated currents using glutamate as activator.

2.3. *Solutions and drugs*

The composition of the standard bath solution (ORi) was (in mM): 90 NaCl, 2 KCl, 2 $CaCl_2$ and 5 MOPS (adjusted to pH 7.4 with Tris). All standard chemicals were purchased from Sigma-Aldrich Co. (St. Louis, MO, USA). EGCG (CAS 989-51-5, purity ≥98%) and (-)-epicatechin (CAS 490-46-0, purity ≥98%) were obtained from Sigma-Aldrich Co. (St. Louis, MO, USA) or Seebio Biotech Co. Ltd. (Shanghai, China), respectively, and tiagabine (TGB, CAS 145821-59-6) from Biotrend Chemicals (Zurich, Switzerland). These chemicals were dissolved in dimethyl sulfoxide (DMSO) to prepare 10 mM stock solution, and diluted before the experiment to the desired concentration with the respective test solution. The final concentration of DMSO was always below 1%. The range of concentrations used in the experiments was based on preliminary screening to cover the range for 50% effects.

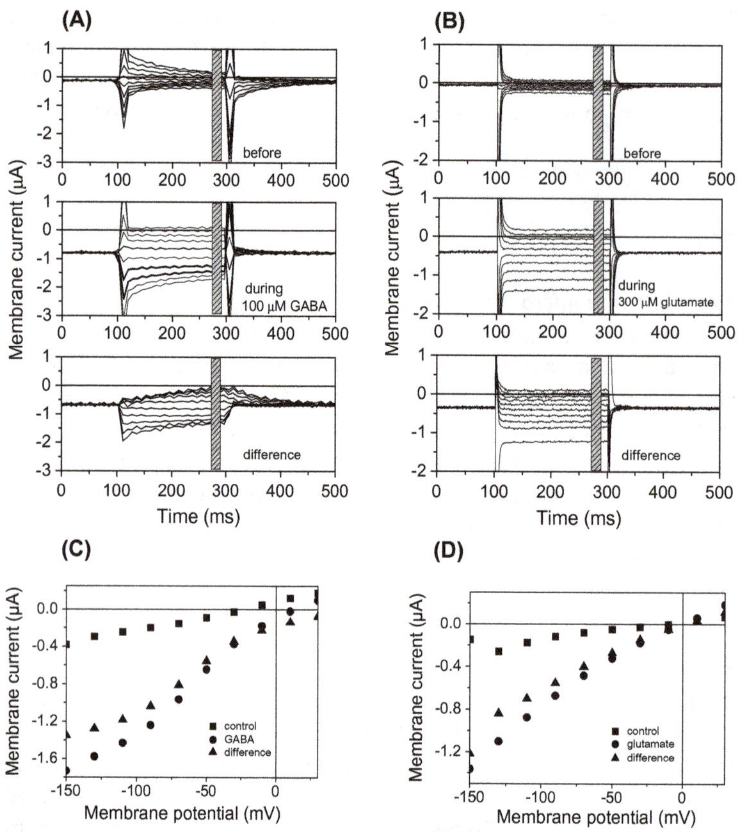

Figure 1. (A, B) Original current traces in response to rectangular voltage pulses. Pulses were applied from −150 to +30 mV in 20 mV increments. Top traces are in the absence, middle traces in the presence of 100 µM GABA (A) or 300 µM glutamate (B). Lower traces show the differences of currents in the presence and absence of drug, respectively. The hatched bars indicate the time period of 20 ms used to determine mean steady-state currents. (C, D) Steady-state current–voltage dependencies determined from the above traces.

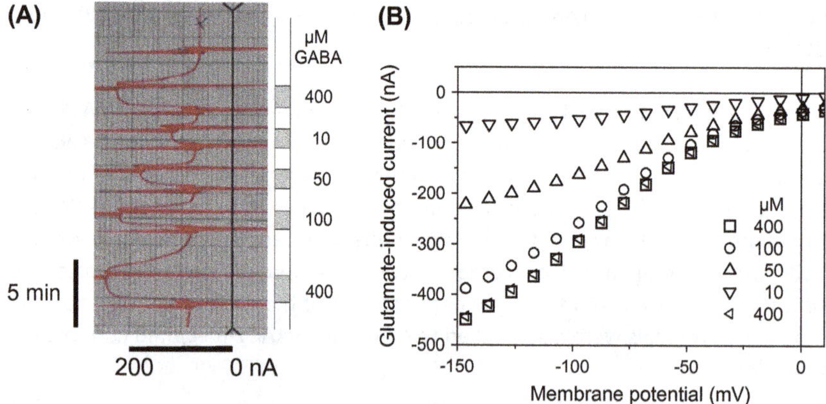

Figure 2. Protocol of a typical experiment. (A) Chart record of holding current at −60 mV illustrating the inward-directed current responses to different concentrations of GABA (deflections to the left). (B) Current–voltage dependencies of the respective GABA-induced steady-state currents determined as the difference of the currents in the presence and absence of GABA.

2.4. Data analysis

To estimate the concentration $K_{1/2}$ that is needed to obtain a 50% effect, we approximated the concentration dependence by a sigmoidal dependency using a Hill equation:

$$Y = Y_{max} \frac{[X]^n}{[X]^n + K_{1/2}^n} \tag{1}$$

with $[X]$ representing the concentration of activator GABA or glutamate, or the concentration of the tested chemical, $K_{1/2}$ represents the concentration where X produces half-maximum effect, n is the Hill coefficient describing cooperativity of binding of the respective substrate. Averaged data are presented as means ± SEM and were considered as significantly different by Student's t test on the basis of $p < 0.05$.

3. Results

3.1. The effects of catechins on GABA transporter GAT1

As illustrated in Figures 1(A), (C) and 2, application of GABA induces inwardly directed current in oocytes with expressed GAT1. The dependency of current on GABA concentration at −100 mV (Figure 3) can be described by Equation (1); for 50% stimulation a $K_{1/2}$ value of 57.6 µM was obtained. Analysis of the concentration dependence at −60 and −150 mV revealed no significant difference in $K_{1/2}$ values. To test the effect of drugs on GAT1-mediated current an intermediate concentration of 50 µM close to the $K_{1/2}$ value was used in all experiments described below.

Since green tea extracts have been used in Chinese medicine for treatment of epilepsy (Noor et al., 2015; Xie et al., 2012), we tested in particular the dominating EGCG and (-)-epicatechin (see Table 1) with respect to their effects on the GAT1-mediated current. While 40 µM (-)-epicatechin exhibited no significant effect (Table 1), and even 500 µM was ineffective, 40 µM EGCG showed about 35% inhibition (Table 1, and Figure 4(A)). A more detailed analysis of the concentration dependency revealed 50% inhibition at about 100 µM EGCG at −100 mV (Figure 4(B)).

To compare the efficacy of EGCG with another potent GAT1 inhibitor that is successfully applied as antiepileptic drug, we determined the efficacy of TGB in inhibiting the GAT1-mediated current (Figure 4(C)), a $K_{1/2}$ value of 2.3 ± 0.8 µM was obtained (Figure 4(D)). These measurements confirm our earlier results from Eckstein-Ludwig, Fei and Schwarz (1999). TGB is by an order of magnitude more efficient in inhibiting GAT1 than EGCG, but nevertheless, this catechin may serve as a basis for the development of new antiepileptic drugs.

3.2. The effects of catechins on glutamate transporter EAAC1

Similar experiments as described above for the effects of the catechins on the GAT1-mediated currents were also performed for EAAC1-mediated currents, which were determined as glutamate-activated currents in oocytes with expressed EAAC1. The dependence of EAAC1-mediated current on glutamate concentration at −100 mV (Figure 5) can be described by Equation (1) with a $K_{1/2}$ value of 82 µM. Similar to the activation of GAT1 by GABA, analysis of glutamate dependency of EAAC1-mediated current at −60 and −150 mV revealed no significant difference in $K_{1/2}$ values. For analysing drug effects, EAAC1-mediated current was determined as current activated by 100 µM glutamate.

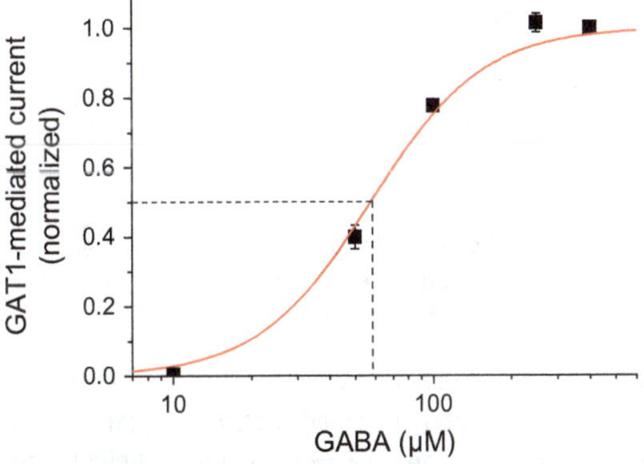

Figure 3. Dependence of GAT1-mediated current at −100 mV on GABA concentration. One corresponds to −415.7 ± 41.2 nA. Data point represent averages ± SEM from N = 5–11 oocytes. The line represents a fit of Equation (1) to the data with $K_{1/2}$ = 57.6 ± 3.8 µM (n = 2).

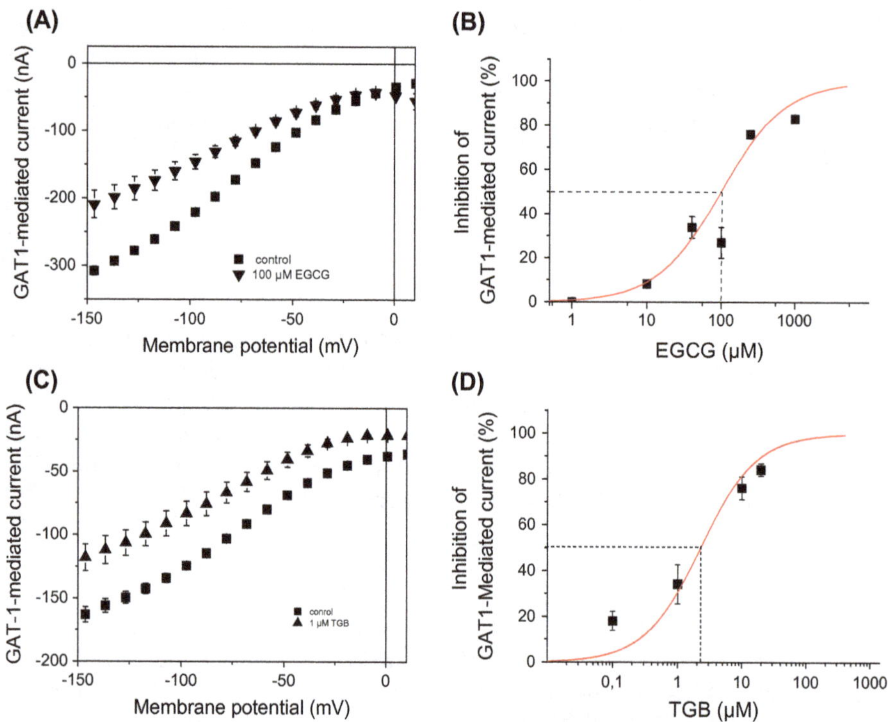

Figure 4. Effect of EGCG (A, B) and of TGB (C, D) on GAT1-mediated current (A, C) Voltage dependence of GAT1-mediated current (activated by 50 μM GABA) in the absence and presence of 100 μM EGCG (A: *N* = 3) and 1 μM TGB (B: *N* = 5). Dependence of the GAT1-mediated, steady-state current at −100 mV on EGCG (B) and TGB (D) concentration. 100% corresponds to −220 ± 11 nA (B) and −148 ± 17 (D). Data point represent averages ± SEM from *N* = 3 (B) and *N* = 5 (D) oocytes. The solid line represents a fit of Equation (1) to the data with $K_{1/2}$ values of 99 ± 14 μM (*n* = 1) (B) and 2.3 ± 0.8 μM (*n* = 1) (D).

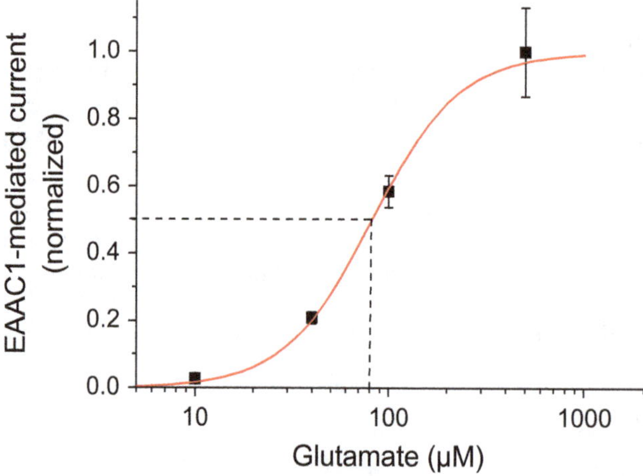

Figure 5. Dependence of EAAC1-mediated current at −100 mV on glutamate concentration. One corresponds to 570 ± 75 nA. Data point represent averages ± SEM from *N* = 7 oocytes. The line represents a fit of Equation (1) to the data with $K_{1/2}$ = 82.0 ± 2.9 μM (*n* = 1.9).

The effects of the two catechins on EAAC1 were opposite to those on GAT1. EGCG did not exhibit any effect on the EAAC1-mediated current, while (-)-epicatechin clearly affected the transporter. In fact, EAAC1-mediated current at −100 mV even became stimulated by about 80% by 50 μM (-)-epicatechin (see Figure 5(A) and compare values in Table 1). A more detailed analysis of the concentration dependency revealed 50% of maximum stimulation at about 5 μM of the (-)-epicatechin (Figure 5(B)). The stimulatory effects seemed to saturate at about 80%.

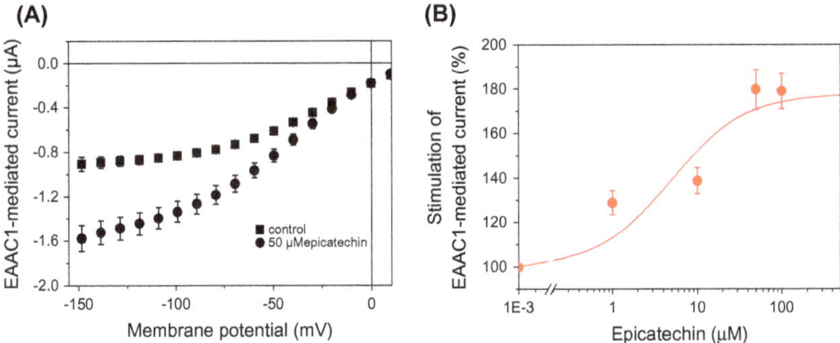

Figure 6. Effect of epicatechin on EAAC1-mediated current (A) Voltage dependence of EAAC1-mediated current (activated by 100 µM glutamate) in the absence and presence of 50 µM epicatechin. (B) Dependence of the EAAC1-mediated current at −100 mV on (-)-epicatechin concentration. 100% corresponds to 407 ± 92 nA. Data point represent averages ± SEM from N = 5 oocytes. The solid line represents a fit of Equation (1) to the data with $K_{1/2}$ = 4.9 + 2.6 µM and maximum stimulation of 79 + 9% (n = 1).

4. Discussion

To investigate modulations of activity of GABA transporter GAT1 and glutamate transporter EAAC1, we determined the currents induced by 50 µM GABA and 100 µM glutamate, respectively. These concentrations were close to the $K_{1/2}$ values of 57.6 µM for GAT1 and 82 µM for EAAC1.

Our results have shown that catechins extracted from green tea may indeed contribute to reported anti-epileptic effects (D'avila et al., 2011; Noor et al., 2015; Xie et al., 2012) by interfering with neurotransmitter transporters. The established anti-epileptic drug TGB exerts it effect as a specific inhibitor of GAT1. The catechin EGCG also inhibits the GABA-uptake activity of GAT1, which would favour inhibitory synaptic activity. Epicatechin, on the other hand, stimulates glutamate-uptake activity of EAAC1, which would reduce excitatory synaptic activity via reducing glutamate concentration in the synaptic cleft. Since EAAC1 also mediates uptake of cysteine, the stimulation may also promote neuroprotection via glutathione metabolism (Aoyama & Nakaki, 2013).

To achieve micromolar concentrations of EGCG or epicatechin in the human plasma, which are necessary for inhibition of GAT1 or EAAC1, respectively, a consumption of 50–60 cups of green tea would be necessary (Chow et al., 2005). Although it is not realistic to treat neuronal disorder like epilepsy by drinking tea, concentrated extracts may have stabilizing potency via the inhibition of GAT1 and stimulation of EAAC1. In addition, bioavailability of catechins is restricted due to metabolism (Yong Feng, 2006), and catechins may therefore form the basis of drug with higher bioavailability for modulating neurotransmitter transporters.

5. Conclusion

Our data demonstrate that catechins extracted from green tea can interfere with the activity of neurotransmitter transporters, EGCG inhibits GAT1 and (-)-epicatechin stimulates EAAC1, which may form the basis for the development of new drugs that exert their effects through interference with the neurotransmitter transporters.

Acknowledgements
We gratefully acknowledge the technical assistance from Guohui Chen and Beibei Xing. DNA for EAAC1 and GAT1 were kindly provided by Dr Jian Fei.

Funding
TE and YW thank for financial travel support from "Vereinigung der Freunde und Förderer der Goethe-Universität Frankfurt" to work in the laboratory in Shanghai. We are also grateful for support from Green Valley Holding Co. the Shanghai Key Laboratory of Acupuncture Mechanism and Acupoint Function [grant number 14DZ2260500], the National Basic Research Program of China [grant number 973 Program 2012CB518502], and the National Natural Science Foundation of China [grant number 81403489] to Y. Xu.

Competing Interests
The authors declare no competing interest.

Author details
Y.X. Wang[1,2,3]
E-mail: wangyixinavg@gmail.com
T. Engelmann[1,2,3]
E-mail: timengelmann@aol.com
Y.F. Xu[1,2]
E-mail: annxu21@hotmail.com
W. Schwarz[1,2,3]

E-mail: schwarz@biophysik.org
[1] Shanghai Research Center for Acupuncture and Meridians, Shanghai, China.
[2] Shanghai Key Laboratory of Acupuncture Mechanism and Acupoint Function, Fudan University, Shanghai, China.
[3] Institute for Biophysics, Goethe-University Frankfurt, Frankfurt am Main, Germany.

References

Adachi, N., Tomonaga, S., Tachibana, T., Denbow, D. M., & Furuse, M. (2006). (-)-Epigallocatechin gallate attenuates acute stress responses through GABAergic system in the brain. *European Journal of Pharmacology, 531*, 171–175.

Aoyama, K., & Nakaki, T. (2013). Neuroprotective properties of the excitatory amino acid carrier 1 (EAAC1). *Amino Acids, 45*, 133–142. http://dx.doi.org/10.1007/s00726-013-1481-5

Bettuzzi, S., Brausi, M., Rizzi, F., Castagnetti, G., Peracchia, G., & Corti, A. (2006). Chemoprevention of human prostate cancer by oral administration of green tea catechins in volunteers with high-grade prostate intraepithelial neoplasia: A preliminary report from a one-year proof-of-principle study. *Cancer Research, 66*, 1234–1240. http://dx.doi.org/10.1158/0008-5472.CAN-05-1145

Bjørn-Yoshimoto, W. E., & Underhill, S. M. (2016). The importance of the excitatory amino acid transporter 3 (EAAT3). *Neurochemistry International, 98*, 4–18. http://dx.doi.org/10.1016/j.neuint.2016.05.007

Borden, L. A., Dhar, T. G. M., Smith, K. E., Weinshank, R. L., Branchek, T. A., & Gluchowski, C. (1994). Tiagabine, SK&F 89976-A, CI-966, and NNC-711 are selective for the cloned GABA transporter GAT-1. *European Journal of Pharmacology: Molecular Pharmacology, 269*, 219–224. http://dx.doi.org/10.1016/0922-4106(94)90089-2

Cabrera, C., Artacho, R., & Giménez, R. (2006). Beneficial effects of green tea—A review. *Journal of the American College of Nutrition, 25*, 79–99. http://dx.doi.org/10.1080/07315724.2006.10719518

Chen, D., Wan, S. B., Yang, H., Yuan, J., Chan, T. H., & Dou, Q. P. (2011). EGCG, green tea polyphenols and their synthetic analogs and prodrugs for human cancer prevention and treatment. *Advances in Clinical Chemistry, 53*, 155–177. http://dx.doi.org/10.1016/B978-0-12-385855-9.00007-2

Chou, C. W., Huang, W. J., Tien, L. T., & Wang, S. J. (2007). (-)-Epigallocatechin gallate, the most active polyphenolic catechin in green tea, presynaptically facilitates Ca^{2+}-dependent glutamate release via activation of protein kinase C in rat cerebral cortex. *Synapse, 61*, 889–902.

Chow, H.-H. S., Hakim, I. A., Vining, D. R., Crowell, J. A., Ranger-Moore, J., Chew, W. M., ... Alberts, D. S. (2005). Effects of dosing condition on the oral bioavailability of green tea catechins after single-dose administration of Polyphenon E in healthy individuals. *Clinical Cancer Research, 11*, 4627–4633. http://dx.doi.org/10.1158/1078-0432.CCR-04-2549

D'avila, B. F., Esteves Lopez, M. C., Patriarcha, F. A., & Araujo Restini, C. B. (2011). Effect of green tea (Camellia sinensis) non epileptic seizures induced by pentylenetetrazol (PTZ) in rats. *Pharmacologia, 2*, 362–368.

Eckstein-Ludwig, U., Fei, J., & Schwarz, W. (1999). Inhibition of uptake, steady-state currents, and transient charge movements generated by the neuronal GABA transporter by various anticonvulsant drugs. *British Journal of Pharmacology, 128*, 92–102. http://dx.doi.org/10.1038/sj.bjp.0702794

Gu, Q. B., Du, H. M., Ma, C. H., Fotis, H., Wu, B., Huang, C. G., & Schwarz, W. (2010). Effects of α-Asarone on the glutamate transporter EAAC1 in xenopus oocytes. *Planta Medica, 76*, 595–598. http://dx.doi.org/10.1055/s-0029-1240613

Hossain, S. J., Hamamoto, K., Aoshima, H., & Hara, Y. (2002). Effects of tea components on the response of GABAA receptors expressed in xenopus oocytes. *Journal of Agricultural and Food Chemistry, 50*, 3954–3960. http://dx.doi.org/10.1021/jf011607h

Jensen, K., Chiu, C. S., Sokolova, I., Lester, H. A., & Mody, I. (2003). GABA transporter-1 (GAT1)-deficient mice: Differential tonic activation of GABAA versus GABAB receptors in the hippocampus. *Journal of Neurophysiology, 90*, 2690–2701. http://dx.doi.org/10.1152/jn.00240.2003

Kandel, E. R., Schwartz, J. H., & Jessell, T. M. (2000). *Principles of neural science* (4th ed.). New York, NY: McGraw Hill.

Lane, M. C., Jackson, J. G., Krizman, E. N., Rothstein, J. D., Porter, B. E., & Robinson, M. B. (2014). Genetic deletion of the neuronal glutamate transporter, EAAC1, results in decreased neuronal death after pilocarpine-induced status epilepticus. *Neurochemistry International, 73*, 152–158. http://dx.doi.org/10.1016/j.neuint.2013.11.013

Löscher, W. (1998). New visions in the pharmacology of anticonvulsion. *European Journal of Pharmacology, 342*, 1–13. http://dx.doi.org/10.1016/S0014-2999(97)01514-8

Mähler, A., Mandel, S., Lorenz, M., Ruegg, U., Wanker, E. E., Boschmann, M., & Paul, F. (2013). Epigallocatechin-3-gallate: A useful, effective and safe clinical approach for targeted prevention and individualised treatment of neurological diseases? *EPMA Journal, 4*, 1–17.

Mandel, S., Weinreb, O., Amit, T., & Youdim, M. B. H. (2004). Cell signaling pathways in the neuroprotective actions of the green tea polyphenol (-)-epigallocatechin-3-gallate: Implications for neurodegenerative diseases. *Journal of Neurochemistry, 88*, 1555–1569. http://dx.doi.org/10.1046/j.1471-4159.2003.02291.x

Mereles, D., & Hunstein, W. (2011). Epigallocatechin-3-gallate (EGCG) for clinical trials: More pitfalls than promises? *International Journal of Molecular Sciences, 12*, 5592–5603.

Noor, N. A., Mohammed, H. S., Khadrawy, Y. A., Aboul Ezz, H. S., & Radwan, N. M. (2015). Evaluation of the neuroprotective effect of taurine and green tea extract against oxidative stress induced by pilocarpine during status epilepticus. *The Journal of Basic & Applied Zoology, 72*, 8–15. http://dx.doi.org/10.1016/j.jobaz.2015.02.001

Pu, L., Xu, Y. F., & Schwarz, W. (2015). Regulation of membrane transporters by delta-opioid receptors. In Y. Xia (Ed.), *Neural function of the delta-opioid receptor* (pp. 349–361). Cham: Springer International Publishing.

Pu, L., Xu, N. J., Xia, Q., Gu, Q. B., Ren, S. L., Fucke, T., ... Schwarz, W. (2012). Inhibition of activity of GABA transporter GAT1 by δ-opioid receptor. *Evidence-Based Complementary and Alternative Medicine, 2012*, ID818451.

Schachter, S. C. (2001). Pharmacology and clinical experience with tiagabine. *Expert Opinion on Pharmacotherapy, 2*, 179–187. http://dx.doi.org/10.1517/14656566.2.1.179

Schwarz, W., & Gu, Q. B. (2003). Cellular mechanisms in acupuncture points and affected sites. In Y. Xia, G. H. Ding, & G.-C. Wu (Eds.), *Current research in acupuncture* (pp. 37–51). New York, NY: Springer Science+Business Media.

Steffgen, J., Koepsell, H., & Schwarz, W. (1991). Endogenous L-glutamate transport in oocytes of Xenopus laevis. *Biochimica et Biophysica Acta (BBA)-Biomembranes, 1066*, 14–20.

Tu, Y. Y. (2011). The discovery of artemisinin (qinghaosu) and gifts from Chinese medicine. *Nature Medicine, 17*, 1217–1220. http://dx.doi.org/10.1038/nm.2471

Xia, P., Pei, G., & Schwarz, W. (2006). Regulation of the glutamate transporter EAAC1 by expression and activation of δ-opioid receptor. *European Journal of Neuroscience, 24*, 87–93. http://dx.doi.org/10.1111/ejn.2006.24.issue-1

Xie, T., Wang, W. P., Mao, Z. F., Qu, Z. Z., Luan, S. Q., Jia, L. J., & Kan, M. C. (2012). Effects of epigallocatechin-3-gallate on pentylenetetrazole-induced kindling, cognitive impairment and oxidative stress in rats. *Neuroscience Letters, 516*, 237–241. http://dx.doi.org/10.1016/j.neulet.2012.04.001

Yang, Z. J., Bao, G. B., Deng, H. P., Du, H. M., Gu, Q. B., Pei, G., ... Xia, P. (2008). Interaction of δ-opioid receptor with membrane transporters: Possible mechanisms in pain suppression by acupuncture. *Journal of Acupuncture and Tuina Science, 6*, 298–300. http://dx.doi.org/10.1007/s11726-008-0298-3

Yong Feng, W. (2006). Metabolism of green tea catechins: An overview. *Current Drug Metabolism, 7*, 755–809. http://dx.doi.org/10.2174/138920006778520552

Zhou, Y., & Danbolt, N. C. (2013). GABA and glutamate transporters in brain. *Front Endocrinol (Lausanne), 4*, 165. doi:10.3389/fendo.2013.00165

12

The rough fur (*ruf*) mutation in mice is an allele of myelin protein zero-like 3 (*Mpzl3*)

Kenneth M. Palanza[1], Legairre A. Radden II[1], Mohammed A. Rabah[1], Tu V. Nguyen[1], Audra C. Kohm[1], Malcolm E. Connor[1], Morgan M. Ricci[1], Jachius J. Stewart[1], Sidney Eragene[1] and Thomas R. King[1*]

*Corresponding author: Thomas R. King, Department of Biomolecular Sciences, Central Connecticut State University, 1615 Stanley Street, New Britain, CT 06053, USA

E-mail: kingt@ccsu.edu

Reviewing editor

Jurg Bahler, University College London, UK

Additional information is available at the end of the article

Abstract: *Background:* The recessive rough fur mutation (*ruf*)—named for the unkempt, greasy appearance of the hair coat in homozygotes—has previously been mapped on mouse Chromosome 9. However, the assignment of *ruf* to a particular gene is needed to facilitate a complete molecular analysis of the mutant phenotype. *Results:* To establish a more refined location for *ruf* (as a basis for positional cloning) DNA isolated from a large backcross family was typed for microsatellite and single-nucleotide markers on Chromosome 9. This analysis restricted the location of *ruf* between sites that flank only four genes known to be expressed in skin, one of which—*Mpzl3*, for myelin protein zero-like 3—generates a similar hair phenotype in mice homozygous for engineered and spontaneous null-alleles. A cross between *ruf* mutants and mice heterozygous for the *Mpzl3*^{rc} mutation (which controls a recessive phenotype called rough coat) produced offspring that displayed matted, damp-looking fur, indicating that *ruf* is a mutant allele of *Mpzl3*. However, sequence analysis of the *Mpzl3* promoter, exons and splice junctions revealed no mutant-specific DNA defect. *Conclusion:* The results presented indicate that *ruf* is a mutant allele of *Mpzl3*. With a genetic assignment in hand, the rough fur variant can now be more fully characterized to advance our understanding of *Mpzl3*'s role in normal skin development and function, hepatic triglyceride synthesis, weight regulation, energy and glucose homeostasis; and to model-related human disorders.

ABOUT THE AUTHORS

The research group led by Thomas R. King at Central Connecticut State University (New Britain, CT, USA) aims to identify the genetic basis of certain spontaneous hair variants in mice and rats. Identification of such causative gene defects promises to make new probes available for the molecular analysis of the mammalian integument, and to facilitate the identification of human conditions that these animal mutants might model. Mohammed A. Rabah, Tu V. Nguyen, Audra C. Kohm, Morgan M. Ricci, Jachius J. Stewart, and Sidney Eragene were undergraduates when they contributed to this project; Kenneth M. Palanza, Legairre A. Radden II, and Malcolm E. Connor were graduate students in the Department of Biomolecular Sciences.

PUBLIC INTEREST STATEMENT

The rough fur mouse variant has been known for more than 30 years, but its detailed analysis has been hampered by a lack of molecular probes. Here, we used a positional approach to identify a small number of gene candidates, and complementation testing to assign *ruf* to the *Mpzl3* gene. This work places the rough fur mutant among a series of other known *Mpzl3* variants that, together, comprise a living resource that promises to advance a complete molecular analysis of *Mpzl3*'s complex role in mammalian physiology, and offers a model system for the study of orthologous human conditions that may be characterized by alopecia, over-active sebaceous glands, or altered glucose homeostasis and weight regulation.

Subjects: Developmental Biology; Genetics; Molecular Biology

Keywords: positional candidate approach; complementation testing; intraspecific backcross mapping; hair morphology; rough coat (*rc*)

1. Introduction

The recessive rough fur mutation in mice (abbreviated *ruf*) generates a hair coat in homozygotes that looks unkempt and greasy by weaning age (see Figure 1) and persists thereafter. This mutation arose spontaneously in the C3H/HeJ inbred strain at The Jackson Laboratory (Bar Harbor, ME, USA) in 1985, and its initial description (including frequently broken hairs and mild hyperkeratosis) and linkage assignment to Chromosome (Chr) 9 were published by Sweet et al. (1990). Light and electron microscopic evaluation of the pelage skin from mutants by Park et al. (2001) showed large lipid deposits on the surface of the epidermis and in abnormally large sebaceous glands, suggesting that the rough fur mutant may be an animal model for a human disorder characterized by hyperlipidogenesis, but no particular disorder was named.

To facilitate assignment of the mutant rough fur phenotype to a specific genetic cause—which would provide molecular probes for the further analysis of this interesting phenotype and help to identify an orthologous human disorder—here we have fine-mapped the *ruf* mutation with respect to various microsatellite and single-nucleotide polymorphisms on mouse Chr 9. This analysis identified a small set of co-localizing, skin-expressed candidate genes, and *ruf* has been found (by complementation testing) to be an allele of one of these, *Mpzl3* (for myelin protein zero-like 3).

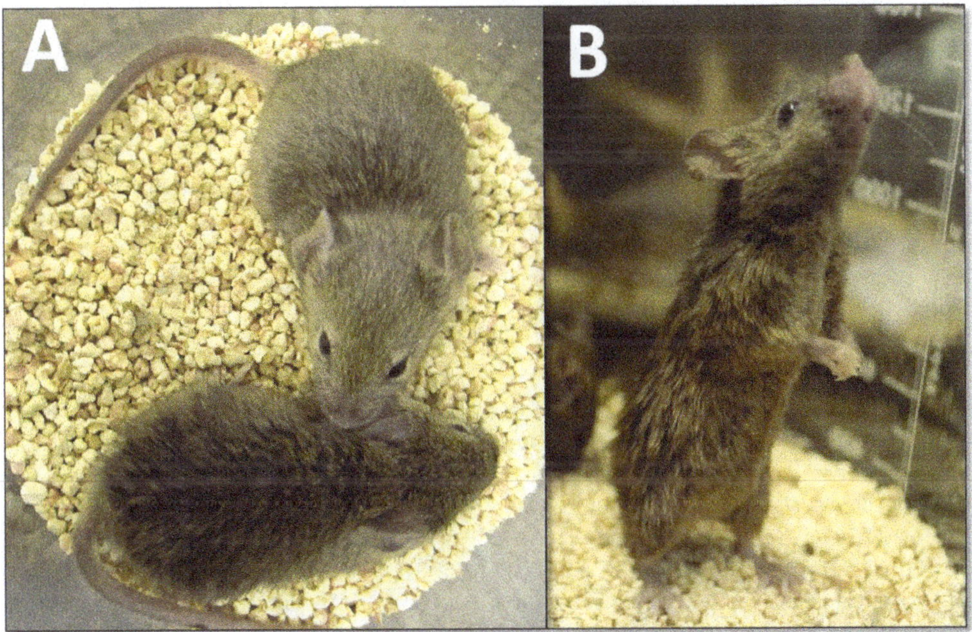

Figure 1. The rough fur phenotype. (A) A 60-day-old mutant (*ruf/ruf*) female is shown (bottom) with a phenotypically normal (heterozygous) sister (top). (B) The same mutant female is shown, standing upright.

2. Materials and methods

Animals were housed and fed according to Federal guidelines, and the Institutional Animal Care and Use Committee at Central Connecticut State University (CCSU) approved of all procedures involving mice (Animal Protocol Applications 151 and 153). Mice from the standard inbred strains C57BL/6 J (Jax Stock Number 664) and C3H/HeJ (Jax Stock Number 659), inbred C3H/HeJ-*ruf*/J mice (Jax Stock Number 1544, segregating for *ruf*), and C57BL/6 J-*Mpzl3*^{rc}/J mice (Jax Stock Number 540, segregating for rough coat, *Mpzl3*^{rc}) were obtained from The Jackson Laboratory (Bar Harbor, ME, USA). Both mutant strains were derived from cryopreserved stocks that were imported and then maintained at CCSU in testcrosses. Both mutants were reliably identified by the presentation of matted, greasy hair at weaning age.

Genomic DNA was isolated from 2 mm tail-tip biopsies taken from two- to three-week-old mice using Nucleospin® Tissue kits distributed by Clontech Laboratories, Inc. (Mountain View, CA, USA), as directed. The polymerase chain reaction (PCR) was performed in 13 µl reactions using the Titanium PCR kit from Clontech Laboratories, as directed. Oligonucleotide primers for PCR were designed and synthesized by Integrated DNA Technologies, Inc. (Coralville, IA, USA), based on sequence information available online (Ensembl Mouse Genome Server, 2017; Mouse Genome Database [MGD], 2017). To score PCR product sizes for seven different microsatellite markers on Chr 9 (Dietrich et al., 1996), reactions plus 2 µl loading buffer were electrophoresed through 3.5% NuSieve® agarose (Lonza, Rockland, ME, USA) gels. Gels were stained with ethidium bromide and photographed under ultraviolet light (see Supplementary Figure S1 for typical results). In addition, four DNA markers based on single nucleotide polymorphisms (SNPs) previously reported to differ between strains C3H/He and C57BL/6 (MGD, 2017) were also scored. These markers (herein designated *SNPA-D*) are described in detail in Supplementary Tables S1 and S2 (and typical results are shown in Supplementary Figure S2). For DNA sequence analysis, about 1.5 µg of individual PCR amplimers were purified and concentrated into a 30 µl volume using QIAquick® PCR Purification kits (Qiagen Sciences, Germantown, MD, USA) prior to primer-extension sequence analysis performed by SeqWright DNA Technology Services (Houston, TX, USA) or by the Keck Foundation Resource Laboratory at Yale University (New Haven, CT, USA).

The *Mpzl3*^{rc} mutation is a G to A transition at position 9:45066394 that is predicted by Cao et al. (2007) to cause the substitution of a conserved, basic arginine at position 100 with a polar glutamine. This point mutation also destroys an *AlwI* target site. Carriers of the recessive *Mpzl3*^{rc} allele were thus identified by endonuclease treatment of a 370 bp PCR product directed by forward (5′ GACAGCCAAGGGAAGAGAAG 3′) and reverse (5′GGAGTCACAGCATGAGAAGAG 3′) primers that flank Exon 3, which contains the *rc* mutation (see Supplementary Figure S3).

Total RNA was isolated from skin samples taken from 1-month-old mutant and control mice using the Nucleospin® RNA Midi kit by Macherey-Nagel (Bethlehem, PA, USA). From these total RNA samples, poly-A^{+} mRNA was purified using the NucleoTrap® mRNA kit (also by Macherey-Nagel), and cDNA was generated using the SMARTer® RACE 5′/3′ kit (Clontech Laboratories, Inc.). To detect *Mpzl3*-specific sequences, primer pairs that flanked exon-junction boundaries [Exon 2 (5′-GTGCGCGTATCGTCCTCTCCTTG-3′) and Exon 4 (5′-TCTTATAGCCAGACCTGCTCCTCTTCTG-3′), or Exon 4 (5′-GCTGTCTTCTGTGGCCCTTCTCTC-3′) and Exon 6 (5′-GTCATGAGTCCCTTCTGCGAGTTTCC-3′)] were used to direct standard PCR amplifications of these cDNAs. To detect mouse β actin (*Actb*) transcripts, a primer pair within Exon 4 (5′-CCCAGCCATGACGTAGCCATCCA-3′ and 5′-GAAGCTGTAGCCACGCTCGGTCAG-3′) was used. The resulting products were visualized in 3% NuSieve® agarose gels. For primer-extension sequencing, these products were purified and concentrated (as described above) and shipped to SeqWright DNA Technology Services or the Keck Foundation Resource Laboratory at Yale University.

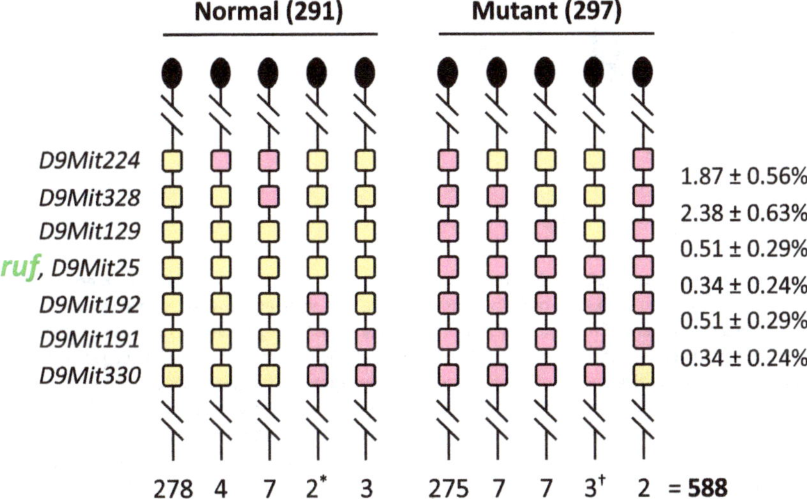

Figure 2. Segregation of alleles of *ruf* and 7 dimorphic microsatellite markers among 588 progeny from an intraspecific mouse backcross. Heterozygous F_1 mice (C3H/HeJ-*ruf*/J x C57BL/6 J) were backcrossed to homozygous C3H/HeJ-*ruf*/J mutants. The resulting progeny were scored for their rough fur phenotype, and a DNA sample from each mouse was typed for the microsatellite markers shown at the left. Only the Chr 9 haplotype inherited from the F_1 parent is shown (with a knob at the top of the haplotype representing the centromere), and the number of mice inheriting that haplotype is shown below it. Very similar numbers of heterozygous and mutant progeny (as expected for a testcross, $p > 0.8$) suggest that the mutant phenotype is both fully penetrant and fully viable. The two phenotypically normal recombinants marked with an asterisk show that the *ruf* locus must lie centromeric to *D9Mit192*. The three mutant recombinants marked with a dagger show that *ruf* must lie telomeric to *D9Mit129*. Genetic distances in percentage recombination are shown to the right (± 1 Standard Error).

3. Results

The linkage of *ruf* on mouse Chr 9, about 3.5% recombination centromeric to *Apoa1*, was reported by Sweet et al. (1990), but its location has not been defined with respect to other molecular markers. To further refine *ruf*'s chromosomal location, we produced a 588-member backcross (N_2) family by breeding (C57BL/6 J x C3H/HeJ-*ruf*/J)F_1 females with C3H/HeJ-*ruf*/J mutant males. These N_2 mice were typed for their hair phenotype, and DNA isolated from each mouse was characterized for various PCR-scorable, microsatellite markers (Dietrich et al., 1996; and see Figure S1) that lie near and centromeric to *Apoa1*. This haplotype analysis (summarized in Figure 2) positioned *ruf* between markers *D9Mit129* and *D9Mit192*, very near marker *D9Mit25* (C3H/HeJ- or C57BL/6 J-derived alleles of which were never meiotically separated from *ruf* or *ruf*$^+$ alleles, respectively, in this backcross panel).

To more precisely locate *ruf* in this interval, the five mice recombinant between *D9Mit129* and *D9Mit192* (see Figure 3(A)) were typed for four single-nucleotide polymorphisms (SNPs) that lie within that interval (and are described in detail in Supplementary Tables S1 and S2, with typical results displayed in Supplementary Figure S2). This analysis, summarized in Figure 3(B), positioned the three crossovers that fell centromeric to *ruf* between *D9Mit129* and *SNP A*; and the two crossovers that fell telomeric to *ruf* were located between *SNP C* and *D*. These results show that the *ruf* mutation must lie between *D9Mit129* and *SNP D*, very near *SNP A, B*, and *C*, which (like *D9Mit25*) were never separated from *ruf* in this backcross panel. This 1.4-Mb span contains 51 genes or processed transcripts, only four of which (*Cbl*, for Casitas B-lineage lymphoma; *Arcn1*, for archain 1; *Kmt2a*, for lysine (K)-specific methyltransferase 2A; and *Mpzl3*, for myelin protein zero-like 3) are known to be expressed in skin (MGD, 2017)(see Figure 3(B) and 3(C)). Each of these candidate genes has been knocked out in mice (*Cbl*, Thien et al., 2003; *Arcn1*, Xu et al., 2010; *Kmt2a*, McMahon et al., 2007), but only mutants homozygous for *Mplz3* null alleles (Czyzyk et al., 2013; Leiva et al., 2014) presented hair and skin anomalies that are similar to the rough fur phenotype.

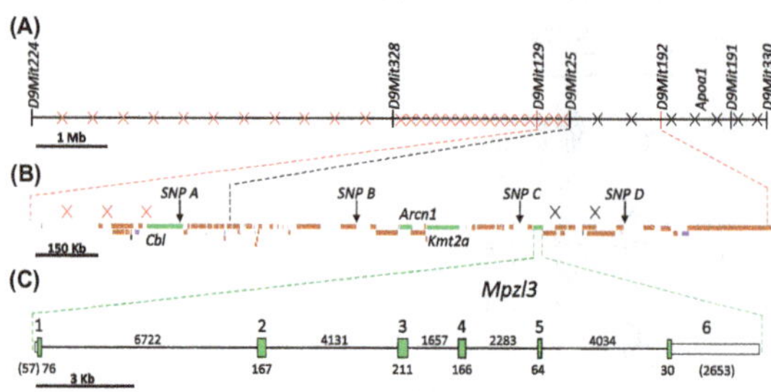

Figure 3. Physical maps of the *ruf* region on mouse Chr 9. (A) The relative positions of the seven microsatellite markers typed within the backcross panel are shown (with a 1 Mb scale bar). The crossovers found in each marker-defined interval are shown as a red "x" for crossovers that fell centromeric of *ruf*, and as a blue "x" for crossovers that fell telomeric to *ruf*. As described in Figure 2, the *ruf* locus must be located between markers *D9Mit129* and *D9Mit192*. (*D9Mit25* was never meiotically separated from *ruf* in this backcross panel.) (B) The relative positions of the four SNP markers typed within the backcross panel are shown for the 1.7 Mb region that is flanked by *D9Mit129* and *D9Mit192* (with a 0.15 Mb scale bar). Three crossovers that fell between *D9Mit129* and *SNP A*, and cenomeric to *ruf*, are depicted by red x's. Two crossovers that fell between *SNP C* and *SNP D*, and telomeric to *ruf*, are depicted by blue x's. (*SNP A, B,* and *C* never separated from *ruf* in this backcross panel.) These data limit the location of *ruf* to a 1.4 Mb span between *D9Mit129* and *SNP D*, where 49 protein coding genes (orange and green boxes), 2 processed transcripts (blue boxes), 11 RNA genes, (purple boxes), and 3 pseudogenes (grey boxes) are located. Only 4 of these protein-coding genes (green boxes) are known to be expressed in skin: *Cbl, Arcn1, Kmt2,* and *Mpzl3*. (C) The *Mpzl3* gene is expanded to show the 6 exons it comprises (with a 3 kb scale bar). Tall green boxes represent coding regions and shorter white boxes indicate the 5′ and 3′ untranslated regions. The number below each coding segment or above each intron is its length in base pairs; the base-pair lengths of noncoding segments are in parentheses. All of the *Mpzl3* exons and more than 3,400 bp upstream of the + 1 site were sequenced, but no mutant-specific changes were found.

To determine if *ruf* might be an allele of *Mpzl3*, we crossed *ruf/ruf* mutants with mice heterozygous for the spontaneous rough coat mutation (Dickie, 1966; Ruvinsky et al., 2002), which has been shown to be an allele of *Mpzl3* by Cao et al. (2007) and is now designated *Mpzl3*^{rc}. Because the progeny that inherited both *ruf* and a recessive *Mpzl3*^{rc} allele showed matted, greasy hair (while siblings that inherited *ruf* and *Mpzl3*$^{+}$ did not)(see Figure 4), we conclude that these recessive mutations "fail to complement", indicating that *ruf* is a variant allele of *Mpzl3*.

Seeking a *ruf*-specific defect, we sequenced all 6 exons of the *Mpzl3* gene (including both coding and untranslated regions—see Figure 3(C)), at least 20 base pairs of all intronic splice junctions, and more than 3,400 bp from the promoter region in genomic DNA from *ruf/ruf* mutants and from control C57BL/6 J and C3H/HeJ mice, but found no sequence differences among any of these templates (data not shown). Next, we isolated poly-A^{+} mRNA from mutant (*ruf/ruf*) and control (C3H/HeJ and +/*ruf* heterozygous) skin, and amplified sequences between Exons 2 and 6 of *Mpzl3* (see Figure 5). Mutant and control product lengths were the same, and sequence analysis of these amplimers (data not shown) verified the normal splicing of *Mpzl3*, Exons 2–6, in mutant (and control) skin.

Random retroviral insertions have been estimated to be the cause of 10–15% of all spontaneous mutations in mice (Maksakova et al., 2006), and more than half of these have occurred on the C3H/He genetic background—where the *ruf* variant originated. On that basis, we set out to test noncoding regions in and around *Mpzl3* for the presence of a large DNA insertion (such as an endogenous retroviral element, ERV) that might be specifically associated with the *ruf* mutation. To do this, we designed primer pairs for PCR that would generate amplimers of about 800–1,200 bp using wild type C3H/HeJ genomic DNA templates. A primer pair that gives the expected amplimer size in reactions including wild type C3H/HeJ template DNA, but fails to amplify a product from C3H/HeJ-*ruf/ruf* templates might flank a mutation-specific ERV. We have recently used this "scanning" approach to identify the novel insertion of an ERV element of the IAP family, sub type IΔ1, into the *Gata3* gene in mice (Connor & King, 2016) that is specific to (and the likely cause of) the juvenile alopecia mutation (abbreviated *jal*) which also occurred spontaneously on the C3H background, and which, like *ruf*, showed no DNA defect in coding regions (Ramirez et al., 2013). However, our scan of *Mpzl3* revealed no large, *ruf*-specific insertion in any of the introns, in more than 4,900 bp downstream of the polyadenylation signal, nor in more than 5,000 bp upstream of the +1 site.

Characterization of other spontaneous (*Mpzl3*^{rc}, Hayashi et al., 2004; Cao et al., 2007) or engineered, loss-of-function alleles (*Mpzl3*$^{tm1(KOMP)Mbp}$, Czyzyk et al., 2013; *Mpzl3*^{tm1Tacz}, Leiva et al., 2014) has revealed a complex array of physiological roles for *Mpzl3* that includes regulation of body mass, accumulation of adipose tissue, and progressive alopecia. We therefore measured these attributes

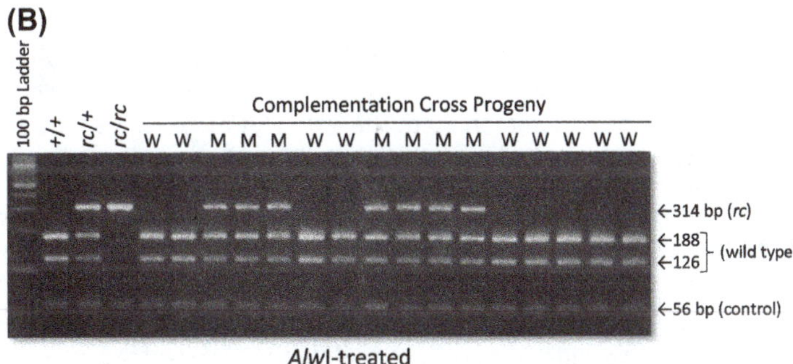

Figure 4. The recessive *ruf* and *Mpzl3*^{rc} mutations fail to complement in doubly heterozygous mice. A cross between *Mpzl3*$^{+}$/*Mpzl3*^{rc} heterozygotes and *ruf*/*ruf* mutants yielded 17 phenotypically normal and 14 progeny with matted hair, suggesting that *ruf* is a mutant allele of *Mpzl3*. (A) Representative, 30-day-old progeny from this cross are shown: the mouse on the lower left is phenotypically normal, while the mouse toward the upper right displays damp-looking, matted fur. (B) The *Mpzl3*^{rc} status of all complementation-test progeny was verified by a DNA test that reveals the *rc*-specific missense mutation (see Supplementary Figure S3). Lanes that used template DNA from mice with a wild-type phenotype are marked "W", and those using template DNA from mice with a mutant phenotype are marked "M". All 31 complementation-cross progeny were genotyped (typical results are shown), and all phenotypic mutants were proven to carry the defective *Mpzl3*^{rc} allele and one copy of the *ruf* allele. All phenotypically normal mice carried one copy of *ruf*, but no copy of *Mpzl3*^{rc}.

in a set of 48 N_2 mice (half male, half female; half mutant and half control heterozygotes) from the backcross described above, over the course of 28 weeks. While mutant (*ruf*/*ruf*) and control (+/*ruf*) males gained mass at the same rate, on average (data not shown), mutant females weighed, on average, less than wild-type female controls at all time points, although this difference was not statistically significant (Figure 6(A)). For example, in week 28, the average mass was 30.6 ± 5.2 g for control females and 25.7 ± 4.6 g for mutant females (± 1 standard deviation). At week 28, these animals were euthanized, abdominal fat was dissected away from other visceral organs, weighed, and recorded as the percentage of total body mass. Again, mutants and controls showed no discernible difference among males, on average (data not shown). By contrast, female mutants showed more than threefold less adipose tissue at 28 weeks of age than heterozygous controls, although this difference was not statistically significant (Figure 6(B)). Each week during this 28-week study, the percentage of furless body surface (to the nearest 5%) was also estimated. No control mice showed any hair loss prior to week 18, and, by the end of the study, only a few (5/24) showed a modest degree of alopecia (affecting less than 50% of the body surface). By contrast, alopecia appeared in the mutant cohort as early as week 7, and, by the end of the study, most mutants (20/24) showed generally larger bald patches, which affected from 5 to 90% of the body surface (see Figure 6(C)).

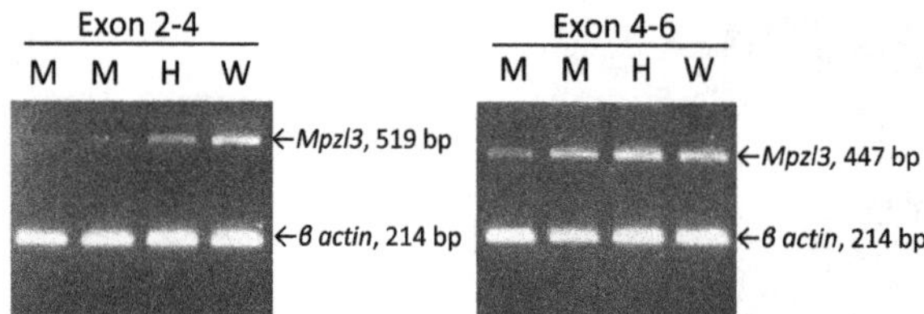

Figure 5. Analysis of *Mpzl3* splice junctions in control and mutant (*ruf/ruf*) cDNA. Poly-A⁺ RNA from skin was reverse transcribed into cDNA and amplified using forward and reverse primers that anneal with Exon 2 and Exon 4 (left panel), or Exon 4 and Exon 6 (right panel) of *Mpzl3*. Primers that anneal within Exon 4 of the mouse β actin (*Actb*) gene were used together with *Mzpl3* primers to provide an internal loading control. Templates from *ruf/ruf* mutants (labeled M above), a *ruf/+* heterozygote (labeled *H*), and a wild type +/+ control (labeled *W*) all yielded amplimers of the size and sequence expected for normal splicing of Exons 2 through 6. While the intensity of *Mpzl3*-specific products amplified from mutant templates may appear less abundant than products amplified from control skin samples, it should be noted that the RT-PCR method used here was not quantitative.

4. Discussion

We have taken a positional-candidate approach to assign the classic rough fur mutation in mice to the *Mpzl3* gene. Although we could find no corresponding DNA defect in the *Mpzl3* gene in rough fur mice, the failure of the recessive *rc* allele of *Mpzl3* to complement the *ruf* defect in compound heterozygotes confirms allelism. We therefore recommend that the rough fur mutation be formally renamed "myelin protein zero-like 3; rough fur", with the abbreviation $Mpzl3^{ruf}$.

We are not the first team to perform a DNA sequence analysis of the *Mpzl3* gene in rough fur mutants. Cao et al. (2007) reported no sequence variation in *Mpzl3* coding regions in a *ruf/ruf* DNA sample compared to one from the parental C3H/HeJ strain. Here, we verified that result, but, with reconstituted *ruf/ruf* mutants that Cao et al. lacked, we were able to test directly for non-complementation and therefore draw a different conclusion (that *ruf* is an allele of *Mpzl3*) than Cao and co-workers did in 2007. While our sequence analysis extended beyond the coding regions, more work is needed to determine the precise nature of the $Mpzl3^{ruf}$ defect. We speculate that this defect might, in some way, impact the expression, processing, or stability of the $Mpzl3^{ruf}$ transcript, although we find that the mutant transcript appears to be normally spliced. Quantitative and qualitative evaluation of *Mpzl3* transcripts and protein in the epidermis of mutant versus wild-type C3H mice should help to refine this array of possibilities. It is notable that, while we sequenced over 3,400 bp upstream of $Mpzl3^{ruf}$ (and found no defect), others have suggested that the regulatory region for *Mpzl3* may extend 12 kb upstream of the transcriptional start site (Boxer, Barajas, Tao, Zhang, & Khavari, 2014; Sen et al., 2012).

In any case, the spontaneous *ruf* allele now joins $Mpzl3^{rc}$ and at least two targeted null alleles of *Mpzl3* (Czyzyk et al., 2013; Leiva et al., 2014), providing a collection of phenotypically similar, but mutationally distinct defects. Side-by-side comparison of the phenotypes controlled by each different allele (on similar genetic backgrounds), might reveal important information relating gene structure to function. For example, the spontaneous *ruf* allele may not show the same pleiotropic effect on adipose accumulation (at least in males) as reported for the targeted *Mpzl3* alleles (Czyzyk et al., 2013; Leiva et al., 2014; Tang et al., 2010), but diet and background would need to be controlled and similar assays performed to verify this impression.

The study of spontaneous and targeted mouse models has emphasized *Mpzl3*'s essential role in skin (where it impacts dermal adipose, hair follicle and sebaceous gland development), hepatic triglyceride synthesis, main adipose deposition, energy expenditure and glycemic control (see Cao et al., 2007; Czyzyk et al., 2013; Hayashi et al., 2004; and Leiva et al., 2014). Consistent with these phenotypes, *Mpzl3* has recently been found to encode an adhesion protein that is highly induced in epidermal differentiation, and is primarily located in the mitochondria (Bhaduri et al., 2015). MPZL3 has been found to interact with numerous mitochondrial proteins including ferrodoxin reductase (FDXR), an enzyme that, together with MPZL3, is necessary for induction of reactive oxygen species (ROS) required for epidermal differentiation (Bhaduri et al., 2015). While no humans with MPZL3 deficiencies have yet been reported, based on the described series of spontaneous-mutant and knock-out mouse models it may be predicted that such conditions might present with altered metabolic

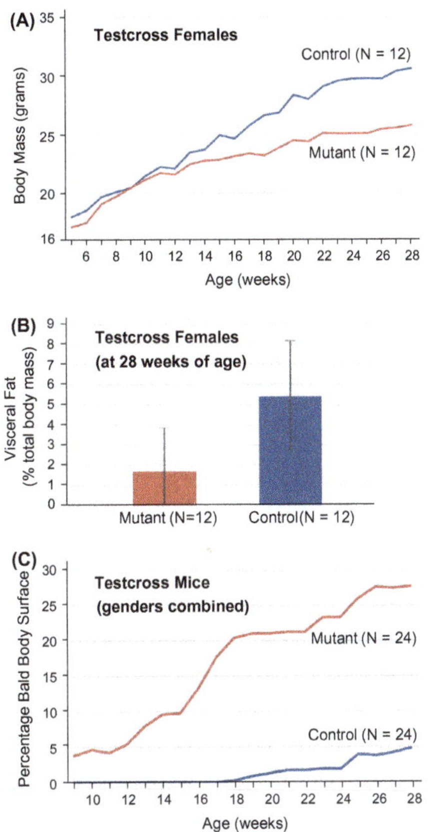

Figure 6. Adiposity and progressive hair loss among N_2 mice segregating for rough fur. Backcross mice (24 *ruf/ruf* and 24 *+/ruf*) from the mapping panel described in Figure 2 were assessed weekly for body mass (in grams) and for the degree of baldness (estimated as the percentage of hairless body surface). (A) While heterozygous control and mutant males showed no difference in body mass at any point in the 28 week study (data not shown), control females (blue line) showed a markedly (but not significantly) higher mass versus mutant females (red line) at all time points. (B) At the end of the survey, visceral fat was dissected from all N_2 subjects, weighed, and expressed as percentage of total body mass. While males showed no difference in fat stores between wild and mutant types (data not shown), control females displayed, on average, three times more visceral fat (5.39 ± 2.76%) than mutants (1.68 ± 2.18%), but this difference was not statistically significant. (C) The estimated size of bald patches (expressed as the percent of the body surface) was averaged among heterozygous control (blue) and mutant (red) N_2 mice. Mutants (males and females combined) show a steady, progressive increase in baldness that begins earlier and is more extensive than the hair loss seen among non-mutant control subjects (which was similar to the incidental alopecia typically displayed by C3H/HeJ mice).

physiology (such as decreased adiposity and increased food intake in response to metabolic demand), sebaceous gland hypertrophy (as in acne vulgaris) and hair loss (as in alopecia areata). Notably, one inherited skin condition in humans, seborrhea-like dermatitis with psoriasiform elements (OMIM #610227, Online Mendelian Inheritance in Man [OMIM], 2017), results from mutation of *ZNF750* (Birnbaum et al., 2006), which encodes a transcription factor that binds to the *MPZL3* promoter to activate its transcription (Bhaduri et al., 2015). The readily-available *Mpzl3* mutants can clearly serve as useful models to advance our understanding of seborrhoeic dermatitis in general and of ZNF750 dysfunction in particular (see Wikramanayake et al., 2016), even in the absence of a *Zfp750* knock-out mouse.

Other tissues—including the brain; the tongue, esophagus, stomach, colon and gall bladder; the trachea and lungs; lymph nodes and white blood cells; the thyroid; and the prostate—also display high levels of myelin protein zero-like 3 expression in both mouse and man (Fagerberg et al., 2014; Mouse ENCODE Consortium, 2014), and the role of *Mpzl3* in the development, maintenance and function of these tissues certainly deserves further investigation. We anticipate that mouse variants with distinct defects in *Mpzl3*—a set which now includes *Mpzl3^{ruf}*—will help to facilitate such studies.

5. Conclusions

The recessive *ruf* mutation in mice maps within a 1.4 Mb interval on mouse Chr 9, and has been shown by complementation testing to be a mutant allele of the *Mpzl3* gene. While further study will be needed to pinpoint the molecular defect that is the basis of this mutant phenotype, this assignment places *ruf* among a series of distinct mutant alleles of *Mpzl3*. This set of mouse variants will surely be valuable as a model system for the in-depth, molecular analysis of the various, pleiotropic functions controlled by *Mpzl3*, and for the study of some inherited human disorders where those functions are impaired.

Abbreviations

Chr	chromosome
CCSU	Central Connecticut State University
PCR	polymerase chain reaction
RT-PCR	reverse transcriptase PCR
SNP	single nucleotide polymorphism
ERV	endogenous retrovirus
IAP	intracisternal A particle
ROS	reactive oxygen species

Author's contributions

TRK conceived of the study, carried out all procedures involving mice, and drafted the manuscript. Student coauthors led the various aspects of the study (as listed below), including experimental design, data acquisition and interpretation. Specifically: KMP and MAR performed the genetic mapping analysis; LAR performed the complementation analysis; TVN, MMR, SE, and JJS sequenced the *Mpzl3^{ruf}* promoter, exons and splice junctions; MEC and MMR scanned *Mpzl3^{ruf}* for the presence of large insertion elements; TVN conducted the analysis of *Mpzl3^{ruf}* mRNA; AK performed the time-course analysis of body mass, adipose accumulation and progressive hair loss. All authors read, edited, and approved the final manuscript.

Acknowledgements

The authors thank undergraduate Erin Casey and high school interns Renukanandan Tumu, Brandy Robinson, Tasheika Telfer and Tamara Rodriguez for help with marker typing; and Mary Mantzaris for excellent animal care.

Funding

This work was supported by a small research grant from the Connecticut State Colleges and Universities System.

Competing Interests

The authors declare no competing interest.

Author details

Kenneth M. Palanza[1]
E-mail: k.palanza@my.ccsu.edu
Legairre A. Radden[1]
E-mail: laradden@my.ccsu.edu
Mohammed A. Rabah[1]
E-mail: rabahm@my.ccsu.edu
Tu V. Nguyen[1]
E-mail: rnguyen@my.ccsu.edu
Audra C. Kohm[1]
E-mail: kohm@my.ccsu.edu
Malcolm E. Connor[1]
E-mail: malcolm.connor@my.ccsu.edu
Morgan M. Ricci[1]
E-mail: morganricci@my.ccsu.edu
Jachius J. Stewart[1]
E-mail: jachius.stewart@my.ccsu.edu
Sidney Eragene[1]
E-mail: eragene.s@my.ccsu.edu
Thomas R. King[1]
E-mail: kingt@ccsu.edu
ORCID ID: http://orcid.org/0000-0001-6332-5484
[1] Department of Biomolecular Sciences, Central Connecticut State University, 1615 Stanley Street, New Britain, CT 06053, USA.

Cover image

Source: Authors

References

Bhaduri, A., Ungewickell, A., Boxer, L. D., Lopez-Pajares, V., Zarnegar, B. J., & Khavari, P. A. (2015). Network analysis identifies mitochondrial regulation of epidermal differentiation by MPZL3 and FDXR. *Developmental Cell*, 35, 444–457. https://doi.org/10.1016/j.devcel.2015.10.023

Birnbaum, R. Y., Zvulunov, A., Hallel-Halevy, D., Cagnano, E., Finer, G., Ofir, R., ... Birk, O. S. (2006). Seborrhea-like dermatitis with psoriasiform elements caused by a mutation in ZNF750, encoding a putative C2H2 zinc finger protein. *Nature Genetics, 38*, 749–751. https://doi.org/10.1038/ng1813

Boxer, L. D., Barajas, B., Tao, S., Zhang, J., & Khavari, P. A. (2014). ZNF750 interacts with KLF4 and RCOR1, KDM1A, and CTBP1/2 chromatin regulators to repress epidermal progenitor genes and induce differentiation genes. *Genes & Development, 28*, 2013–2026. https://doi.org/10.1101/gad.246579.114

Cao, T., Racz, P., Szauter, K. M., Groma, G., Nakamatsu, G. Y., Fogelgren, B., ... Csiszar, K. (2007). Mutation in *Mpzl3*, a novel (corrected) gene encoding a predicted (corrected) adhesion protein, in the rough coat (*rc*) mice with severe skin and hair abnormalities. *Journal of Investigative Dermatology, 127*, 1375–1386. https://doi.org/10.1038/sj.jid.5700706

Connor, M. E., & King, T. R. (2016). The spontaneous juvenile alopecia (*jal*) mutation in mice is associated with the insertion of an IAP element in the *Gata3* gene. *Cogent Biology, 2*, 1264691.

Czyzyk, T. A., Andrews, J. L., Coskun, T., Wade, M. R., Hawkins, E. D., Lockwood, J. F., ... Statnick, M. A. (2013). Genetic ablation of myelin protein zero-like 3 in mice increases energy expenditure, improves glycemic control, and reduces hepatic lipid synthesis. *American Journal of Physiology - Endcrinology and Metabolism, 305*, E282–E292. https://doi.org/10.1152/ajpendo.00228.2013

Dickie, M. M. (1966). Rough coat. *Mouse News Letter, 34*, 30.

Dietrich, W. F., Miller, J., Steen, R., Merchant, M. A., Damron-Boles, D., Husain, Z., ... O'Connor, T. J. (1996). A comprehensive genetic map of the mouse genome. *Nature, 380*, 149–152. https://doi.org/10.1038/380149a0

Ensembl Mouse Genome Server. (2017). *Mouse genome sequencing consortium: The European Bioinformatics Institute (EBI) and Welcome Trust Sanger Institute (WTSI)*. Release 89. Retrieved July 2017, from https://www.ensembl.org

Fagerberg, L., Hallström, B. M., Oksvold, P., Kampf, C., Djureinovic, D., Odeberg, J., ... Uhlén, M. (2014). Analysis of the human tissue-specific expression by genome-wide integration of transcriptomics and antibody-based proteomics. *Molecular & Cellular Proteomics, 13*, 397–406. https://doi.org/10.1074/mcp.M113.035600

Hayashi, K., Cao, T., Passmore, H., Jourdan-Le Saux, C., Fogelgren, B., Khan, S., ... Csiszar, K. (2004). Progressive hair loss and myocardial degeneration in rough coat mice: Reduced lysyl oxidase-like (LOXL) in the skin and heart. *Journal of Investigative Dermatology, 123*, 864–871. https://doi.org/10.1111/j.0022-202X.2004.23436.x

Leiva, A. G., Chen, A. L., Devarajan, P., Chen, Z., Damanpour, S., Hall, J. A., ... Wikramanayake, T. C. (2014). Loss of *Mpzl3* function causes various skin abnormalities and greatly reduced adipose depots. *Journal of Investigative Dermatology, 134*, 1817–1827. https://doi.org/10.1038/jid.2014.94

Maksakova, I. A., Romanish, M. T., Gagnier, L., Dunn, C. A., van de Lagemaat, L. N., & Mager, D. L. (2006). Retroviral elements and their hosts: Insertional mutagenesis in the mouse germ line. *PLoS Genetics, 2*, e2. https://doi.org/10.1371/journal.pgen.0020002

McMahon, K. A., Hiew, S. Y., Hadjur, S., Veiga-Fernandes, H., Menzel, U., Price, A. J., ... Brady, H. J. (2007). *Mll* has a critical role in fetal and adult hematopoietic stem cell self-renewal. *Cell Stem Cell, 1*, 338–345. https://doi.org/10.1016/j.stem.2007.07.002

Mouse ENCODE Consortium. (2014). A comparative encyclopedia of DNA elements in the mouse genome. *Nature, 20*, 355–364.

Mouse Genome Database (MGD). (2017). *Mouse genome database group: The mouse genome informatics website*. Bar Harbor ME: The Jackson Laboratory. Retrieved July 2017, from https://www.informatics.jax.org

Online Mendelian Inheritance in Man (OMIM). (2017). *McKusick-Nathans institute of genetic medicine*. Baltimore, MD: Johns Hopkins University. Retrieved July 2017, from https://omim.org

Park, Y.-G., Hayasaka, S., Takagishi, Y., Inouye, M., Okumoto, M., & Oda, S. (2001). Histological characteristics of the pelage skin of rough fur mice (C3H/HeJ-*ruf/ruf*). *Experimental Animals, 50*, 179–182. https://doi.org/10.1538/expanim.50.179

Ramirez, F., Feliciano, A. M., Adkins, E. B., Child, K. M., Radden II, L. A., Salas, A., ... King, T. R. (2013). The juvenile alopecia mutation (*jal*) maps to mouse Chromosome 2, and is an allele of GATA binding protein 3 (*Gata3*). *BMC Genetics, 14*, 40. https://doi.org/10.1186/1471-2156-14-40

Ruvinsky, I., Chertkov, O., Borue, X. V., Agulnik, S. I., Gibson-Brown, J. J., Lyle, S. R., & Silver, L. M. (2002). Genetics analysis of mouse mutations Abnormal feet and tail and rough coat, which cause developmental abnormalities and alopecia. *Mammalian Genome, 13*, 675–679. https://doi.org/10.1007/s00335-002-2191-6

Sen, G. L., Boxer, L. D., Webster, D. E., Bussat, R. T., Qu, K., Zarnegar, B. J., ... Khavari, P. A. (2012). ZNF750 is a p63 target gene that induces KLF4 to drive terminal epidermal differentiation. *Developmental Cell, 22*, 669–677. https://doi.org/10.1016/j.devcel.2011.12.001

Sweet, H. O., Oda, S. I., Taylor, B. A., Rowe, L., Davisson, M. T., Cook, S., ... Sundberg, J. (1990). Rough fur (*ruf*). *Mouse Genome, 86*, 236–237.

Tang, T., Li, L., Tang, J., Li, Y., Lin, W. Y., Martin, F., ... de Sauvage, F. J. (2010). A mouse knockout library for secreted and transmembrane proteins. *Nature Biotechnology, 28*, 749–755. https://doi.org/10.1038/nbt.1644

Thien, C. B., Scaife, R. M., Papadimitriou, J. M., Murphy, M. A., Bowtell, D. D., & Langdon, W. Y. (2003). A mouse with a loss-of-function mutation in the c-Cbl TKB domain shows perturbed thymocyte signaling without enhancing the activity of the ZAP-70 tyrosine kinase. *The Journal of Experimental Medicine, 197*, 503–513. https://doi.org/10.1084/jem.20021498

Wikramanayake, T. C., Borda, L. J., Kirsner, R. S., Wang, Y., Duffort, S., Reyes-Capo, A., ... Perez, V. L. (2016). Loss of MPZL3 function causes seborrheic dermatitis-like phenotype in mice. *Experimental Dermatology, 26*, 736–738.

Xu, X., Kedlaya, R., Higuchi, H., Ikeda, S., Justice, M. J., Setaluri, V., & Ikeda, A. (2010). Mutation in archain 1, a Subunit of COPI coatomer complex, causes diluted coat color and Purkinje cell degeneration. *PLoS Genetics, 6*, e1000956. https://doi.org/10.1371/journal.pgen.1000956

13

Acute and chronic toxicities assessment of arsenic (III) to catfish, *Silurus lanzhouensis* in China

Zongqiang Lian[1,2] and Xudong Wu[1,2*]

*Corresponding author: Xudong Wu, Ningxia Fisheries Research Institute, Yinchuan 750001, China; Ningxia Engineering Research Center for Fisheries, Yinchuan 750001, China

E-mail: 838873204@qq.com

Reviewing editor: Mahmud Hossain, Dhaka University, Bangladesh

Additional information is available at the end of the article

Abstract: We evaluated the lethality, uptake, depuration, accumulation, and effects of waterborne arsenic in Lanzhou catfish (*Silurus lanzhouensis*). The 96-h LC_{50} and safe concentrations (SC) for Lanzhou catfish were 12.88 and 1.288 mg/L, respectively. We evaluated the effects of chronic exposure to 0 mg/L (C), 1.288 mg/L (T1), 0.5 mg/L (T2), and 0.1 mg/L (T3) and measured depuration rates post-exposure. As accumulated in the target organs in the following order of concentration: gill > muscle > brain > liver, which is consistent with the variation in k_1. The values for k_1 and C_{Amax} declined with a decrease in arsenic concentration in the different target organs, whereas the reverse was true for BCF. The $C_{L,50}(t)$ values decreased initially and then approached equilibrium status after 30 of exposure. The gill tissue had the highest depuration rates, followed by muscle, brain, and liver. The treatment groups exposed to lower arsenic concentrations treats had lower k_2 values in the target organs, but higher depuration half-lives ($t_{1/2}$) at lower arsenic concentrations. Our results demonstrate that the target organs of Lanzhou catfish are capable of regulating arsenic toxicity by way of internal regulation mechanisms, and the rate of arsenic uptake and depuration over time are concentration- and tissue-dependent.

Subjects: Environmental Sciences; Fisheries Science; Agriculture and Food; Aquaculture; Environmental & Ecological Toxicology; Environment & Health

Keywords: *Silurus lanzhouensis*; arsenic; bioconcentration; histopathology

1. Introduction

Arsenic (As) is a heavy metal that is relatively common in the environment as a consequence of both anthropogenic and natural processes. Arsenic is classified as a top environmental chemical concern by Petrusevski, Sharma, Schippers, and Shordt (2007). United States Food and Drug Administration

ABOUT THE AUTHOR

Xudong Wu, PhD, is a professor in the Ningxia Engineering Research Center for Fisheries, Yinchuan, China. He is mainly engaged in environmental toxicology, aquatic animal protection genetics, and fish breeding research. Together with her research team members Zongqiang Lian has extensive publications on the environmental toxicology, aquatic animal protection.

PUBLIC INTEREST STATEMENT

Arsenic (As) is a heavy metal and highly toxic to animals, plants, and humans. Arsenic toxicity has been evaluated for some aquaculture species, but until now, little was known about the mechanisms of arsenic toxicity in freshwater fishes, especially the fishes without scales. To assess the lethality, distribution, and kinetics of waterborne As, the study evaluated the uptake of As, its elimination and its distribution in the target organs of Lanzhou catfish. This is the first study to document bioaccumulation and depuration of As in the catfish.

(1993) noted that 90% of total arsenic exposure is derived from fish and other seafood. But one of the major challenges for assessing the potential risk of heavy metals to aquatic organisms is predicting the time-dependent internal effective dose that causes the toxic effects. The response of an organism is not merely determined by the dose–response relationship, but is also affected by the ability for biological regulation or detoxification during long-term exposures (Schuler, Landrum, & Lydy, 2004; Vijver, van Gestel, Lanno, van Straalen, & Peijnenburg, 2004). A number of models have been developed to predict toxicity to aquatic organisms by integrating the mechanistic process between accumulated chemical dose and the induced dynamics of tissue damage and recovery. These include the time-integrated concentration toxicity model (TIC), the concentration–time toxicity model (CT), the uptake–depuration toxicity model (UD), (Liao & Lin, 2001), the damage assessment model (DAM) (Lee, Landrum, & Koh, 2002), and the biotic ligand model (BLM) (Tsai, Chen, Ju, & Liao, 2009). All these models are based on a well-established one-compartment bioaccumulation model.

Lanzhou catfish (*Silurus lanzhouensis*), commonly known as the Yellow River catfish, are native to China and are distributed throughout the Ningxia section of the Yellow River. The species is widely cultured in China and has a high market value. Due to water pollution, overfishing, construction of water conservancy facilities, and other factors, wild Lanzhou catfish populations have declined in abundance recently and species is now listed on the China Species Red List (Wang & Xie, 2009). The declines are largely Because of the value of this species, it has been the focus of considerable research in recent years in China. In recent years, we have discovered and cloned four antitoxicity-related genes specific for Lanzhou catfish. These genes may help us to understand the antitoxicity mechanisms used by aquatic animals and take measures to protect this rare species. At present, to our knowledge, little is known about the uptake and effects of arsenic in Lanzhou catfish and, likewise, little is known about the actual uptake and elimination processes or on the kinetics of arsenic.

We evaluated the uptake of As, its elimination and its distribution in the target organs of Lanzhou catfish exposed to As-contaminated water at environmentally relevant concentrations in the laboratory. This is the first study to document bioaccumulation and depuration of As in the Lanzhou catfish, and to assess the lethality, distribution, and kinetics of waterborne As.

2. Materials and methods

2.1. Fish and water quality

The number of 1200 Lanzhou catfish (*S. lanzhouensis*) (mean body length = 5.437 ± 1.21 cm and mean body weight = 4.32 ± 0.85 g wet wt) were obtained from the Ningxia Fisheries Research Institute, Yinchuan, China. These fish were tested for arsenic contamination and the levels were non detectable. The fish were also visibly free of any lesions, deformities, or disease. This study was performed with the approval of the local ethical committee and all the experiments were performed according to the National Institutes of Health Guide for the Care and Use of Laboratory Animals. All experiments were carried out in 80-L indoor rectangular plastic aquaria filled with 60 L water. The experimental water was filtered and aerated tap water (pH = 7.3 ± 0.27, alkalinity = 186.65 ± 22.4 mg/L, and Cl^{-1} < 13 mg/L, K^+, Na^+ < 24.3 mg/L, Ca^{2+} < 29.5 mg/L, p < 0.01 mg/L, H_2S, Hg, and As not detected). Dissolved oxygen in each tank was maintained at close to saturation (7.21 ± 0.1 mg/L) by aeration. The temperature was maintained at 24.4 ± 0.5°C using submerged heaters in each aquarium.

2.2. Acute toxicity assays

Laboratory bioassays were conducted to determine the 24-, 48-, 72-, and 96-h LC_{50} values for Lanzhou catfish exposed to sodium arsenite ($NaAsO_2$). Thirty fish were randomly selected and transferred into each test aquarium. After a 7-d acclimation period, the catfishes were exposed to one of five logarithmically spaced concentrations of As. All treatments were conducted in triplicate and the control group consisted of catfishes held in the experimental aquaria without the addition of As. The exposure concentrations were chosen based on the results of a preliminary test. The nominal concentrations of arsenic we tested were 0 (control), 6.67, 10, 15, 22.5, and 33.75 mg/L.

The measured arsenic concentrations were 0.67 ± 0.02, 9.96 ± 0.05, 15.06 ± 0.05, 22.55 ± 0.10, and 33.68 ± 0.08 mg/L.

To maintain water quality, the fish were not fed throughout the test and the entire arsenic solution was replaced daily in each tank. Gross mortality of fish in each tank was recorded every hour for the first 12 h and every 2 h thereafter for 96 h. Dead fish were removed every 1–2 h.

2.3. Chronic toxicity

Based on the safe concentration estimated from toxicity testing above, we evaluated the pattern of bioaccumulation and depuration at four arsenic concentrations: 0 mg/L (C), 1.288 mg/L (T1), 0.5 mg/L (T2), and 0.1 mg/L (T3). Thirty fish were stocked into each experimental tank and acclimated to laboratory conditions for 2 weeks before exposure. During the acclimation (14 d) and experimental periods (42 d), the fish were fed with water worms (bait was cultured in our laboratory, Hg and As not detected) once a day for 7 d at a rate of 0.5% of fish biomass. We removed feces every 4 h and collected any uneaten food 1 h after feeding. During the 21-d exposure period, the entire arsenic solution was replaced daily in each tank, the water level was checked in each aquarium every 6 h and distilled water was used to keep levels constant. At the end of the 21-d exposure period, the fish were held for 21 d to evaluate depuration.

To measure the tissue concentrations of arsenic, three fish were sequentially sampled from each aquarium on days 0, 4, 7, 14, 21, 28, 35, and 42. Each fish was dissected and the gill, muscle, brain, and liver of each individual was removed, weighed, cleaned with deionized water, and individually wrapped in a plastic bag and stored frozen (−20°C) until trace element analysis. Dissected tissues samples were lyophilized and then ground to fine powder in a grinder. A ~500 mg sample of the powder was digested with concentrated nitric acid (7 ml) and perchloric acid (1 ml) in an Acid Digestion Bomb (5 min at 160°C, 35 Bar, 1,000 Watts). After cooling, the samples were diluted to 50 ml with deionized water and evaporated with an Electric Hot Plate (100°C) to a 1-ml solution. We then added 5% ascorbic acid (5 ml) and 5% thiocarbamide (5 ml) and made the volume up to 25 ml with 5% hydrochloric acid. Samples were analyzed by an AFS-230E Atomic Fluorescence Spectrometer to determine As concentrations in all tissues. Concentrations were expressed as μg/g dry weight.

2.4. Data analysis

We used a computerized probit analysis program (United States Environmental Protection Agency, 2000) to calculate the LC_{50} values (24-, 48-, 72-, and 96-h) and upper and lower 95% confidence levels. The 96-h safe concentration (1/10th of the 96-h LC_{50} value) was calculated following the method of Zhou and Zhang (1989). The uptake rate constant (k_1) and the depuration rate constant (k_2) of arsenic were calculated using nonlinear regression based on the chronic-toxicity data. The organ-specific bioconcentration factor (BCF) was calculated using the formula $BCF = k_1/(k_2 + g)$ (Falusi & Olanipekun, 2007), the maximal content in the organism (C_{Amax}) at the theoretic equilibrium was estimated using the formula $C_{Amax} = BCF \times C_W$. Depuration half-life ($t_{1/2}$) was calculated as $t_{1/2} = \ln 2/k_2$.

We used an AUC-based TIC toxicity model to estimate the internal lethal body burden [$C_{L,50}(t)$] associated with 50% mortality (Liao et al., 2004). All data are presented as the mean ± SE. We tested for differences between groups using a one-way ANOVA ($\alpha = 0.05$) followed by Duncan's multiple range test.

3. Results

3.1. Arsenic acute-toxicity analysis

The LC_{50} values for the 24-, 48-, 72-, 96-, 120-, and 144-h exposures to arsenic ranged from 31.72 to 11.54 mg/L (Table 1). The $LC_{50}(t)$ decreased as the duration of exposure increased. The safe concentration for a 96-h exposure to sodium arsenite was 1.288 mg/L in Lanzhou catfish.

Table 1. LC_{50} of arsenic in Lanzhou catfish at selected time intervals			
Time (h)	LC_{50} (mg/L)	STD	r^2
24	31.72	0.55	0.91
48	20.89	−0.62	0.87
72	13.28	−0.2	0.97
96	12.88	0.28	0.93
120	12.23	0.23	0.95
144	11.54	0.24	0.93

Note: Values obtained using a trimmed Spearman-Karber method with 95% confidence.

Figure 1. Changes in As concentrations in the brain, gill, liver, and muscle during the exposure phase and recovery phase. Values represent the mean S.E. of triplicate subsamples. Values at a given time that are significantly different are indicated by different letters ($p < 0.05$).

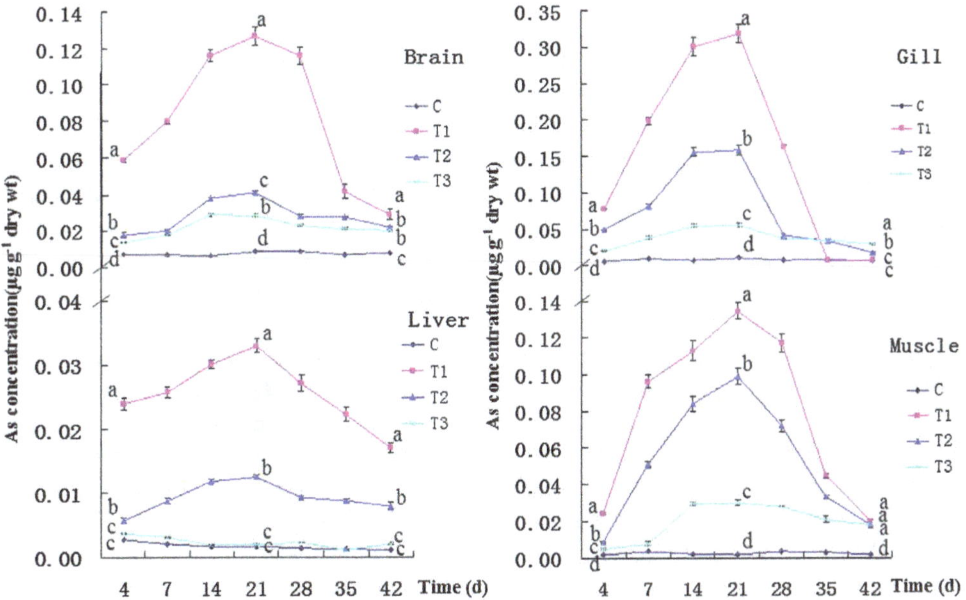

3.2. As bioaccumulation and depuration

During the exposure phase, arsenic concentrations in the gill, muscle, brain, and liver varied significantly with time ($F_{3,41}$ = 8.91, 7.29, 5.94, and 18.75, $p < 0.01$), between treatments ($F_{3,41}$ = 35.09, 44.37, 502.28, and 43.95, $p < 0.01$), and with the interaction between treatment and time ($F_{1,3}$ = 20.04, 28.79, 11.07, and 313.79; p = 0.02, 0.01, 0.05, and 0), respectively. After a 21-d exposure, As concentrations were much higher in the gill than in other target organs of Lanzhou catfish at each of the exposure concentrations ($F_{(T1)3,79} = 555.66$, $F_{(T2)3,79} = 707.24$, $F_{(T3)3,79} = 559.28$, $p < 0.01$). The order of accumulation was: gill > muscle > brain > liver. With the exception of the liver in T3 ($F_{3,8} = 1767.69$, $p > 0.05$), there was a significant difference in arsenic concentrations in each target organ between all three arsenic exposure treatment groups and the control at the end of the exposure period (Figure 1).

During the recovery interval, arsenic concentrations in the gill, muscle, brain, and liver varied significantly with time ($F_{3,41}$ = 12.07, 12.76, 16.09, 9.44, $p < 0.01$), between treatments ($F_{3,41}$ = 3.84, 152.78, 33.91, and 200.53; $p < 0.01$), and with the interaction between treatment and time ($F_{1,3}$ = 3.08, 12.72, and 78.60; p = 0.18, 0.04, 0.003) in brain, muscle, and liver, except for the gill ($F_{1,3} = 2.89$; $p = 0.19$). After 21 d of recovery, arsenic concentrations had returned to normal levels in the gill of T1 ($F_{3,8} = 922.7$, $p < 0.01$) and liver of T3 ($F_{2,6} = 414.62$, $p < 0.01$) (Figure 1).

Table 2. Kinetic parameters of arsenic for Lanzhou catfish

Organ	Treats	k_1 (mL/g d)	k_2 (d^{-1})	BCF (mL/g)	C_{Amax}(mg/kg)	$t_{1/2}$ (d)
Gill	T1	0.1473	0.2105	0.70	0.90	3.3
	T2	0.0723	0.0943	0.77	0.38	7.3
	T3	0.0219	0.0276	0.79	0.08	25.1
Muscle	T1	0.0612	0.0952	0.64	0.83	7.3
	T2	0.0536	0.0831	0.65	0.32	8.3
	T3	0.0175	0.0263	0.67	0.07	26.4
Brain	T1	0.0426	0.0781	0.55	0.70	8.9
	T2	0.0156	0.0270	0.58	0.29	25.7
	T3	0.0099	0.0164	0.60	0.06	42.3
Liver	T1	0.0054	0.0304	0.18	0.23	22.8
	T2	0.0042	0.0197	0.21	0.11	35.2
	T3	0.0012	0.0061	0.20	0.02	113.6

3.3. Target organ As kinetics

The uptake rate constants (k_1), the depuration rate constants (k_2), and C_{Amax} values, declined with a decrease in arsenic concentrations in the target organs (Table 2). However, the BCF values and the depuration half-lives ($t_{1/2}$) increased with a decrease in arsenic concentrations in the target organs (Table 2). Thus, the rates of arsenic uptake and depuration over time are concentration- and tissue-dependent.

The gill had higher k_1 and BCF values than the remaining target organs in T1, T2, and T3. The k_1 values ranged from 0.0219 to 0.1473 mL/g d and the BCF values ranged from 0.70 to 0.79 mL/g. Thus, As tends to accumulate primarily in the gills of Lanzhou catfish. The k_1 values for the liver ranged from 0.0012 to 0.0054 mL/g d and the BCF values ranged from 0.18 to 0.20 mL/g, which were the lowest among the organs. Thus, the liver had the lowest capacity to uptake arsenic. The k_2 values for the target organs ranged between 0.0061 and 0.2105 d^{-1}, and were in the order gill > muscle > brain > liver. The depuration half-lives ($t_{1/2}$) ranged from 3.3 to 22.8 d in T1, from 7.3 to 35.2 d in T2, and 25.1 to 113.6 d in T3. Among the target organs, the liver had the highest $t_{1/2}$ value, indicating that it will take a longer time to eliminate arsenic than the other organs.

3.4. Prediction of the internal lethal body burden

The optimal fit of the TIC toxicity model to the $LC_{50}(t)$ data are listed in Table 1. The regression coefficients ($r^2 = 0.98$, $p < 0.05$) demonstrate the quality of fit for the TIC toxicity model (Figure 2). Based on the TIC toxicity model, the internal lethal body burden [$C_{L,50}(t)$] for each target organ at the site of action that causes 50% mortality was predicted assuming an exposure to 1.288 mg/L waterborne arsenic. The $C_{L,50}(t)$ values decreased initially and then approached an equilibrium after 30-d exposure (Figure 3). The equilibrium $C_{L,50}(t)$ values in the gill, muscle, brain, and liver were 56.21, 50.93, 42.35, and 10.98 μg/g, respectively.

4. Discussion

Sures (2003) speculated that teleosts are better adapted to resist metal and non-metal poisoning than are other vertebrates. Arsenic toxicity has been evaluated for some aquaculture species, but until now, little was known about the mechanisms of arsenic toxicity in catfish. We designed a toxicological model experiment to estimate the time–concentration-dependent tolerances of Lanzhou catfish to As toxicity and determine the mechanism of arsenic toxicity. The 96-h LC50 of arsenic to Lanzhou catfish was 12.88 mg/L, which is similar to that for rainbow trout *Oncorhynchus mykiss* (15.3 mg/L) (Tišler & Zagorc-Končan, 2002) and Medaka *Oryzias latipes* (14.6 mg/L) (Suhendrayatna, Ohki, Nakajima, & Maeda, 2002), but lower than for tilapia *O. mossambicusis* (28.68 mg/L: Liao et al., 2004; and 26.55 mg/L: Hwang & Tsai, 1993) and zebrafish *Danio rerio* (28.1 mg/L) (Tišler &

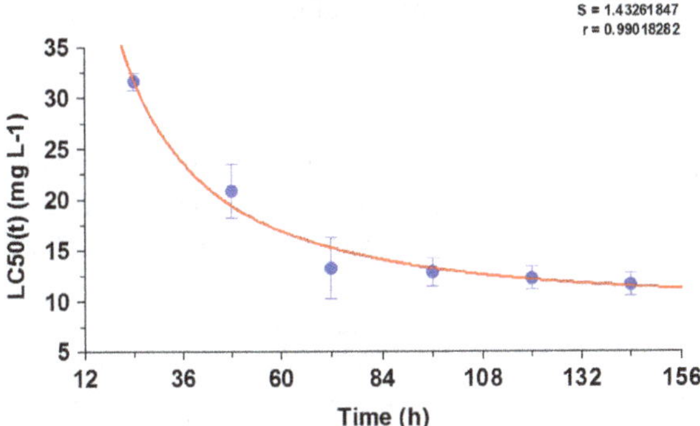

Figure 2. Optimal fit of the AUC-based TIC acute toxicity model to the LC50(t) data (mean ± 95% confidence interval) listed in Table 2.

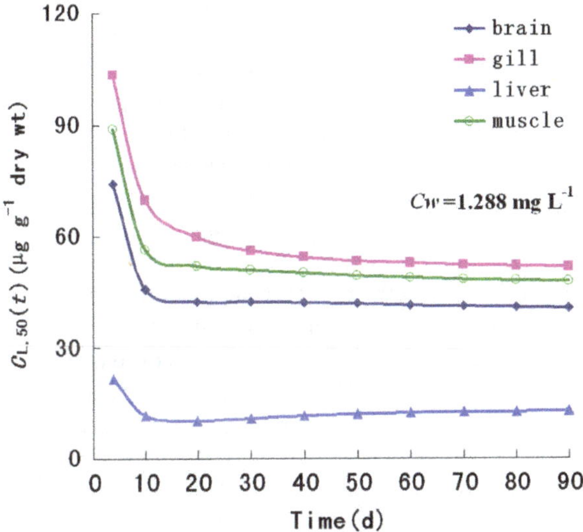

Figure 3. Predicted internal effect concentration–time response relationships in target organs of Lanzhou catfish when exposed to 1.288 mg/L waterborne arsenic, at the site of action that causes 50% mortality.

Zagorc-Končan, 2002), but higher than for milk fish *Chanos chanos* (7.29 mg/L) (Chou, Liao, Lin, & Cheng, 2006). In general, the LC50 values of teleost species for arsenic are between 7 and 29 mg/L, depending on age, species, and environment conditions. We hypothesized that Lanzhou catfish would be more susceptible to arsenic uptake and toxicity than fish with larger scales, thereby making them a more suitable bioindicator for studying the accumulation and transformation of arsenic in freshwater organisms.

The TIC toxicity concept has a significant use in evaluating metal toxicity (Liao et al., 2004). We used a TIC toxicity model to predict the internal lethal body burden for each target organ at the site of action that causes 50% mortality [$C_{L,50}(t)$ values] in Lanzhou catfish when exposed to 1.288 mg/L waterborne arsenic concentration. The $C_{L,50}(t)$ values decreased initially and then approached an equilibrium after 30 d of exposure, suggesting that the target organs of Lanzhou catfish are capable of regulating arsenic toxicity by way of internal regulation. The AUC-based TIC model successfully predicted the trend in $C_{L,50}(t)$ values, which were consistent with the organ-specific toxic kinetic parameters.

As accumulated in the target organs of Lanzhou catfish in the following order of concentration: gill > muscle > brain > liver. Additionally, the sigmoid pattern of accumulation was steeper in profile in the gill than for the other target organs. The gills are an important site for the accumulation of many metals and organic pollutants. Furthermore, the gills are played a major role in regulating metal toxicity in fish, primarily by regulating against metal ion disruption (Daglish & Nowak, 2002).

This explains why the gill is more sensitive and of value as a biomarker of exposure to waterborne arsenic in Lanzhou catfish.

Metals typically accumulate preferentially in the liver rather than muscle tissue (Liao et al., 2004; Storelli & Marcotrigiano, 2000). However, higher values have been reported in the muscle tissue in marine turtles (Storelli, Ceci, & Marcotrigiano, 1998) and in fish from the North Sea (De Gieter et al., 2002). In our study, the liver had the lowest capacity for arsenic accumulation, which may be a function of rapid arsenic methylation in this tissue. Inorganic arsenic methylation is the primary mechanism by which the body removes arsenic a process that occurs in the liver (Yamauchi & Yamamura, 1985). Thus, the rate of arsenic uptake was organ-specific and time-dependent in Lanzhou catfish. The capacity of the different target organs to accumulate arsenic varied depending on the exposure concentration. At the end of the recovery period, the majority of arsenic was cleared in individuals exposed to lower concentrations, except in the muscle. Interestingly, levels returned rapidly to normal during the recovery period in the gill of animals exposed to high concentrations of arsenic. This may have been a function of serious tissue damage, leading to increased cell membrane permeability, and excretion of arsenic by simple diffusion. Consistent with our observations, Calamari, Gaggino, and Pacchetti (1982) reported a marked drop in tissue metal concentrations after the exposure was terminated. Taken together, these observations suggest that fish are able to regulate the concentrations of metals in their organs over time through the processes of absorption, excretion, detoxification, and storage (Chen, Yu, & Liu, 2001).

Subathra and Karuppasamy (2008) and Kalay and Canli (2000) demonstrated that the rate of Cu accumulation was higher than the rate of Cu elimination in fish. A similar pattern was reported by Liao et al. (2004) for arsenic in tilapia. In contrast, our results suggest that the opposite is true in Lanzhou catfish. k_2 was higher than k_1, suggesting that Lanzhou catfish can more easily eliminate As in their body tissue. Thus, we speculate that Lanzhou catfish have a more effective mechanism for arsenic metabolism than other teleosts.

The k_1 and k_2 values decreased with an increase in waterborne arsenic concentrations. In contrast, Dang, Zhong, and Wang (2009) concluded that k_1 and k_2 are not significantly affected by the concentration and exposure duration. Luoma and Rainbow (2005) suggested that the assumption of a constant uptake rate might only be applicable in natural conditions or up to concentrations that are an order of magnitude higher than those seen in nature. The uptake rate is likely subject to saturation kinetics at very high concentrations because most of the metal ions pass through the gill membranes via channels or other facilitated diffusion processes (Green, Mirza, Wood, & Pyle, 2010). Tsai et al. (2012) observed a decrease in the values of k_1 and BCF with an increase in waterborne copper concentrations. Kraemer, Campbell, and Hare (2008) showed that fish can alter their ability to uptake and eliminate metals over a longer time period or at higher concentrations in a field situation. Tsai, Huang, Chen, and Liao (2012) suggested that if the tissue arsenic concentration exceeds the metal influx threshold (CIT) during short-term exposure conditions, the detoxification rate (k_{dex}) will increase with the waterborne metal concentration, whereas if the reverse is true, the value of k_{dex} will tend to be zero. The authors also demonstrated that BCF decreased as waterborne arsenic concentrations (C_w) increased ($C_w > 5.1$ μg/mL). McGeer et al. (2003) comprehensively reviewed theoretical and experimental data describing BCF for Cu, Zn, Cd, Pb, and other metals, and found an inverse relationship between BCF and aqueous exposure concentrations. Liao et al. (2003) noted that tilapia held in the field had a higher tendency to accumulate As (BCF = 143–421) than did fish from the same population in a 7-d lab bioaccumulation assay (BCF = 1.04–4.19). The lab group was exposed waterborne As concentrations that were ~30 times higher than the field group. Our results also showed an inverse relationship between BCF and exposure concentration. These observations are consistent with a number of studies that have developed metal bioaccumulation models for fish species. The relatively high value of BCF obtained from exposure to lower metal concentrations may result from the active regulation or acclimation by the organism to metals.

Among the various tissues (gill, liver, brain, and muscle) we evaluated for depuration of arsenic in Lanzhou catfish, the half-life was lowest for the gill, followed by brain, muscle, and liver. According to Anderson and Spear (1980), the gills of pumpkin seed sunfish exhibit monophasic elimination of accumulated Cu. Metals that are accumulated in the gills may either be transferred back to water (adsorbed metals) or transferred to other tissues, particularly the liver, for detoxification (Subathra & Karuppasamy, 2008). This may explain why metals have a shorter biological half-life in the gills than in the liver. Together, these observations support the view that higher BCF values indicate that species or organ is more effective at the removal of pollutants (Kara & Zeytunluoglu, 2007). Muscle tissue is thought to be particularly inefficient at metal elimination. For example, muscle Cu concentrations of *Mystus vittatus* were not significantly lowered during a 30-d depuration period (Subathra & Karuppasamy, 2008). This lack of elimination by fish muscle explains why 90% of total arsenic in food comes from fish and other seafood.

Within an individual fish, the kinetics of metal accumulation and release are expected to be very complex because physical and chemical parameters, water temperature, salinity, diet, fish species, and many other parameters may affect the rate of metal accumulation and release in aquatic animals (Lemus & Chung, 1999).

5. Conclusion

Our results demonstrate that the target organs of Lanzhou catfish are capable of regulating arsenic toxicity by way of internal regulation mechanisms, and the rate of arsenic uptake and depuration over time are concentration- and tissue-dependent.

Funding

This work was supported by the External Cooperation Projects of Ningxia.

Competing Interests

The authors declare no competing interest.

Author details

Zongqiang Lian[1,2]
E-mail: lianzq04@163.com
Xudong Wu[1,2]
E-mail: 838873204@qq.com
[1] Ningxia Fisheries Research Institute, Yinchuan 750001, China.
[2] Ningxia Engineering Research Center for Fisheries, Yinchuan 750001, China.

References

Anderson, P. D., & Spear, P. A. (1980). Copper pharmacokinetics in fish gills—1 kinetics in pumpkinseed sunfish, lepomis gibbosus, of different body sizes. *Water Research, 14*, 1101–1105. https://doi.org/10.1016/0043-1354(80)90159-1

Calamari, D., Gaggino, G. F., & Pacchetti, G. (1982). Toxicokinetics of low levels of Cd, Cr, Ni and their mixture in long-term treatment on *Salmo gairdneri* Rich. *Chemosphere, 11*, 59–70. https://doi.org/10.1016/0045-6535(82)90094-7

Chen, C. M., Yu, S. C., & Liu, M. C. (2001). Use of Japanese medaka (*Oryzia latipes*) and tilapia (*Oreochromis mossambicus*) in toxicity tests on different industrial effluents in Taiwan. *Archives of Environmental Contamination and Toxicology, 40*, 363–370. https://doi.org/10.1007/s002440010184

Chou, B. Y., Liao, C. M., Lin, M. C., & Cheng, H. H. (2006). Toxicokinetics/toxicodynamics of arsenic for farmed juvenile milkfish *Chanos chanos* and human consumption risk in BFD-endemic area of Taiwan. *Environment International, 32*, 545–553. https://doi.org/10.1016/j.envint.2006.01.004

Daglish, R. W., & Nowak, B. F. (2002). Rainbow trout gills are a sensitive biomarker of short-term exposure to waterborne copper. *Archives of Environmental Contamination and Toxicology, 43*, 98–102. https://doi.org/10.1007/s00244-002-1184-5

Dang, F., Zhong, H., & Wang, W. X. (2009). Copper uptake kinetics and regulation in a marine fish after waterborne copper acclimation. *Aquatic Toxicology, 94*, 238–244. https://doi.org/10.1016/j.aquatox.2009.07.011

De Gieter, M., Leermakers, M., Van Ryssen, R., Noyen, J., Goeyens, L., & Baeyens, W. (2002). Total and toxic arsenic levels in north sea fish. *Archives of Environmental Contamination and Toxicology, 43*, 406–417. https://doi.org/10.1007/s00244-002-1193-4

Falusi, B. A., & Olanipekun, E. O. (2007). Bioconcentration factors of heavy metals in tropical crab (*carcinus sp*) from River Aponwe, Ado-Ekiti, Nigeria. *Journal of Applied Sciences and Environmental Management, 11*, 51–54.

Green, W. W., Mirza, R. S., Wood, C. M., & Pyle, G. G. (2010). Copper binding dynamics and olfactory impairment in fathead minnows (*Pimephale promelas*). *Environmental Science & Technology, 44*, 1431–1437. https://doi.org/10.1021/es9023892

Hwang, P. P., & Tsai, Y. N. (1993). Effects of arsenic on osmoregulation in the tilapia *Oreochromis mossambicus* reared in seawater. *Marine Biology, 117*, 551–558. https://doi.org/10.1007/BF00349765

Kalay, M., & Canli, M. (2000). Elimination of essential (Cu, Zn) and nonessential (Cd, Pb) metals from tissues of a freshwater fish Tilapia zilli. *Turkish Journal of Zoology, 24*, 429–436.

Kara, Y., & Zeytunluoglu, A. (2007). Bioaccumulation of toxic metals (Cd and Cu) by *Groenlandia densa* (L.) Fourr. *Bulletin of Environmental Contamination and Toxicology, 79*, 609–612. https://doi.org/10.1007/s00128-007-9311-7

Kraemer, L. D., Campbell, P. G. C., & Hare, L. (2008). Modeling cadmium accumulation in indigenous yellow perch (*Perca flavescens*). *Canadian Journal of Fisheries and Aquatic Sciences, 65*, 1623–1634. https://doi.org/10.1139/F08-081

Lee, J. H., Landrum, P. F., & Koh, C. H. (2002). Prediction of time-dependent PAH Toxicity in *Hyalella azteca* using a damage assessment model. *Environmental Science & Technology, 36*, 3131–3138. https://doi.org/10.1021/es011202d

Lemus, M. J., & Chung, K. S. (1999). Effect of temperature on copper toxicity, accumulation and purification in tropical fish juveniles Petenia kraussii. *Caribbean Journal of Science, 35*, 64–69.

Liao, C. M., & Lin, M. C. (2001). Acute toxicity modeling of rainbow trout and silver sea bream exposed to waterborne metals. *Environmental Toxicology, 16*, 349–360. https://doi.org/10.1002/(ISSN)1522-7278

Liao, C. M., Chen, B. C., Singh, S., Lin, M. C., Liu, C. W., & Han, B. C. (2003). Acute toxicity and bioaccumulation of arsenic in tilapia (*Oreochromis mossambicus*) from a blackfoot disease area in Taiwan. *Environmental Toxicology, 18*, 252–259. https://doi.org/10.1002/(ISSN)1522-7278

Liao, C. M., Tsai, J. W., Ling, M. P., Liang, H. M., Chou, Y. H., & Yang, P. T. (2004). Organ-specific toxicokinetics and dose? Response of arsenic in tilapia *Oreochromis mossambicus*. *Archives of Environmental Contamination and Toxicology, 47*, 502–510. https://doi.org/10.1007/s00244-004-3105-2

Luoma, S. N., & Rainbow, P. S. (2005). Why is metal bioaccumulation so variable? Biodynamics as a unifying concept. *Environmental Science & Technology, 39*, 1921–1931. https://doi.org/10.1021/es048947e

McGeer, J. C., Brix, K. V., Skeaff, J. M., DeForest, D. K., Brigham, S. I., Adams, W. J., & Green, A. (2003). Inverse relationship between bioconcentration factor and exposure concentration for metals: Implications for hazard assessment of metals in the aquatic environment. *Environmental Toxicology and Chemistry, 22*, 1017–1037. https://doi.org/10.1002/etc.v22:5

Petrusevski, B., Sharma, S., Schippers, J. C., & Shordt, K. (2007). *Arsenic in drinking water* (Thematic Overview Paper 17). IRC International Water and Sanitation Centre.

Schuler, L. J., Landrum, P. F., & Lydy, M. J. (2004). Time-dependent toxicity of fluoranthene to freshwater invertebrates and the role of biotransformation on lethal body residues. *Environmental Science & Technology, 38*, 6247–6255. https://doi.org/10.1021/es049844z

Storelli, M. M., & Marcotrigiano, G. O. (2000). Total organic and inorganic arsenic from marine turtles (*Caretta caretta*) beached along the Italian Coast (South Adriatic Sea). *Bulletin of Environmental Contamination and Toxicology, 65*, 732–739. https://doi.org/10.1007/s0012800184

Storelli, M. M., Ceci, E., & Marcotrigiano, G. O. (1998). Distribution of heavy metal residues in some tissues of *Caretta caretta* (Linnaeus) specimen beached along the Adriatic Sea (Italy). *Bulletin of Environmental Contamination and Toxicology, 60*, 546–552. https://doi.org/10.1007/s001289900660

Subathra, S., & Karuppasamy, R. (2008). Bioaccumulation and depuration pattern of copper in different tissues of *Mystus vittatus*, related to various size groups. *Archives of Environmental Contamination and Toxicology, 54*, 236–244. https://doi.org/10.1007/s00244-007-9028-y

Suhendrayatna, Ohki, A., Nakajima, T., & Maeda, S. (2002). Studies on the accumulation and transformation of arsenic in freshwater organisms I. Accumulation, transformation and toxicity of arsenic compounds on the Japanese Medaka, *Oryzias latipes*. *Chemosphere, 46*, 319–324. https://doi.org/10.1016/S0045-6535(01)00084-4

Sures, B. (2003). Accumulation of heavy metals by intestinal helminths in fish: An overview and perspective. *Parasitology, 126*, S53–S60. https://doi.org/10.1017/S003118200300372X

Tišler, T., & Zagorc-Končan, J. (2002). Acute and chronic toxicity of arsenic to some aquatic organisms. *Bulletin of Environmental Contamination and Toxicology, 69*, 421–429.

Tsai, J. W., Chen, W. Y., Ju, Y. R., & Liao, C. M. (2009). Bioavailability links mode of action can improve the long-term field risk assessment for tilapia exposed to arsenic. *Environment International, 35*, 727–736. https://doi.org/10.1016/j.envint.2009.01.014

Tsai, J. W., Ju, Y. R., Huang, Y. H., Deng, Y. S., Chen, W. Y., Wu, C. C., & Liao, C-M. (2012). Toxicokinetics of tilapia following high exposure to waterborne and dietary copper and implications for coping mechanisms. *Environmental Science and Pollution Research, 20*, 3771–3780.

Tsai, J. W., Huang, Y. H., Chen, W. Y., & Liao, C. M. (2012). Detoxification and bioregulation are critical for long-term waterborne arsenic exposure risk assessment for tilapia. *Environmental Monitoring and Assessment, 184*, 561–572. https://doi.org/10.1007/s10661-011-1988-8

United States Environmental Protection Agency. (2000). *Understanding and accounting for method variability in whole effluent toxicity applications under the national pollutant discharge elimination system* (EPA833-R-00-003).

United States Food and Drug Administration. (1993). *Guidance document for arsenic in shellfish*. Washington, DC: Author. Retrieved from http://www.cfsan.fda.gov/%7Efrf/guid-as.html

Vijver, M. G., van Gestel, C. A., Lanno, R. P., van Straalen, N. M., & Peijnenburg, W. J. (2004). Internal metal sequestration and its ecotoxicological relevance: A review. *Environmental Science & Technology, 38*, 4705–4712. https://doi.org/10.1021/es040354g

Wang, S., & Xie, Y. (2009). *China species red list, Vol. 2: Vertebrates part 1* (pp. 294–295). Beijing: Higher Education Press.

Yamauchi, H., & Yamamura, Y. (1985). Metabolism and excretion of orally administrated arsenic trioxide in the hamster. *Toxicology, 34*, 113–121. https://doi.org/10.1016/0300-483X(85)90161-1

Zhou, Y. X., & Zhang, Z. S. (1989). *Biological Aquatic toxicity test method* (pp. 11–27). Beijing: China Agriculture Press.

14

Recovery of proliferative cells up to 15- and 49-day postmortem from bovine skin stored at 25°C and 4°C, respectively

Brian Walcott[1] and Mahipal Singh[1*]

*Corresponding author: Mahipal Singh, Animal Biotechnology Program, Agricultural Research Station, Fort Valley State University, Fort Valley, GA 31088, USA

E-mail: singhm@fvsu.edu

Reviewing editor: Rajni Hatti Kaul, Lunds Universitet, Sweden

Additional information is available at the end of the article

Abstract: The objective of this study was to assess the time limits within which proliferative cells can be recovered in bovine skin stored separately at 4 and 25°C after animal death. In the first experiment, skin explants ($n = 110$; 2–3 mm^2) from 11 animals stored at 4°C were cultured weekly up to 7 weeks in Dulbecco's modified Eagle medium (DMEM) supplemented with 10% fetal bovine serum (FBS), 50 units/mL of penicillin, 50 µg/mL of streptomycin, and 2.5 µg/mL of fungizone. The presence/absence of fibroblast-like cell outgrowth around explants was scored. Out of 640 explants cultured, 567 (87%) adhered to dish surface, of which 333 (58.73%) exhibited outgrowth including 16.67% explants from 49 days postmortem tissues. Similarly, in the second experiment, when the tissues were stored at 25 ± 2°C prior to culturing on alternate days up to 17 days, 204 (48%) explants exhibited outgrowth that included 19.15% from 15 dpm tissues. The number of explants exhibiting outgrowth was inversely proportional to postmortem time interval in both temperatures studied. Secondary cultures established from outgrowth for selected time points showed stable chromosomes, normal GFP gene expression, and comparable growth morphology to fresh tissue-derived cells. The cells lasted in

ABOUT THE AUTHORS

Mahipal Singh is associate professor of Animal Biotechnology at the Fort Valley. He teaches undergraduate and graduate students and has an active research program in animal biotechnology. His research interests include genetic engineering of livestock, stem cells, and animal cloning. The group utilizes molecular biological, reproductive, surgical and cell culture methods in their research. Recent focus of his laboratory has been to understand the time limits of cell/tissue survival after the death of an animal in various temperature regimes and their survival mechanisms. The interest is to recover live stem cells from postmortem tissues and their potential for cloning dead animals and to use for cellular therapies in humans. Recently, his laboratory has undertaken collaborative studies on understanding dynamics of peri-implantation phase of ruminant reproduction as relates to the pregnancy recognition genes. Brian Walcott is a former MS graduate student and is currently pursuing MD degree program.

PUBLIC INTEREST STATEMENT

Death is an ultimate destination of life. Or is it? When we die, what happen to the individual cells, of which, we are made up of? Are they all dead as the blood circulation stops and the oxygen, which is our life line, is no more transported to the cells? Or are they still alive for some time, and if yes, for how long, are the questions that always puzzled scientists. This research paper addresses some of these questions and demonstrates that individual cells in mammalian tissues are alive for much longer time than was previously thought. This study further demonstrates that these live cells retain normal characters and can be stored frozen for long time. This research highlights the huge potential of utilization of postmortem tissues, for the recovery of stem cells, for cellular therapies in human and veterinary medicine and preservation of germplasm to meet the climatic challenges to feed the expanding world by cloning food animals.

culture for more than 20 passages. These results suggest that live and usable cells can be recovered from bovine skin tissues up to about 2 weeks postmortem, if skin is stored at 25°C, and about three times more (>6 weeks), if stored at 4°C.

Subjects: General Science; Biology; Biotechnology; Biodiversity & Conservation

Keywords: bovine; fibroblast; skin stem cells; cell culture; postmortem cell recovery; cryopreservation; animal cloning

1. Introduction

Angus is a beef-producing breed of cattle, which is said to have been developed from native cattle of Aberdeen shire and Angus in Scotland and thus is known as Aberdeen Angus in most parts of the world (Briggs & Briggs, 1980). They are naturally polled (hornless) animals and are found in black or red colors. They were introduced in United States in 1873 by George Grant. He brought four Angus bulls without cows from Scotland for crossbreeding with native cattle herds. Subsequently, many more animals of both sexes were imported by US farmers, due to their desirable qualities. Black Angus is the most common breed today. There are 320,362 heads of Angus cattle registered in 2015 in the United States (http://www.angus.org/pub/faqs.aspx).

Cloning of animals including Angus by nuclear transfer is an important milestone in agricultural biotechnology. Almost every livestock species have now been cloned (https://en.wikipedia.org/wiki/List_of_animals_that_have_been_cloned). Cloning requires fusion of desired somatic cells (nuclear donor) with an enucleated oocyte cytoplasm. Therefore, the preservation of somatic cells/tissues from elite animals has been suggested, so as to utilize these resources globally. This will help to meet the climatic and/or other challenges in future by cloning of these animals to supply the global demand for protein. Additionally, cloning of animals with desirable meat qualities after postmortem carcass evaluation is gaining importance among producers. Recent studies show effective preservation of postmortem tissues and production of live animals from these tissues by SCNT *aka* cloning (Ogura, Inoue, & Wakayama, 2013). However, for the success of a cloning experiment, nuclear integrity of donor cells is a requirement (Hoshino et al., 2009). *In vitro* culture of cells is one of the methods to ensure nuclear integrity (Mastromonaco, Perrault, Betts, & King, 2006). *In vitro* culture of cells from both live and dead animal tissues preserved at subzero temperatures has been reported (Erker et al., 2010; Hoshino et al., 2009; Palmer et al., 2001; Viel, McManus, Cady, Evans, & Brewer, 2001; Wakayama et al., 2008) and the cells have been used to clone the animals even after many years of their death (Hoshino et al., 2009; Wakayama et al., 2008). However, in all these cases, the tissues were preserved within few hours of animal death. Delays in preservation may compromise cloning efficiency and quality of cloned offspring. There are only limited studies to show as how long the live and culturable cells can be recovered from postmortem tissues in mammalian species and, how different these cells are, compared to fresh tissue-derived cells. Literature survey shows that neural stem/progenitor cells were cultured from postmortem rat brains, stored at 4°C for a week (Xu, Kimura, Matsumoto, & Ide, 2003). Fibroblasts were recovered in postmortem skin tissues stored at 4°C up to 14 days in rabbits and pigs (Silvestre, Saeed, Cervera, Escribá, & García-Ximénez, 2003), 12 days in goat, and sheep (Silvestre, Sánchez, & Gómez, 2004), and up to 41 days in goats (Okonkwo & Singh, 2015). Muscle stem cells have been shown to survive up to 17 days postmortem in human beings and up to 16 days postmortem in mice (Latil et al., 2012). To our knowledge, there are only two studies in bovine; in first study, fibroblasts were cultured up to 12 days from skin stored at 4°C (Silvestre et al., 2004), and in another muscle and cartilage, cells were cultured up to 9 days of postmortem tissue storage at 4°C (Caputcu, Akkoc, Cetinkaya, & Arat, 2012). Here, we show *in vitro* culture of fibroblast-like cells up to 49 days in bovine steer skin tissues, if the tissue is stored at 4°C, and up to 15 days, if stored at 25 ± 2°C after the animal death. We further show that these cells are cytogenetically stable with normal karyotype, have comparable growth profile, can be recultured post-cryopreservation, and transfected to express GFP gene from a plasmid, suggesting their potential utility in postmortem cryopreservation for animal cloning and/or cell therapy.

2. Materials and methods

2.1. Sample preparation and explant culture

The tissue samples were procured from the USDA-inspected university slaughter house. The animals were not slaughtered for these experiments but were routine food animals for human consumption, and thus, there were no ethical violations. Animal ears of slaughtered animals were excised from the head and brought to the laboratory. Ear skin was cleaned first with tape water and then with 70% alcohol swabs, wrapped in clean sterile paper towels, and stored in plastic bags in the laboratory, either at room temperature (25 ± 2°C) or in refrigerator set at 4°C. After different days of postmortem storage, small skin samples were excised aseptically from inner side of the ear and chopped into ten small explants (2–3 mm^2 size) and adhered onto two 60-mm-diameter dishes (Falcon, BD Biosciences, Oxnard, CA) for each animal individually. Explants were cultured in DMEM supplemented with 10% FBS, 50 units/mL of penicillin, 50 µg/mL of streptomycin and 2.5 µg/mL of fungizone at 37°C, 5% CO_2 in a humidified environment. Media were changed once a week and the dishes were observed for any microbial or fungal contamination, explant dislodging, and for the outgrowth of fibroblast-like cells under an inverted microscope. Dislodged explants and contaminated dishes were removed as soon as observed. The presence or absence of the outgrowth around each explant was recorded. Any outgrowth containing a cluster of more than 50 cells was considered positive.

2.2. Establishment of secondary cultures and cryopreservation of cells

Primary outgrowing cells on dish surface were trypsinized at 70–90% confluence and secondary cultures established as described (Singh & Ma, 2014). Briefly, the cells in dishes were washed twice with 3.0 mL of the balanced salt solution without calcium and magnesium (Gibco, Carlsbad, CA) and incubated with 2.0 mL of 0.125% trypsin for 5–10 min at 37°C. The trypsinized cells were neutralized with 5 vol. of growth media, counted to assess cell viability using Trypan Blue Dye Exclusion Method (Strober, 2001), and pelleted at 200 × g for 7 min. The cells were resuspended in Synth-a-Freeze® (Life Technologies Corp., Carlsbad, CA) media, aliquoted into cryogenic storage vials (1.0×10^5 cells/vial) and frozen at −80°C o/n using Nalgene™ Cryo 1°C Freezing Container (Nalgene, Rochester, NY). The vials were transferred to liquid nitrogen tank after 24 h and stored till used. Representative cryovials were thawed and postfreezing cell-viability percentage was determined at least after a week of storage. To establish secondary cultures and to expand them, the frozen vials were quickly thawed at 37°C, mixed slowly with 10 vol. of the media, pelleted at 200 × g for 7 min, dissolved in growth media, and cultured in appropriate (T25 or T75) culture flasks as described (Singh & Ma, 2014).

2.3. Transfection and GFP expression

Transfection of 42 dpm cell line (p5) was performed as per the vender instructions using a transfectamine 3000 kit from Invitrogen (Life Technologies Inc., CA). GFP containing plasmid pcDNA3.1-GFP:NT was prepared using Endo-Free plasmid kit (Qiagen) and the plasmid DNA was quantified using Nanodrop 2000 (Fisher Scientific, Wilmington, DE, USA). GFP gene expression was observed under UV light in EVOS Cell Imaging System (Thermo Fisher Scientific, Atlanta, GA) using GFP filter and the appropriate images captured using their software.

2.4. Karyotype analysis

Bovine cells from 42 dpm cell line at p5 were processed for karyotyping using GTL Banding technique as previously established (Meisner & Johnson, 2008) at Cell-line Genetics (Madison, WI; www.clgenetics.com). Chromosome assignments were made according to the Atlas of Mammalian Cytogenetics (O'Brien, Menninger, & Nash, 2006).

2.5. Cell passaging

Bovine cell cultures were splitted (1:3) after each subculture at 70–80% confluence until the cells stopped growing as described earlier (Freshney, 2000). In brief, the spent culture media from the cell culture dish were aspirated and washed twice with phosphate-buffered saline (PBS). The cells on surface were treated with 1.0 mL of trypsin-EDTA and incubated for 5–10 min at 37°C till the cells detached completely from the dish surface. Once the cells detached as observed under microscope, the complete media were added to the dish. The volume was made to 10 mL, and the cells were harvested at 200 × g for 7 min. The pellet was dissolved in 0.5–1.0 mL of complete media, the cells counted if needed, and otherwise splitted into one-third and cultured in new dishes. This procedure was repeated till cells stopped growing.

3. Results

3.1. Effect of postmortem time interval on in vitro culture of cells from tissues stored at 4°C

In order to test the time limit, within which proliferative cells can be recovered after animal death, we analyzed a total of 640 skin explants for *in vitro* culture. These explants were excised from 11 individual animal ears that were stored at 4°C. A subset of explants (*n* = 110) was cultured after different storage time intervals, i.e. 0, 7, 14, 21, 28, 35, 42 and 49 days postmortem. The outgrowing cells in cultured dishes were observed weekly until they reached > 50% confluence (Figure 2, panel B). The presence or absence of cell outgrowth around each adhered explant for each time interval was recorded. We observed outgrowth around tissues in all the time points studied including 16.67% of 49 dpm tissues (Table 1). Out of 640 explants cultured, 567 (87.23%) adhered to dish surface, of

Table 1. Outgrowth of fibroblast-like cells around skin explants stored at 4°C for different days postmortem

Steer number	Date of culture initiation	# of explants exhibiting outgrowth/# of explants adhered to dish surface									
		0 dpm	7 dpm	14 dpm	21 dpm	28 dpm	35 dpm	42 dpm	49 dpm	Total	%
S1	25 February 2014	10/10	07/07	01/01	05/05	n/d	n/d	n/d	n/d	23/23	100
S2	25 February 2014	10/10	10/10	08/08	10/10	08/08	07/07	n/d	n/d	53/53	100
S3	28 May 2014	10/10	10/10	06/06	03/10	–	–	–	n/d	29/36	80.56
S7	11 September 2014	10/10	10/10	03/05	00/10	00/10	00/10	00/10	n/d	23/65	35.38
S8	11 September 2014	10/10	04/10	05/05	00/10	00/10	00/10	00/10	00/10	19/75	25.33
S9	11 September 2014	10/10	10/10	05/10	01/05	00/05	n/d	n/d	n/d	26/40	65.00
S10	6 January 2015	05/05	05/05	10/10	10/10	04/05	05/10	00/10	00/10	39/65	60.00
S11	6 January 2015	10/10	10/10	06/10	03/05	03/05	01/10	04/10	05/10	42/70	60.00
S12	6 January 2015	10/10	08/10	08/10	07/10	00/10	00/10	n/d	n/d	33/60	55.00
S13	21 January 2015	10/10	n/d	00/10	n/d	07/10	n/d	00/10	n/d	17/40	42.50
S14	21 January 2015	10/10	n/d	06/10	n/d	09/10	n/d	04/10	n/d	29/40	72.50
Total		105/105	74/82	58/85	39/75	31/73	13/57	08/60	05/30	**333/567**	
%		100	90.24	68.24	52.00	42.47	22.81	13.33	16.67	**58.73**	

Notes: Eleven Black Angus steers (2–3 years of age) were used in this study. For each time point, 10 explants (2–3 mm^2) were cultured onto two 60-mm dishes (5 explants/dish) for each animal. n/d, not done; –, both dishes contaminated during culture and thus removed from study and not included in analysis.

which 333 (58.73%) exhibited fibroblast-like outgrowth. We never observed outgrowth around tissues dislodged from dish surface within first 3 days of culture initiation. Outgrowth ranged from 13.33 to 100% in tissues stored for various time intervals and show a reduction trend with increasing postmortem time interval (Figure 1, panel A). Percent outgrowth in tissues from different animals varied and ranged from 25 to 100% (Figure 1, panel B).

3.2. *Effect of postmortem time interval on in vitro culture of cells from tissues stored at 25 ± 2°C*

Similar to tissues stored at 4°C, we also studied effect of room temperature (25 ± 2°C) storage on outgrowth. Skin explants (n = 100) from 10 animals were cultured after 0, 3, 6, 9, 13, 15 and 17 days of postmortem. Out of 425 explants adhered, 204 (48%) exhibited outgrowth including 19.15% of 15 dpm tissues (Table 2). Outgrowth ranged from 12.25 to 100% and exhibited a decreasing trend with increasing postmortem time interval as was observed for tissues stored at 4°C (Figure 1, panel C). Outgrowth in tissues from different animals varied and ranged from 14 to 93.33% (Figure 1, panel D).

3.3. *Cryopreservation, postfreezing cell morphology, and gene expression*

The dishes from selected time points were trypsinized at around 70% confluence to recover the primary cells and cryopreserved until used for further analysis. Secondary cultures from selected cell lines (0 dpm and 42 dpm) were then established by serial passaging. Secondary cultures of these cells grow much faster, as compared to the primary outgrowth, and reach 70–90% confluence in 5–7 days. The cells cryopreserved in DMEM with 10% DMSO retained >80% postfreezing cell viability and normal cell morphology, i.e. elongated, fibrous, and bipolar, which are typical characteristics of fibroblast cells (Figure 2, panel C–F). In order to test whether the postmortem tissue-derived cells have normal transcriptional machinery, we transfected 42 dpm cell line (p5) with pcDNA3.1-GFP: NT plasmid DNA vector. The transfected cells exhibited GFP gene expression when observed under UV light in an inverted fluorescent microscope (Figure 2, panel G).

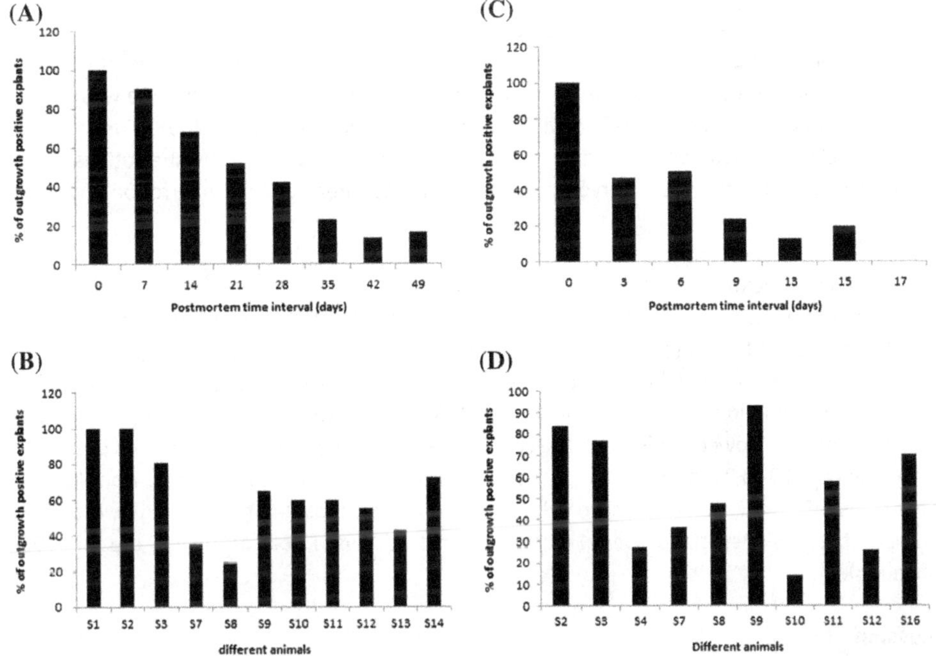

Figure 1. Recovery of fibroblast-like cells from stored tissues at two different temperatures: (A) outgrowth after different days of storage at 4°C; (B) outgrowth in individual animal tissues stored at 4°C; (C) outgrowth after different days of storage at 25 ± 2°C; (D) outgrowth in individual animal tissues stored at 25 ± 2°C.

Notes: Percentages in the figure were calculated by dividing number of outgrowth-positive explants by explants that adhered to dish surface and by multiplying by 100.

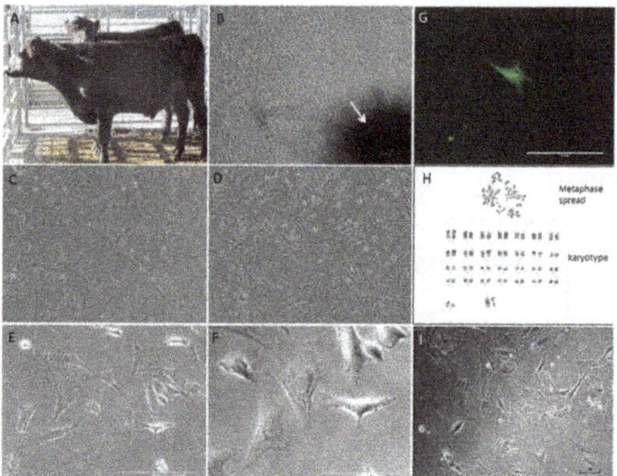

Figure 2. Postmortem recovery of fibroblast-like cells in cattle: (A) a sample of Black Angus just prior to slaughter; (B) migration of primary outgrowth around skin explant (arrow-marked dark-shaded area is explant) adhered on petridish surface after 10 days of culture (light microscopy, ×100 magnification); (C) & (D) low- (1×10^5 cells) and high (2×10^5 cells)-density postcryopreservation cultures of 42 dpm cell population (p5) after 24 h of culture (×100 magnifications); (E) & (F) ×200 and ×400 magnifications, respectively, of the cells in panel D; (G) GFP gene expression after 72 h of transfection in 42 dpm (p5) cultures (×200 magnification); (H) chromosome analysis at p5 level in 42 dpm fibroblast cell line (a representative metaphase cell spread and karyotype with 30 pairs of XX chromosomes are shown); (I) high passage (p25, on day 53) culture of 42 dpm cell line (×100 magnification but contrast correction of 56% to show better cell morphology).

Notes: The light microscopic images (B), (C), (D), and (I) were captured using a TS100-inverted microscope and DSL2 camera (Nikon), whereas (E) and (F) were captured using an EVOS microscope with a bright field filter. GFP image (panel G) was captured using EVOS fluorescent microscope using a GFP filter.

3.4. Cytogenetic stability

In order to determine if there was any genetic change in the cells derived from stored postmortem tissues, we performed a cytogenetic analysis on a cell line derived from tissues stored at 4°C until 42 days postmortem. The diploid number of chromosomes determined for this cell line was 60, XX, which is consistent with earlier studies in bovine (Iannuzzi, King, & Di Berardino, 2009). Furthermore, 20 G-banded metaphase cells were also analyzed to see if there were any genetic aberrations. All 20 cells demonstrated normal female karyotype without any apparent genetic aberration (Figure 2, panel H).

3.5. Proliferative life span

In order to determine as how long the postmortem tissue-derived cell populations can last in culture, we chose three cell lines, (a) 42 dpm tissue derived (4°C stored tissue), (b) 15 dpm tissue derived (25°C stored tissues) and, (c) 0 dpm (control; fresh tissue derived). We observed that these cell lines proliferated well in initial passages giving about 70% confluence in 5–7 days. However, after passage 15 the growth slowed exhibiting 70% confluence in 2–3 weeks. The cells stopped proliferating around passage 23 to 27. An example of the cell morphology on day 53 of culture from 42 dpm cell line is shown (Figure 2, panel I). These cells at later passages look flat, thinner, highly branched and comparatively bigger sized (see Figure 2, panel C/D vs I, all of which are at 100x magnification).

4. Discussion

The development of animal cloning technology and induction of pluripotent stem cells (iPSCs) from somatic cells in last two decades has revolutionized not only the agricultural biotechnology but also biomedical and veterinary field, benefitting both agriculture and human health. However, for the proper utilization of somatic cells, it is essential that they retain some ability of proliferation ensuring the integrity of genetic material within the cell. Biopsies taken from live animals or immediately after animal death, cultured within few hours, are preferred somatic cells for cloning. However, in many instances, it is not possible to set up cultures within few hours. It is not precisely known as how long

Table 2. Outgrowth of fibroblast-like cells around skin explants stored at room temperature (25 ± 2°C) for different days postmortem

Steer number	Date of culture initiation	# of explants with outgrowth/# adhered onto dish surface							Total	%
		0 dpm	3 dpm	6 dpm	9 dpm	13 dpm	15 dpm	17 dpm		
S2	25 February 2014	09/09	n/d	n/d	09/09	04/04	09/09	00/06	31/37	83.78
S3	28 May 2014	10/10	03/05	07/10	03/05	–	–	–	23/30	76.67
S4	10 June 2014	10/10	03/10	00/10	00/05	00/10	00/03	n/d	13/48	27.08
S7	11 September 2014	10/10	03/10	09/10	00/10	00/10	00/10	n/d	22/60	36.67
S8	11 September 2014	10/10	02/10	07/10	00/10	–	–	–	19/40	47.5
S9	11 September 2014	10/10	10/10	08/10	–	–	–	–	28/30	93.33
S10	6 January 2015	05/05	02/05	00/10	00/10	00/10	00/10	n/d	7/50	14.00
S11	6 January 2015	10/10	03/10	09/10	02/05	02/05	00/05	n/d	26/45	57.78
S12	6 January 2015	10/10	00/05	03/10	01/10	00/10	00/10	n/d	14/55	25.46
S16	24 February 2015	10/10	09/10	02/10	n/d	n/d	n/d	n/d	21/30	70.00
Total		94/94	35/75	45/90	15/64	06/49	09/47	00/06	204/425	
%		100	46.67	50.00	23.441	12.25	19.15	00.00	48.00	

Notes: For each time point, 10 explants (2–3 mm²) were cultured onto two 60 mm dishes (5 explants/dish) for each animal. A total of 10 Black Angus steer tissues were used. n/d, not done; –, both dishes contaminated during culture and thus removed from study.

the individual cells in a biological tissue are alive and retain culture potential, after it is excised from mammalian body. Therefore, the objective of this study was to determine the postmortem survival limits of bovine tissues stored at two different temperatures, i.e. 25 and 4°C. To study this phenomenon, a simple *in vitro* culture procedure, previously optimized for goat and sheep skin tissue (Singh & Ma, 2014) was utilized. Here, tiny skin explants are adhered onto plastic dish surfaces and are cultured in fibroblast culture media. The outgrowing cells are measured as cell survival data points. Using this procedure, a comprehensive *in vitro* culture study was conducted on 11 Black Angus Steer tissues for various time intervals postmortem.

Results show outgrowth of cells around bovine explants up to 49 days of postmortem storage at 4°C and up to 15 days after storage at 25 ± 2°C. In an earlier study, bovine skin fibroblasts were cultured only up to 12 days of postmortem storage at 4°C (Silvestre et al., 2004), while muscle and cartilage were cultured only up to 9 days of postmortem storage at 4°C (Caputcu et al., 2012). To our knowledge, this is the first report of *in vitro* culture of mammalian cells from postmortem tissues stored for such a long period of 49 days at 4°C and 15 days at 25°C, after animal death. We observed that the numbers of explants exhibiting outgrowth decreased with increasing postmortem storage time interval in both the temperatures studied (Figure 1). It is likely that the number of viable adult skin stem cells decreased in tissues with an increased postmortem time interval, which could explain the reduction with increasing postmortem time interval. These results are in agreement with a previous similar study in goats (Singh, Ma, & Sharma, 2012). Furthermore, it was interesting to learn that such a long postmortem time lapse did not affect cytogenetic stability of cells, since all 20 metaphase cell spreads analyzed had normal chromosomes. Postmortem stored tissue-derived cell populations were also competent for transfection and expressed GFP gene in their cytoplasm from a plasmid DNA, suggesting that the transcriptional and translational machinery of the cells are intact. These cell populations were recultured after cryopreservation in liquid nitrogen. Results show that they last in *in vitro* cultures for more than 20 passages suggesting that postmortem stored tissue-derived cells can be utilized for long-term storage and could benefit animal agriculture and biomedicine in future.

5. Conclusions

This study show that, (a) skin tissue of bovine stored at 25 ± 2°C and 4°C has potential for *in vitro* culture up to 15- and 49-day postmortem, respectively, (b) the number of explants with outgrowth decreases with increasing postmortem time interval, and (c) that the cell populations derived from tissues stored over a month of postmortem are genetically stable, have similar transcriptional and translational machinery, and can be easily re-cultured *in vitro* after cryopreservation. Future studies should reveal the potential of these cells for reprogramming, cloning of animals and for cell therapies.

Acknowledgements

This work is part of MS thesis of BW. We would like to thank NSF for graduate assistantship to BW through an educational grant. X. Ma and N. Degala are thankfully acknowledged for technical assistance. Part of this work was presented in the Society for In Vitro Biology Meeting held at San Diego, CA, June 13, 2016 (abstract No: A3000).

Funding

This work was partly supported by a USDA-NIFA Capacity Building Teaching [grant number 2011-38821-30910] to MS.

Competing Interests

The authors declare no competing interest.

Author details

Brian Walcott[1]

E-mail: bwalcott1@live.com

Mahipal Singh[1]

E-mail: singhm@fvsu.edu

[1] Animal Biotechnology Program, Agricultural Research Station, Fort Valley State University, Fort Valley, GA 31088, USA.

References

Briggs, H. M., & Briggs, D. M. (1980). *Modern breeds of livestock* (4th ed.). New York, NY: Macmillan publishing Company.

Caputcu, A. T., Akkoc, T., Cetinkaya, G., & Arat, S. (2012). Tissue cryobanking for conservation programs: Effect of tissue type and storage time after death. *Cell Tissue Bank, 14*, 1–10.

Erker, L., Azuma, H., Lee, A. Y., Guo, C., Orloff, S., Eaton, L., ... Grompe, M. (2010). Therapeutic liver reconstitution with murine cells isolated long after death. *Gastroenterology, 139*, 1019–1029. https://doi.org/10.1053/j.gastro.2010.05.082

Freshney, R. I. (2000). *Culture of animal cells: A manual of basic techniques*. Hoboken, NJ: Wiley-Liss.

Hoshino, Y., Hayashi, N., Taniguchi, S., Kobayashi, N., Sakai, K., Otani, T., ... Saeki, K. (2009). Resurrection of a bull by cloning from organs frozen without cryoprotectant in a −80°C freezer for a decade. *PLoS ONE, 4*, e4142. https://doi.org/10.1371/journal.pone.0004142

Iannuzzi, L., King, W. A., & Di Berardino, D. (2009). Chromosome evolution in domestic bovids as revealed by chromosome banding and fish-mapping techniques. *Cytogenetic Genome Research, 126*, 49–62. https://doi.org/10.1159/000245906

Latil, M., Rocheteau, P., Châtre, L., Sanulli, S., Mémet, S., Ricchetti, M., ... Chrétien, F. (2012). Skeletal muscle stem cells adopt a dormant cell state post mortem and retain regenerative capacity. *Nature Communications, 3*, 903. https://doi.org/10.1038/ncomms1890

Mastromonaco, G. F., Perrault, S. D., Betts, D. H., & King, W. A. (2006). Role of chromosome stability and telomere length in the production of viable cell lines for somatic cell nuclear transfer. *BMC Developmental Biology, 6*, 41. https://doi.org/10.1186/1471-213X-6-41

Meisner, L. F., & Johnson, J. A. (2008). Protocols for cytogenetic studies of human embryonic stem cells. *Methods, 45*, 133–141. https://doi.org/10.1016/j.ymeth.2008.03.005

O'Brien, S. J., Menninger, J. C., & Nash, W. G. (2006). *Atlas of mammalian chromosomes* (pp. 653–655). Hoboken, NJ: Wiley. https://doi.org/10.1002/0471779059

Ogura, A., Inoue, K., & Wakayama, T. (2013). Recent advancements in cloning by somatic cell nuclear transfer. *Philosophical Trans Royal Society of London B Biological Sciences, 368*, 20110329.

Okonkwo, C., & Singh, M. (2015). Recovery of fibroblast-like cells from refrigerated goat skin up to 41 d of animal death. *In Vitro Cellular & Developmental Biology - Animal, 51*, 463–469. https://doi.org/10.1007/s11626-014-9856-9

Palmer, T. D., Schwartz, P. H., Taupin, P., Kaspar, B., Stein, S. A., & Gage, F. H. (2001). Cell culture. Progenitor cells from human brain after death. *Nature, 411*, 42–43. https://doi.org/10.1038/35075141

Silvestre, M. A., Saeed, A. M., Cervera, R. P., Escribá, M. J., & García-Ximénez, F. (2003). Rabbit and pig ear skin sample cryobanking: Effects of storage time and temperature of the whole ear extirpated immediately after death. *Theriogenology, 59*, 1469–1477. https://doi.org/10.1016/S0093-691X(02)01185-8

Silvestre, M. A., Sánchez, J. P., & Gómez, E. A. (2004). Vitrification of goat, sheep, and cattle skin samples from whole ear extirpated after death and maintained at different storage times and temperatures. *Cryobiology, 49*, 221–229. https://doi.org/10.1016/j.cryobiol.2004.08.001

Singh, M., & Ma, X. (2014). In vitro culture of fibroblast-like cells from sheep ear skin stored at 25–26°C for 10 days after animal death. *International Journal of Biology, 6*, 96–102.

Singh, M., Ma, X., & Sharma, A. (2012). Effect of postmortem time interval on *in vitro* culture potential of goat skin tissues stored at room temperature. *In Vitro Cellular and Developmental Biology- Animal, 48*, 478–482. https://doi.org/10.1007/s11626-012-9539-3

Strober, W. (2001). Trypan blue exclusion test of cell viability. *Current protocols in immunology*. Appendix 3: Appendix 3B. doi:10.1002/0471142735.ima03bs111

Viel, J. J., McManus, D. Q., Cady, C., Evans, M. S., & Brewer, G. J. (2001). Temperature and time interval for culture of postmortem neurons from adult rat cortex. *Journal of Neuroscience Research, 64*, 311–321. https://doi.org/10.1002/(ISSN)1097-4547

Wakayama, S., Ohta, H., Hikichi, T., Mizutani, E., Iwaki, T., Kanagawa, O., & Wakayama, T. (2008). Production of healthy cloned mice from bodies frozen at −20 C for 16 years. *Proceedings of the National Academy of Sciences, 105*, 17318–17322. https://doi.org/10.1073/pnas.0806166105

Xu, Y., Kimura, K., Matsumoto, N., & Ide, C. (2003). Isolation of neural stem cells from the forebrain of deceased early postnatal and adult rats with protracted post-mortem intervals. *Journal of Neuroscience Research, 74*, 533–540. https://doi.org/10.1002/(ISSN)1097-4547

15

Echolocation calls of *Natalus primus* (Chiroptera: Natalidae): Implications for conservation monitoring of this species

Lida Sanchez[1,2]*, Christian R. Moreno[1] and Emanuel C. Mora[1,3]

*Corresponding author: Lida Sanchez, Faculty of Biology, Department of Human and Animal Biology, Havana University, Havana, Cuba; Division of Ecology and Evolutionary Biology, Graduate School of Life Sciences, Tohoku University, Sendai 980-8578, Japan

E-mail: liditasanchez89@gmail.com

Reviewing editor: Hynek Burda, Universitat Duisburg-Essen, Germany

Additional information is available at the end of the article

Abstract: *Natalus primus* constitutes one of the most vulnerable mammalian species of Cuba. Until now, only one extant population is known to live in one single cave in the westernmost part of Cuba, within the Guanahacabibes National Park. Over multiple trips, we recorded ultrasonic vocalizations from several individuals of this species. We found short, high frequency-modulated multiharmonic calls for *N. primus*; these could be used to identify this species in acoustic inventories conducted in Cuba. Identifying *N. primus* through their echolocation calls will allow conducting passive acoustic monitoring, constituting a noninvasive approach to study this vulnerable species without causing disturbances on its roosts and foraging areas.

Subjects: Environment & Agriculture; Bioscience; Environmental Studies & Management

Keywords: acoustic monitoring; echolocation; Guanahacabibes; *Natalus primus*

1. Introduction

Natalus primus (Cuban greater funnel-eared bat) is the largest species within Natalidae. Few studies had addressed issues from its natural history, most of them referring taxonomy (Tejedor, Silva-Taboada, & Rodríguez-Hernández, 2004; Tejedor, Tavares, & Silva-Taboada, 2005). This bat species

ABOUT THE AUTHORS

The author has been enrolled in bat bioacoustics for more than 5 years, learning about acoustic recordings and characterization of animal vocalizations. The author is interested in long monitoring programs on target species (endangered mammals), in order to know more about their ecology and contribute to their conservation. Acoustics constitutes a convenient non-harmful technique to assess the presence of some of these species, which sometimes avoid conventional trapping techniques. The research presented here was one of the first results of a conservation project dedicated to a vulnerable species: Cuban greater funnel-ear bat. The author and his co-workers intended to monitor this species through acoustic surveys, recognizing it through several acoustic variables and the frequency pattern of its echolocation calls. Scarce studies report information about the natural history of this species. The author's intentions are to use bioacoustics as a non-harmful technique to unravel information regarding the foraging sites and hunting strategies from this species.

PUBLIC INTEREST STATEMENT

Natalus primus is a vulnerable bat species reported only in a single cave in the westernmost part of Cuba. Scarce research has been conducted in this species, making difficult to design conservation plans to protect it. Approaching this species by means of acoustics might constitute a non-harmful way to investigate more about its natural history. In this research, we describe the features of the vocalizations emitted by this species and discussed how these sounds and its characterization might be used to identify *N. primus* among other bat species when conducting acoustic surveys. This could be useful to inform about the places selected by this species to perform its foraging activity as well as the foraging range relative to its roost. Other caves in the surrounding areas could be explored as well to look for the presence of *N. primus* using bioacoustics.

was thought to be extinct but was recently rediscovered in a cave in the westernmost part of Cuban archipelago (Tejedor, Silva-Taboada, & Rodríguez-Hernández, 2004), although fossil records suggest the species resided across the entire archipelago at one time (Silva-Taboada, 1979). This species was categorized as vulnerable (Dávalos & Mancina, 2010), regarding the unique extant population of this species in this remote area in Cuba. In spite of this, no direct conservation plans are taking over this species, neither a long- or short-term monitoring to check its population status. More studies regarding its natural history and environmental requirements are needed, in order to draw more accurate conservation plans for this bat species.

Bioacoustics, the study of sounds produced by animals, has been widely used for species identification on highly vocal species such as bats (Hughes et al., 2011), cetaceans (Oswald, Barlow, & Norris, 2003), frogs (Brandes, Naskrecki, & Figueroa, 2006), and birds (Anderson et al., 1996). Acoustic surveys constitute a good complement for traditional capturing methods (Kuenzi & Morrison, 1998), allowing the detection of otherwise elusive species and causing almost no disturbance in their habitats (Laiolo, Vögeli, Serrano, & Tella, 2007). Acoustic studies have helped to determine habitat selection (Russ & Montgomery, 2002) and identification of cryptic bat species (Davidson-Watts, Walls, & Jones, 2006). Description of species vocal repertoire opens the possibility of using this powerful tool in studies regarding ecology, population dynamics, and conservation (Laiolo, 2010). Several variables measured from these vocalizations have been used to distinguish species, such as duration, mean frequency, and inter-pulse intervals (Fukui, Agetsuma, & Hill, 2004). Recent studies also had included the frequency pattern of vocalizations to recognize species, using several methods such as cross-correlation and neural networks (Gaston & O'Neill, 2004; Mellinger & Clark, 1997). Using bioacoustics could constitute a successful way to make a noninvasive approach to study this vulnerable bat species.

We report here the first description of the echolocation calls of *N. primus*. This description together with the frequency pattern of its vocalizations will allow the identification of this species in future acoustic surveys.

2. Materials and methods

2.1. Study site and capture of individuals

We conducted our study in the site where *N. primus* was rediscovered: Cave La Barca (N 21°50′36.2″; W 084°45′57.5″) (Tejedor, Tavares, & Rodríguez-Hernández, 2005), located within Guanahacabibes National Park. This is a hot cave composed by five main galleries with 500 m of linear extension and multiple entrances (Tejedor, Tavares, & Rodríguez-Hernández, 2005). We visited this cave five times from February to September 2015, accorded to the validity of the permissions requested to enter the National Park and handle the animals (Permission number: PO2014/96 and PO2015/16). To know more about features of this cave, we measured two climatic variables, such as temperature and relative humidity, using HOBO data loggers (Onset Computer Corp. Bourne, MA) in June 2015 for three consecutive days at four locations. We took climatic measurements outside the entrance of the cave and in three galleries in the cave: in the most inner (heat Gallery), in an intermediate one and a gallery with a skylight.

Bats were captured with hand and mist nets; these were located very close to the walls at low heights (<1.5 m above the floor) in the heat gallery of this cave. Individuals were captured and placed in cloth bags to prevent bite and injuries among individuals. This is a very delicate species with broad wings and long tail membranes and becomes stressed very easily, thus, gentle handling is required. For each individual, we recorded morphological variables such as forearm length and body weight, sex, and reproductive condition.

2.2. Acoustic recordings

We recorded their echolocation calls by allowing the bats to freely fly within a small gallery in the cave where no other bats were already roosting. This gallery was similar to a narrow hall (rectangular shape), allowing to close the entrance of it with long cloth sheets to avoid the animals to escape.

Each individual was allowed to fly in this gallery and when two flight paths (full sequence of calls approaching the microphones) were obtained, individuals we allowed to fly out of the gallery by moving the cloth sheets. We used an array of microphones to obtain good quality recordings (obtaining vocalizations from different angles) and the construction of flight trajectories for this species. The array was arranged in a t-shape form, using four ultrasonic condenser microphones (Avisoft CMPA/CM16) coupled to an amplifier, connected to a laptop running Avisoft USGH software (Avisoft Bioacoustics, Germany). Microphone's gain was set in order to have a good signal to noise ratio in our recordings. We selected the channel containing calls with the highest amplitude to build the flight trajectories. We did not have flight paths containing echolocation calls from more than one individual, due that each individual was released inside the recording gallery after the previous one flew out. Flight trajectories were build using the software Bat3D (González, 2012) based on the time delays from the arrival of every emitted sound to each microphone, taking as reference the microphone located in the center of the array.

2.3. Call analysis

We conducted call analysis using two different softwares in order to characterize differently the echolocation calls. We intended to have a characterization from the whole call (all harmonics included) and from each harmonic separately. The channel containing calls with the highest amplitude was selected to make the call analysis, and only those calls above 20 dB relative to background noise, were analyzed. We used the automated detection feature of callViewer18 (Skowronski & Fenton, 2008) to characterize each harmonic independently from the echolocation calls. This is a custom echolocation sound analysis program written with MATLAB software (The MathWorks, Nadick, MA, USA). Detection parameter settings used with this software were: minimum link length = 10, Window size = 0.3, frame rate = 10,000, chunk size = 1, minimum energy = 12 dB, Echo filter hold = 6 dB, and window type = Blackman.

Call variables measured for each harmonic from every call were: duration (ms), minimum and maximum frequencies (kHz), and frequency of most energy (kHz). With these values, we calculated other variables such as bandwidth (kHz) (the difference between maximum and minimum frequencies) and slope rate (kHz/ms) (bandwidth divided by duration). We also used automatic measurement feature from Avisoft SASLab Pro software (Avisoft Bioacoustics, Germany) to characterize entire call. Here, signals were characterized—10 dB from the maximum amplitude peak in the power spectrum of each signal. Call variables measured were the same as in callViewer18.

3. Results

Fifteen adult individuals from *N. primus* were captured in the innermost parts of La Barca cave (e.g. Figure 1(A)). Their unique flight pattern resembling a butterfly at low heights bordering the walls of the caves allowed to distinguish this species from the rest of the bats inhabiting this cave. Forearm length ranged from 46.9 to 50 mm and body weight ranged from 7 to 9.5 g. The two pelage-color morphs were also observed from these captured individuals. Females were found lactating in September (Figure 1(B)). To confirm the precise period in which this species gave birth their

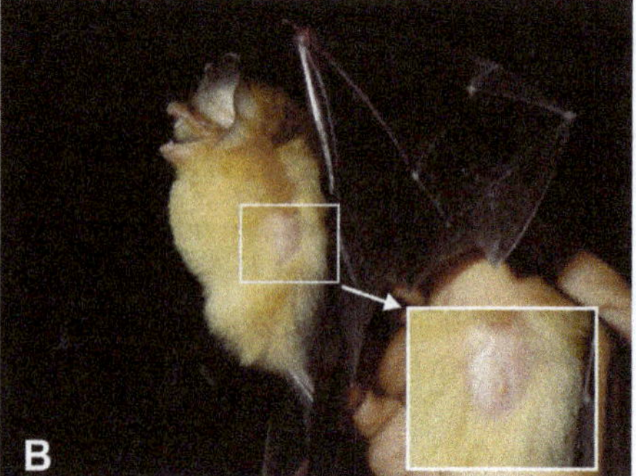

Figure 1. Individual of *Natalus primus* on the wing (A) A lactating female of *N. primus* captured in September 2015, Guanahacabibes National Park, Pinar del Rio, Cuba (B).

Photography: Emanuel C. Mora.

newborns was difficult. The inner gallery of cave La Barca in which most of the species concentrate is a guano swamp, making very difficult to explore it by foot. None of the captures females were carrying their babies.

We successfully recorded a total of 349 echolocation calls from 13 individuals. Two individuals were released after taking their morphological measurements because they were very stressed and we decided not to record their vocalizations. Flight trajectories were very similar among individuals (Figure 2(A)), each one of it contained more than 30 echolocation calls. These flight trajectories allowed to make a good selection of the flight sequences to be analyzed in the acoustic characterization, we show waveforms of some of these (Figure 2(B)). Those trajectories in which the animals were bordering the walls or flying in a direction to the end of the gallery were not analyzed.

Echolocation calls were downward frequency modulated (FM) of short durations, composed of three harmonics (Figure 2(C)). The first harmonic had in most of the cases a short constant frequency (CF) component located around 73 kHz (maximum frequency from the first harmonic) (Table 1), the second and third harmonic were completely frequency modulated with a major bandwidth, although the third harmonic was very faint in most of our recordings and no acoustic measurements could be done for it. The first harmonic was always detected, but the second harmonic was more evident, with the greatest amplitude, especially during the approach to microphone's array. Spectral overlap among harmonics was clear on some calls but was absent in others. Almost no difference was found in the mean value for duration from the whole call in relation to each harmonic (Table 1). However, for the other call variables, means differed and is understandable, due to the specificities in the location of measurement made on each sound analysis software.

Temperature and relative humidity profiles revealed stable values for several galleries of cave La Barca, especially regarding the relative humidity (Figure 3(B)–(D)), compared to the values outside the cave (Figure 3(A)). Although, a slight temperature fluctuation was observed (±1°C) in the gallery containing skylights (Figure 3(C)). Most of the captured individuals of *N. primus* were in the heat trap, where both of these variables were constant (Figure 3(D)).

4. Discussion

Articles referring the echolocation behavior from the bat species within Natalidae are not numerous. Most of them are reflecting the description of the species vocalizations (Murray, Fraser, Davy, Fleming, & Fenton, 2009), or their presence in different localities (García-Rivera, Montes Espín, Hernández Hernándes, Borroto, & Mancina, 2015). This study presents the description of the

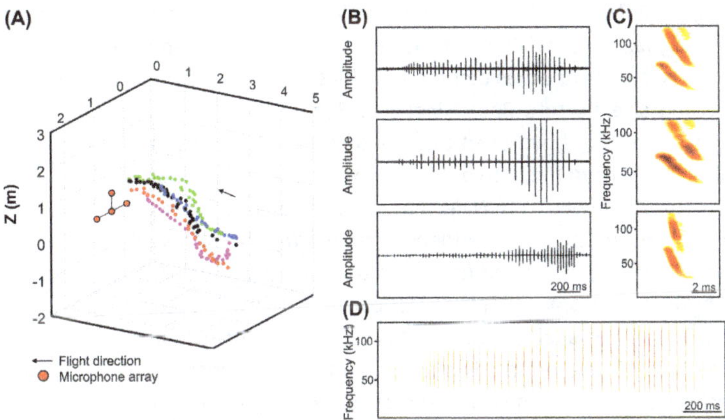

Figure 2. Flight trajectories (A), waveforms from flight paths performed by three individuals (B), and echolocation calls (C) of *Natalus primus* obtained in a gallery in La Barca cave, Guanahacabibes, Pinar del Rio, Cuba, (D) represents the spectrogram from the first waveform shown in (B).

Notes: Colors in A are signaling individual flight trajectories selected from five individuals of *N. primus*. Spectrograms shown in C represent the variation found in these echolocation calls regarding the presence of the short CF component.

Table 1. Echolocation call variables measured for 13 individuals of *N. primus*

N = 349 Harmonics	Duration (ms)	Minimum frequency (kHz)	Maximum frequency (kHz)	Peak frequency (kHz)	Bandwidth (kHz)	Slope modulation (kHz/ms)
1st (*n* = 343)	1.9 ± 0.5	38.8 ± 10.0	73.0 ± 13.1	60.2 ± 10.6	34.1 ± 6.6	34.1 ± 5.6
2nd (*n* = 200)	1.7 ± 0.5	64.0 ± 8.0	114.4 ± 5.5	85.3 ± 6.9	50.3 ± 9.8	29.5 ± 6.0
Whole call	1.7 ± 0.5	46.1 ± 8.9	103.8 ± 13.9	78.5 ± 15.6	57.7 ± 13.9	-

Notes: Arithmetic means, variation coefficient, and standard deviations are given to each call variable in each harmonic and for the entire call. *N* represents the number of echolocation calls obtained from all recordings and *n* represents the number of call harmonics.

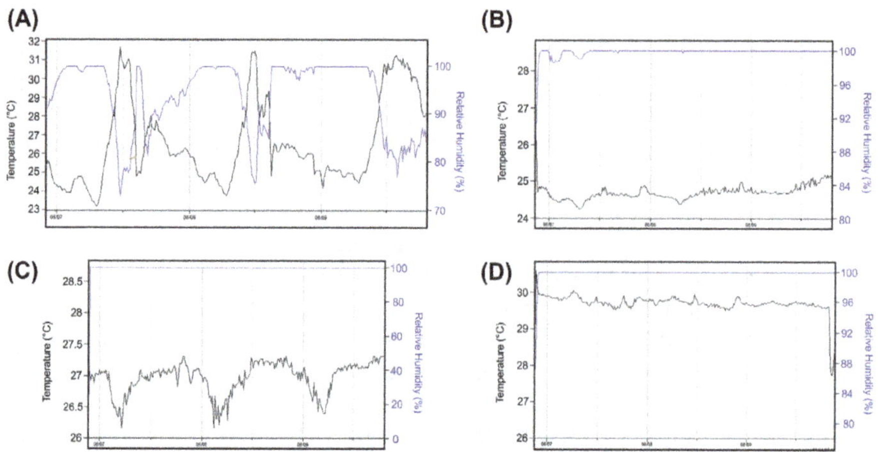

Figure 3. Temperature and relative humidity profiles from different galleries of La Barca cave, Guanahacabibes, Pinar del Rio, Cuba; on 3 days of continuous sampling (*X* axis), reporting 1 value every 30 min. Surroundings outside the cave (A).

Notes: Inner galleries of this cave (B and C). Heat trap (D).

echolocation calls from the largest species of Natalidae, as the first step to undertake more studies about the ecology of this species. In Cuba, specifically most of the bioacoustics and ecological studies had focused on Molossidae (Mora et al., 2011; Mora & Torres, 2008), Phyllostomidae (Macías & Mora, 2006; Macías, Mora, García, & Macías, 2006) and Vespertilionidae (Mora, Rodríguez, Macías, Quiñonez, & Mellado, 2005; Rodríguez & Mora, 2005). The butterfly bat (*Nyctiellus lepidus*), the smallest bat in Cuba is quite numerous in several caves along the archipelago (Silva-Taboada, 1979) and their colonies are of a considerable size (Borroto-Páez & Mancina, 2011). Although, almost no studies regarding aspects of their natural history have been published so far. *N. lepidus* inhabits cave La Barca as well as *N. primus*, within the cave these two species can be easily identified by their flight maneuvers and body size. Echolocation calls from *N. lepidus* are quite distinctive in their frequency pattern from those of *N. primus*, presenting non-overlapped multiharmonic FM calls (Murray et al., 2009) and they lack this short CF component found for *N. primus*. From the 15 captured individuals of *N. primus*, their forearm length and body weight measures concur with those obtained previously for the species (Tejedor et al., 2004). This species presents FM calls, while the CF component of the first harmonic in these (Figure 2) could be a feature that, together with call variables measured, will allow distinguishing this species from the rest of FM bat species coexisting in the same area.

With data presented here, acoustic surveys could be implemented in the surrounding areas from cave La Barca, trying to locate the foraging zones of *N. primus* and what kind of environment they prefer to develop this activity: open, edge or narrow environments. This information may contribute as well to give a right estimation of the species extend of occurrence and offer new areas of occupancy. Still though, more acoustic recordings should be conducted on this species, in order to obtain a more accurate identification of this species on the field by considering possible modifications in their frequency pattern and values of their call variables, as have been detected in other bat species (Mora, Macías, Vater, Coro, & Kössl, 2004; Mora & Macías, 2007).

According to the spectro-temporal call pattern and pulse intervals, this species can be classified as a low duty cycle (LDC) bat. LDC bats can separate their call and echo in time, to avoid self-deafening (Fenton, 2013; Lazure & Fenton, 2011). Frequency-modulated signals of broad bandwidth, as the ones emitted by this species, are well suited for exact target location where range and angle must be precisely measured (Schnitzler & Kalko, 2001). This spectral pattern might be convenient for this species moving at low heights within vegetation (field observations). Individuals of *N. lepidus* were observed to forage at sunset, very close to the beach, at the vegetation edge. Based on the acoustic recordings made in the entrance of cave La Barca, *N. lepidus* is the first species to go out of the cave to forage (data not shown in this article). No individuals from *N. primus* were seen at this time in these locations.

Further studies are needed to better understand the ecology of this species. Priority areas include improving our knowledge about their roosting needs, foraging strategies, and habitat use. Previous studies point that bats from Natalidae prefer roosts (caves and abandoned mines) with stable values of temperature and relative humidity, and with water reservoirs in some galleries, such as the case of *N. stramineus* (Torres-Flores & López-Wilchis, 2010). In our study, we present temperature and relative humidity profiles from three consecutive days, obtained from different galleries in this cave (Figure 3), they show relatively constant values of relative humidity among the three galleries for those sampled days, but temperature remained constant only in the heat trap (Figure 3(D)) which also presents a guano swamp. This might by one of the features preferred for this species in the selection of its roost, but the distance to the seashore could constitute another important one, due that most of the fossil records from this species in Cuba were caves with these characteristics (Silva-Taboada, 1979). Recording temperature and relative humidity in multiple galleries within the cave for a long-term period will be necessary, to make a more conclusive argument regarding these climatic roosting requirements for the species. This species had been seen roosting together with *Mormoops blainvillei* (Tejedor et al., 2004), suggesting that maybe could have similar preferences for roosting sites within the cave. *M. blainvillei* in general sense is a quite fragile species, compared to other cave-dwelling bats species, during handling or conducting experiments outside cave environments, individuals from *M. blainvillei* can dehydrate very easily and even die (author's field observations). This could be one of the reasons explaining the similarity in the roosting sites within the cave among *M. blainvillei* and *N. primus*. Thermal conductance and metabolic rates measurements jointly with long-term measurements of cave temperatures, similar to the ones conducted for *M. blainvillei* and other cave-dwelling bats (Rodríguez-Durán, 1995), will be necessary for *N. primus* to confirm this.

Cave La Barca has been recognized as one of the most diverse caves in Cuba, housing up to 13 bat species (Tejedor, Tavares, Silva-Taboada, 2005). For almost all of them, the echolocation behavior has been addressed in several publications (Macías & Mora, 2006; Macías, Mora, & García, 2006; Macías, Mora, Koch, & Von Helversen, 2005; Mora & Macías, 2011). Our description of *N. primus* echolocation calls (regarding call measurements and frequency pattern) will be useful to identify this species among the rest of the bat species inhabiting Guanahacabibes National Park. Our future intentions are record vocalizations in different areas within this National Park and conduct automated detection to detect the presence of *N. primus* in these locations using their call measurements presented here or spectrogram templates, similar to what have been done already in other species (Bardeli et al., 2010; Mellinger & Clark, 1997).

Acknowledgments

Special thanks to the personnel from Guanahacabibes National Park for all of their support, especially to the forest rangers Jorge Luis Camejo, Jorge Luis Márquez, and Lázaro Izquierdo. We thank Professor Brock Fenton, Matthew Emrich, and Rachel Hamilton for use of acoustic equipment and help with acoustic analysis. We also thank the Rufford Small Grants for Nature Conservation [grant number 16094-1], for supporting the field trips and environmental education activities conducted at the same time with this research. We also thank Humboldt Foundation for funding to purchase equipment. Lastly, thanks to Drs Carlos A. Mancina, Annia Rodriguez-San Pedro and Amanda Adams, and Mr Lainet Garcia-Rivera for their useful comments on this manuscript.

Funding

This work was supported by Rufford Foundation [grant number 16094-1].

Author details
Lida Sanchez[1,2]
E-mail: liditasanchez89@gmail.com
Christian R. Moreno[1]
E-mail: christian.morenoleon@gmail.com
Emanuel C. Mora[1,3]
E-mail: emanuel_mora@yahoo.com
[1] Faculty of Biology, Department of Human and Animal Biology, Havana University, Havana, Cuba.
[2] Division of Ecology and Evolutionary Biology, Graduate School of Life Sciences, Tohoku University, Sendai 980-8578, Japan.
[3] Instituto de Ciencias Biomédicas, Universidad Autónoma de Chile, El Llano Subercaseaux 2801, Santiago de Chile, Chile.

Corrigendum
This article was originally published with errors. This version has been corrected. Please see Corrigendum (https://doi.org/10.1080/23312025.2017.1374910).

Cover image
Source: Emauel C. Mora.

References
Anderson, S. E., Dave, A. S., & Margoliash, D. (1996). Template-based automatic recognition of birdsong syllables from continuous recordings. *The Journal of the Acoustical Society of America, 100*, 1209–1219.

Bardeli, R., Wolff, D., Kurth, F., Koch, M., Tauchert, K. H., & Frommolt, K. H. (2010). Detecting bird sounds in a complex acoustic environment and application to bioacoustic monitoring. *Pattern Recognition Letters, 31*, 1524–1534. doi:10.1016/j.patrec.2009.09.014

Borroto-Páez, R., & Mancina, C. A. (2011). *Mamíferos en Cuba.* Vaasa: Spartacus-Säätiö y Sociedad Cubana de Zoologia.

Brandes, T. S., Naskrecki, P., & Figueroa, H. K. (2006). Using image processing to detect and classify narrow-band cricket and frog calls. *The Journal of the Acoustical Society of America, 120*, 2950–2957. doi:10.1121/1.2355479

Dávalos, L., & Mancina, C. (2010). *Natalus primus.* Gland, VD: IUCN.

Davidson-Watts, I., Walls, S., & Jones, G. (2006). Differential habitat selection by Pipistrellus pipistrellus and Pipistrellus pygmaeus identifies distinct conservation needs for cryptic species of echolocating bats. *Biological Conservation, 133*, 118–127.

Fenton, M. B. (2013). Questions, ideas and tools: Lessons from bat echolocation. *Animal Behaviour, 85*, 869–879. doi:10.1016/j.anbehav.2013.02.024

Fukui, D., Agetsuma, N., & Hill, D. A. (2004). Acoustic identification of eight species of bat (*Mammalia: Chiroptera*) inhabiting forests of Southern Hokkaido, Japan: Potential for conservation monitoring. *Zoological Science, 21*, 947–955. doi:10.2108/zsj.21.947

García-Rivera, L., Montes Espín, R., Hernández Hernándes, N., Borroto, R., & Mancina, C. A. (2015). Murciélagos (*Mammalia: Chiroptera*) del Parque Nacional Alejandro de Humboldt, la ciudad de Baracoa y el Elemento Natural Destacado Yara - Majayara, Cuba. *Poeyana, 500*, 24–32.

Gaston, K. J., & O'Neill, M. A. (2004). Automated species identification: Why not? *Philosophical Transactions of the Royal Society B: Biological Sciences, 359*, 655–667. doi:10.1098/rstb.2003.1442

González, W. (2012). *Localización tridimensional de animales emisores de sonido.* Havana: Havana University.

Hughes, A. C., Satasook, C., Bates, P. J. J., Soisook, P., Sritongchuay, T., Jones, G., & Bumrungsri, S. (2011). Using echolocation calls to identify Thai bat species: *Vespertilionidae, Emballonuridae. Nycteridae and Megadermatidae. Acta Chiropterologica, 13*, 447–455. doi: 10.3161/150811011X624938

Kuenzi, A. J., & Morrison, M. L. (1998). Detection of bats by mist-nets and ultrasonic detectors. *Wildlife Society Bulletin, 26*, 307–311. doi:10.2307/3784055

Laiolo, P. (2010). The emerging significance of bioacoustics in animal species conservation. *Biological Conservation, 143*, 1635–1645. doi:10.1016/j.biocon.2010.03.025

Laiolo, P., Vögeli, M., Serrano, D., & Tella, J. L. (2007). Testing acoustic versus physical marking: Two complementary methods for individual-based monitoring of elusive species. *Journal of Avian Biology, 38*, 672–681. doi:10.1111/j.2007.0908-8857.04006.x

Lazure, L., & Fenton, M. B. (2011). High duty cycle echolocation and prey detection by bats. *The Journal of Experimental Biology, 214*, 1131–1137. doi:10.1242/jeb.048967

Macías, S., & Mora, E. C. (2006). Variability in the echolocation behavior of the big fruit-eating bat *Artibeus jamaicensis parvipes* (Chiroptera: Phyllostomidae) in Cuba. *Revista Biología, 20*, 24–29.

Macías, S., Mora, E. C., & García, A. (2006). Acoustic identification of mormoopid bats: A survey during the evening exodus. *Journal of Mammalogy, 87*, 324–330. doi:10.1644/05-MAMM-A-124R1.1

Macías, S., Mora, E. C., García, A., & Macías, Y. (2006). Echolocation behavior of *Brachyphylla nana* (Chiroptera: Phyllostomidae) under laboratory conditions. *Caribbean Journal of Science, 42*, 114–120.

Macías, S., Mora, E. C., Koch, C., & Von Helversen, O. (2005). Echolocation behaviour of *Phyllops falcatus* (Chiroptera: Phyllostomidae): Unusual frequency range of the first harmonic. *Acta Chiropterologica, 7*, 275–283. doi:10.3161/1733-5329(2005)7[275:EBOPFC]2.0.CO;2

Mellinger, D. K., & Clark, C. W. (1997). Methods for automatic detection of mysticete sounds. *Marine and Freshwater Behavior and Physiology, 29*, 163–181. https://doi.org/10.1080/10236249709379005

Mora, E. C., Ibáñez, C., Macías, S., Juste, J., López, I., & Torres, L. (2011). Plasticity in the echolocation inventory of *Mormopterus minutus* (Chiroptera, Molossidae). *Acta Chiropterologica, 13*, 179–187. doi:10.3161/15081101 1X578723

Mora, E. C., & Macías, S. (2007). Echolocation calls of Poey 's flower bat (*Phyllonycteris poeyi*) unlike those of other phyllostomids. *Naturwissenschaften, 94*, 380–383. doi:10.1007/s00114-006-0198-7

Mora, E. C., & Macías, S. (2011). Short CF-FM and FM-short CF Calls in the echolocation behavior of *Pteronotus macleayii* (Chiroptera: Mormoopidae). *Acta Chiropterologica, 13*, 457–463. doi:10.3161/150811011X624947

Mora, E. C., Macías, S., Vater, M., Coro, F., & Kössl, M. (2004). Specializations for aerial hawking in the echolocation system of *Molossus molossus* (Molossidae, Chiroptera). *Journal of Comparative Physiology A: Neuroethology, Sensory, Neural, and Behavioral Physiology, 190*, 561–574. doi:10.1007/s00359-004-0519-2

Mora, E. C., Rodríguez, A., Macías, S., Quiñonez, I., & Mellado, M. M. (2005). The echolocation behaviour of *Nycticeius Cubanus* (Chiroptera: Vespertilionidae): Inter- and intra-individual plasticity in vocal signatures. *Bioacoustics, 15*, 175–193. doi:10.1080/09524622.2005.9753546

Mora, E. C., & Torres, L. (2008). Echolocation in the large molossid bats *Eumops glaucinus* and *Nyctinomops macrotis. Zoological Science, 25*, 6–13. doi:10.2108/zsj.25.6

Murray, K. L., Fraser, E., Davy, C., Fleming, T. H., & Fenton, M. B. (2009). Characterization of the echolocation calls of bats from exuma. *Bahamas. Acta Chiropterologica, 11*, 415–424. doi:10.3161/150811009X485639

Oswald, J. N., Barlow, J., & Norris, T. F. (2003). Acoustic identification of nine delphinid species in the Eastern Tropical Pacific Ocean. *Marine Mammal Science, 19*, 20–37. doi:10.1111/j.1748-7692.2003.tb01090.x

Rodríguez-Durán, A. (1995). Metabolic rates and thermal conductance in four species of neotropical bats roosting in hot caves. *Comparative Biochemistry and Physiology - Part A: Physiology, 110*, 347–355. doi:10.1016/0300-9629(94)00174-R

Rodríguez, A., & Mora, E. C. (2005). Acoustic identification of Nycticeius cubanus (Gundlach, 1867) and Eptesicus fuscus dutertreus (Gervais, 1837) (*Chiroptera: Vespertilionidae*) in Western Cuba. *Revista Biología, 19*, 25–32.

Russ, J. M. & Montgomery, W. I. (2002). Habitat associations of bats in Northern Ireland: Implications for conservation. *Biological Conservation, 108*, 49–58.

Schnitzler, H.-U., & Kalko, E. K. V. (2001). Echolocation by insect-eating bats. *BioScience, 51*, 557. doi:10.1641/0006-3568(2001)051[0557:EBIEB]2.0.CO;2

Silva-Taboada, G. (1979). *Murciélagos de Cuba*. La Habana: Editorial Academia.

Skowronski, M. D., & Fenton, M. B. (2008). Model-based automated detection of echolocation calls using the link detector. *The Journal of the Acoustical Society of America, 124*, 328–336. doi:10.1121/1.2924122

Tejedor, A. (2011). Systematics of funnel-eared Bats. Chiroptera: Natalidae.

Tejedor, A., Silva-Taboada, G., & Rodríguez-Hernández, D. (2004). Discovery of extant Natalus major (Chiroptera: Natalidae) in Cuba. *Mammalian Biology - Zeitschrift für Säugetierkunde, 69*, 153–162. doi:10.1078/1616-5047-00130

Tejedor, A., Tavares, V. D. C., & Silva-Taboada, G. (2005). A revision of extant greater antillean bats of the genus *Natalus. American Museum Novitates,* 1–22. doi:10.1206/0003-0082(2005)493[0001:AROEGA]2.0.CO;2

Tejedor, A., Tavares, V. D. C., & Rodríguez-Hernández, D. (2005). New records of hot-cave bats from Cuba and the dominican republic. *Boletin de La Sociedad Venezolana de Espeleologia, 39*, 10–15.

Torres-Flores, J. W., & López-Wilchis, R. (2010). Condiciones microclimáticas, hábitos de percha y especies asociadas a los refugios de Natalus stramineus en México. *Acta Zoológica Mexicana, 26*, 191–213.

16

Melatonin bio-synthesizing enzyme genes (*Tph1*, *Aanat1*, *Aanat2*, and *Hiomt*) and their temporal pattern of expression in brain and gut of a tropical carp in natural environmental conditions

Haobijam Sanjita Devi[1], Chongtham Rajiv[1], Gopinath Mondal[1], Zeeshan Ahmad Khan[1], Sijagurumayum Dharmajyoti Devi[1], Thangal Yumnamcha[1], Rupjyoti Bharali[2] and Asamanja Chattoraj[1]*

*Corresponding author: Asamanja Chattoraj, Biological Rhythm Laboratory, Animal Resources Programme, Department of Biotechnology, Institute of Bioresources and Sustainable Development, Government of India, Takyelpat, Imphal 795 001, Manipur, India

E-mails: achattoraj.ibsd@nic.in, asamanja.chattoraj@gmail.com

Reviewing editor: William Wisden, Imperial College London, UK

Additional information is available at the end of the article

Abstract: The study demonstrates the temporal pattern of expression of melatonin bio-synthesizing enzyme genes (*Tph1*, *Aanat1*, *Aanat2*, and *Hiomt*) in the brain and gut of a tropical carp (*Catla catla*) on a daily and seasonal basis under natural photo-thermal conditions. The measurement of melatonin in brain and gut (both *in vivo* and *in vitro*) demonstrated a higher content in gut, with evidence of melatonin bio-synthesizing machinery in both tissues. All melatonin bio-synthesizing gene expressions in these two tissues were negatively correlated (except *Aanat1* in brain and *Hiomt* in the gut) with the water temperature on an annual cycle. The higher expression of *Aanat2* gene, rather than *Aanat1*, signifies the importance of *Aanat2* isoform in the melatonin production in tropical carp. Furthermore, the *Aanat2* and *Hiomt* genes are highly expressed in the gut. Analysis of the rhythm and acrophase of expression of these genes in the brain and gut imply a pineal-independent melatonin synthesizing machinery in these two organs, possibly involving both environmental and endogenous cues for the regulation of melatonin rhythm to synchronize the physiology of the animal.

ABOUT THE AUTHORS

The research group led by Asamanja Chattoraj is mainly focused on the synchronization of the environmental cues with body physiology in relation to the "chronobiotic" molecule melatonin and clock-associated genes. The abiotic parameters of the nature are changing abruptly; fish are the best model to study the variation and possible reason in the alteration of body physiology. The research interest concerns about the relationship between the central and peripheral clock mechanisms and their orientation either through "master-slave" or "orchestra". The findings can lead to develop a sustainable strategy for the maintenance of the species population in a favorable niche as well as tools for their conservation. Moreover, several lifestyle diseases are emerging from the circadian rhythm disruption. We are trying to address this inappropriate harmonization between environmental cues and body functioning with special emphasis on clock-associated genes and "melatonin" using fish as model.

PUBLIC INTEREST STATEMENT

This article provides information about the melatonin bio-synthesizing machinery in the brain and gut of a tropical fish. The production of melatonin, an indole amine, is directly linked to the environmental photo-thermal conditions. This study also gives an idea about the variation and/or changes in the rhythmic expression of the melatonin bio-synthesizing genes under natural environmental conditions. Moreover, a relationship between the brain (central) and gut (peripheral) has also been focused in respect to the production of melatonin. The present finding is an opening toward the brain–gut axis and its synchronization with respect to melatonin in fish.

Subjects: Biology; Molecular Biology; Neurobiology

Keywords: *Tph1*; *Aanat1*; *Aanat2*; *Hiomt*; gut; brain; melatonin; tropical carp

1. Introduction

Melatonin (5-Methoxy-*N*-acetyltryptamine), an indole amine, is primarily synthesized in the pineal gland of vertebrates. It is also synthesized from retina, liver, gut, and brain (Chowdhury & Maitra, 2012; Jimenez-Jorge et al., 2007). Among these, the gut is the richest source of melatonin (Bubenik, 2002) in various groups of vertebrates, including fish (Bubenik & Brown, 1997; Bubenik & Pang, 1997; Munoz, Ceinos, Soengas, & Miguez, 2009). The limited data on brain melatonin in vertebrates suggest the presence of least amount of melatonin in the cerebral structures (Jimenez-Jorge et al., 2007; Uz, Qu, Sugaya, & Manev, 2002). Melatonin is formed from tryptophan by four sequential enzymatic reactions. Tryptophan hydroxylase (TPH) catalyzes the hydroxylation of tryptophan, and decarboxylation is done by aromatic amino acid decarboxylase, forming serotonin, an important neurotransmitter in the functioning of body physiology. It has already been reported that many parts of brain perform these two reactions where serotonin is formed (Appelbaum & Gothilf, 2006). The remaining two steps, *N*-acetylation by arylalkylamine *N*-acetyltransferase (AANAT) and *O*-methylation by hydroxyindole-*O*-methyltransferase (HIOMT), are known to occur predominantly in the pineal gland and retina, and therefore are considered as pineal-/retina-specific enzymes (Appelbaum & Gothilf, 2006). In teleost, the existence of three TPH isoforms (TPH1a, TPH1b, and TPH2), (Lillesaar, 2011), two AANAT subfamilies AANAT1 (AANAT1a and/or AANAT1b) and AANAT2 (Paulin et al., 2015), and two isoforms of HIOMT (Velarde, Cerda-Reverter et al., 2010) has been reported.

The circadian organization and the regulatory system of melatonin are shown to be dependent on environmental photo-thermal conditions resulting in a peak at midnight in the pineal gland of mammals (Reiter, 1991b), fish (Falcón, Migaud, Muñoz-Cueto, & Carrillo, 2010; Iigo et al., 2007), and carp (Seth & Maitra, 2010). In this way, melatonin acts as the internal neurohormonal signal of darkness and plays a role as a "zeitgeber" (Falcón, 1999; Reiter, 1991a; Seth & Maitra, 2011). The involvement of suprachiasmatic nucleus (SCN) in maintaining biological rhythms in mammals is well known (Korf, Schomerus, & Stehle, 1998). Studies on fish and amphibians have given the indication of involvement of brain in photo neuroendocrine system beside the pineal gland and retina (Falcón et al., 2010). The diencephalic cells of *Rana perezi* (including SCN cells) show the expression of *Aanat1* (Isorna et al., 2006) which leads to the hypothesis that some parts of the brain can concentrate photosensitivity (Falcón et al., 2010). However, pinealectomized (brain only) rat demonstrated the reduction of brain melatonin (Jimenez-Jorge et al., 2007). It has also been evidenced that the developing brain (without pineal gland) of rat can produce its own melatonin rhythmically with higher nocturnal content than the diurnal concentration (Jimenez-Jorge et al., 2007). The study on zebrafish showed that circadian rhythms can be initiated and maintained in the absence of SCN and other tissues in the ventral brain, though the SCN may play a deciding role in the regulation of amplitude of rhythms in the absence of environmental cues (Noche, Lu, Goldstein-Kral, Glasgow, & Liang, 2011). Moreover, a recent study on zebrafish also demonstrated the expression pattern of melatonin bio-synthesizing enzymes in the whole brain (Khan et al., 2016) in different photic conditions.

The presence of melatonin in the gut of many vertebrates, including fish, is reported (Bubenik & Brown, 1997; Kulczykowska, Kalamarz, Warne, & Balment, 2006). Based on the high expression levels of its synthesizing enzymes, some studies pointed out its local synthesis in the gut (Fernández-Durán et al., 2007; Stefulj et al., 2001). The study on goldfish informed us of the *gAanat2* mRNA expression in both foregut and hindgut (Velarde, Cerda-Reverter et al., 2010). Rhythmic AANAT activity is also reported in the foregut of the same fish (Nisembaum et al., 2013). A recent study on trout also demonstrated a variation in melatonin content and the expression of *Aanat1*, *Aanat2*, and *Hiomt* in the mucosa and wall of the gut (Munoz-Perez et al., 2016). Stable presence of melatonin in the lower gut in a pinealectomized animal is indicative of local production of this indole amine (Chen,

Fichna, Bashashati, Li, & Storr, 2011). Melatonin level in gut showed a significant difference at the species level in fish, amphibians, and reptiles (Bubenik & Pang, 1997). Recently, a study on the seasonal and diurnal profiles of melatonin (serum, gut, retinal, and pineal) and expression of AANAT in natural and altered photic conditions in same species gave an idea about the presence of gut melatonin in any tropical carp (Mukherjee & Maitra, 2015; Mukherjee, Moniruzzaman, & Maitra, 2014). In fish, there is no difference in the melatonin content in various parts of the gut (Mukherjee & Maitra, 2015). The study on aged mice depicted the protective role of melatonin and its precursor serotonin in the gut (Bertrand, Bertrand, Camello, & Pozo, 2010). Presence of melatonin in the hepatobiliary-gastrointestinal tract (Messner, Huether, Lorf, Ramadori, & Schwörer, 2001) may contribute to reduce pathophysiology in humans (Reiter, Tan, Manchester, & Gitto, 2010) and gut motility in fish through its receptor or bio-synthesizing enzyme (AANAT) (Nisembaum et al., 2013; Velarde, Delgado, & Alonso-Gómez, 2010).

Information about the expression of melatonin bio-synthesizing enzyme genes and its temporal pattern of expression at a transcriptional level along with the concentration, in brain and gut, is not available. Therefore, the present study is the first attempt to show diurnal and seasonal variations and/or rhythms of melatonin bio-synthesizing enzyme genes (*Tph1*, *Aanat1*, *Aanat2*, and *Hiomt*) at the transcriptional level in brain (without pineal) and gut in any tropical carp (*Catla catla*), under natural photo-thermal conditions. Moreover, measurement of melatonin *in vivo* and in culture provided the evidence of the melatonin production unit in these two organs.

2. Materials and methods

2.1. Animals, housing conditions, and sample collection

An economically important surface-dwelling adult major carp of both sexes of Indian subcontinent, *C. catla* (n = 3) weighing between 700 and 900 g was used for the sampling in this study continuously for three years. The fish were collected from local fish farm located in Kodompokpi, Imphal. After transportation to the laboratory, the fish were acclimatized in cemented tanks in identical photo-thermal conditions, as they were, for a week (Chattoraj, Seth, & Maitra, 2009). Artificially prepared balanced food comprising 35% fish meal, 28% mustard oil cake, 28% rice bran, 2% each sunflower and cod liver oils, 5% carboxy methyl cellulose, and multivitamin–multimineral tablets ("Becozyme Forte", Glaxo India Ltd., 25 tabs/kg food) were given two times daily *ad libitum*. Laboratory care of fish and adopted study schedules were in agreement with ethical standards journal (Portaluppi, Smolensky, & Touitou, 2010). The ethical clearance was obtained from the Institutional Animals Ethical Committee constituted as per the recommendations of Committee for the Purpose of Control and Supervision of Experiments on Animals (CPCSEA), Government of India. Samples were collected in preparatory, pre-spawning, spawning, and post-spawning phases in Imphal valley (latitude 24°44′N, longitude 93°58′E), Manipur, with an elevation of 786 m above sea level. This region of the world is under the Indo-Burma biodiversity hotspot. The area is having a seasonal, tropical monsoonal climate, with a cool dry season (~November–April) and warm wet season (~May–October). From winter to summer, the temperature varies from 4 to 32°C and the photoperiod ranges from 11 to 13 h in a year. Animals were anesthetized (0.1% Tricaine, SIGMA; pH 7.00 of the media) and euthanized before tissues were collected. Samplings at dark were carried out in dim red light. The brain tissues without the end vesicle of pineal and whole gut were collected. The previous studies on the morphoanatomy of this carp clearly demonstrated that the end vesicle part of pineal organ is mainly considered as the unit for photoreception and production of melatonin. This end vesicle is very much distant from the dorsal sac and connected by a long pineal stalk (Figure 1 of Dey, Bhattacharya, Maitra, & Banerji, 2003 and Figure 2 of Chattoraj, Bhattacharyya, Dey, Seth, & Maitra, 2006). The gut content was removed from whole gut tissues prior to storage in −80°C freezer. Both the tissues were collected in an identical method and time points (8:00, 12:00, 16:00, 20:00, 24:00, 4:00 h) in preparatory (January; duration of Natural Photoperiod, NP: 11 h 18 min; Water Temperature, WT: maximum 15°C, minimum 14°C), pre-spawning (April; NP: 13 h; WT: maximum 28.3°C, minimum 26.3°C), spawning (August; NP: 13 h; WT: maximum 26°C, minimum 25°C), and post-spawning phases (October; NP: 10 h 50 min; WT: maximum 18.5°C, minimum 17°C) (Table 1). There was no change in the feeding

Table 1. The average day length and water temperature in four seasons from where fish were collected (latitude 24°44′N, longitude 93°58′E)

Seasons	Sunrise (h)	Sunset (h)	Day length (h)	Water temperature (°C)
Preparatory	5.5	17.08	11:18	15.0
Pre-spawning	4.44	17.41	13:00	28.3
Spawning	4.49	17.48	13:00	26.0
Post-spawning	5.35	16.25	10:50	18.5

conditions and timing before the euthanization of fish. Each phase is designated by characterized profile of germ cells in the gonad; serum melatonin varies in its amplitude and pattern (Bhattacharya, Chattoraj, & Maitra, 2007; Dey, Bhattacharya, & Maitra, 2005; Hasan, Moniruzzaman, & Maitra, 2014; Maitra & Chattoraj, 2007) in the mentioned period. Immediately after isolation, tissues were put in TRIzol and frozen at −80°C until further processing.

2.2. Gene expression study

Gene-specific primers were used for quantitative PCR, listed in Table 2. Total RNA was isolated using "PureLink® RNA Mini Kit" (Ambion) as recommended by manufacturer's protocol. RNA quality and quantity were assessed using a spectrophotometer (BioSpec-nano, Shimadzu). RNA integrity was checked by taking 200 ng of total RNA loaded onto 1% denaturing agarose gel electrophoresis, stained with GelRed™ Nucleic Acid Gel Stain (Biotium Inc, USA). Genomic DNA contamination was removed by treating 10 µg of total RNA with "TURBO DNA-free™ Kit (Ambion)". Using "High Capacity cDNA Reverse Transcription Kit" (Applied Biosystems), 2 µg DNase-treated total RNA was reverse transcribed into cDNA following manufacturer's protocol in "ProFlex™ PCR System" (Applied Biosystems). The thermal cycling conditions for cDNA synthesis were 25°C for 10 min, followed by 37°C for 2 h, 85°C for 5 min, and final incubation at 4°C. Furthermore, the cDNA was diluted five times for downstream process. This cDNA was amplified using *Tph1*, *Aanat1*, *Aanat2*, *Hiomt*, and *β-actin* primers followed by clean up using QIAquick PCR Purification Kit (Qiagen) and quantified. A serial dilution of concentration ranges from 0.001 to 100 pg/µl was made and used as template to generate a standard curve. Quantitative PCR was performed in "StepOnePlus Real-Time PCR System" (Applied Biosystems) using GoTaq® qPCR Master Mix (Promega) following manufacturer-recommended protocol. The PCR cycling conditions were as follows: 95°C for 10 min followed by 40 cycles of 95°C for 15 s, and 60°C for 1 min. Primers were blast against NCBI Primer blast tool to check specificity. Amplified products were also assessed in 2% agarose gel to check whether single amplicon was amplified or not. No template control and reverse transcriptase negative control were setup by replacing cDNA with water and DNase-treated total RNA, respectively. After quantitative PCR amplification, standard curve was prepared and sample quantification was done. The quantified amount

Table 2. Details about primer sequences used for RT-qPCR for *Tph1*, *Aanat1*, *Aanat2*, and *Hiomt* in the brain and gut of *Catla catla*

Name of the genes	Primer	Primer sequences	Amplicon (bp)	Sequence ID
Tryptophan hydroxylase 1	qPCR *Tph1*-F	GCGGATCATCCAGGATTC	121	KR024229.1
	qPCR *Tph1*-R	CCTCTTCCTCTGTGAACTC		
Arylalkylamine N-acetyl-transferase 1	qPCR *Aanat1*-F	GGCAGTCCACCGGACCTT	58	KR024226.1
	qPCR *Aanat1*-R	GTAACGCCACAGCAGGATAG		
Arylalkylamine N-acetyl-transferase 2	qPCR *Aanat2*-F	TCTGTGCACCGCCACTGT	54	KR024227.1
	qPCR *Aanat2*-R	GCGCCAAAGCAAGATGGA		
Hydroxyindole-O-methyl-transferase	qPCR *Hiomt*-F	CGCAAGACGAGCGCATCT	58	KR024228.1
	qPCR *Hiomt*-R	TGCGGGAGCTCGTCCTTA		
Beta Actin	qPCR *Actb*-F	GACACAGATCATGTTCGAGA-CCTT	59	JQ991014.1
	qPCR *Actb*-R	CAGCCTGGATGGCAACGT		

was then normalized (Piesiewicz, Kedzierska, Turkowska, Adamska, & Majewski, 2015) using *β-actin* (Kumari, Pathakota, Annam, Kumar, & Krishna, 2015; Munoz-Perez et al., 2016) and the gene expression level was represented as the number of mRNA copies per 10^6 copies of *β-actin* mRNA. The quantified concentrations using standard curve were converted into copy number using the following equation (Whelan, Russell, & Whelan, 2003),

$$\text{Copy number} = \frac{6.02 \times 10^{23}(\text{Copy/mole}) \times \text{DNA amount (g)}}{\text{DNA length (bp)} \times 660\ (\text{g/mol/bp})}$$

2.3. Brain and gut tissue culture and measurement of melatonin
Adult *C. catla* (n = 3) of similar weight were taken. After euthanization, brain and gut tissues were taken out from the fish at 08.00 h (light phase) and 20.00 h (dark phase). Each tissue was incubated in culture medium (M199, SIGMA) for four hours (up to 12.00 h and 24.00 h respectively) and kept immediately at −80°C for further melatonin measurement. The tissues were maintained with 95% O_2/5% CO_2 at 28°C in a CO_2 Incubator (Thermo Scientific). The incubator was fitted with a florescent lamp (300 lux) with automated timer (12 h light: 12 h Dark). The light was on at 05.00 h and off at 17.00 h. The M199 medium was supplemented with 10 mm HEPES, 0.1% BSA, 0.1 mM phenyl-methyl-sulfonyl fluoride (PMSF), penicillin (100 U/ml), and Streptomycin (100 µg/ml) (Chattoraj et al., 2005; Chowdhury et al., 2010; Khan et al., 2016). For *in vivo* melatonin estimation, each tissue was collected in phosphate buffer (PBS; pH 7.4) at 08.00 and 20.00 h from adult fish. Utmost care was given to the identical body weight of the fish for either *in vitro* or *in vivo* assay. For both *in vivo* and *in vitro* measurements of melatonin, 0.3 g of the homogenized tissues were taken in 500 µl of PBS (pH 7.4) and sonicated followed by centrifugation at 3000 rpm until a clear supernatant was obtained. Finally, 50 µl of the supernatant was taken for the measurement of melatonin by a commercially available ELISA kit (Gen Asia Biotech Co., Ltd, China) according to the manufacturer's protocol. The assay was measured at an absorbance of 450 nm wavelength (Thermo, Multiskan spectrum).

2.4. Statistical analysis
Gene expressions shown here are in copy number per 10^6 copies of ***β-actin***. All statistical analyses were performed in Statgraphics Centurion XVI (Statpoint Inc, USA). Seasonal change in expression was done by comparing gene expression level of the sample taken at three different time points during light and dark phases of different seasons. One-way ANOVA followed by Tukey HSD or Kruskal–Wallis test followed by Bonferroni (if the data-set doesn't follow normal distribution even after log transformation) was performed to analyze these seasonal significant changes. Normal distribution of data was assisted using Shapiro–Wilk test.

Diurnal rhythmic changes in expression of these genes were done using COSINOR PERIODOGRAM Version 3.1 (Refinetti, Cornélissen, & Halberg, 2007) based on cosinor-rhythmometry (Nelson, Tong, Lee, & Halberg, 1979). A nonlinear regression curve was fitted to the data using the formula "Y = Mesor + Amplitude Cos (Frequency X + Acrophase)" with GraphPad Prism (Version 6.01, GraphPad Software, Inc., USA). Two-way ANOVA followed by Student–Newman–Keuls was performed to analyze significant changes in the melatonin production *in vitro* and *in vivo* conditions. All statistical analyses were conducted at 0.05 level of significance.

3. Result

3.1. Expression of Tph1 in brain and gut
In the brain, the average transcripts of *Tph1* in all seasons were approximately 125 (Figure 1(A)). The highest expression of *Tph1* in the brain was found in post-spawning while lowest was during the preparatory dark phase (Figure 1(A)). Significant day-night variation in the expression of *Tph1* was only found during preparatory and pre-spawning phases (Kruskal–Wallis Test statistic = 17.9, p-value = 0.01; Figure 1(A)). The Cosinor analysis also revealed the insignificant diurnal rhythm of brain *Tph1* transcript (Figure 3). Highest rhythm percentage was detected in preparatory followed by

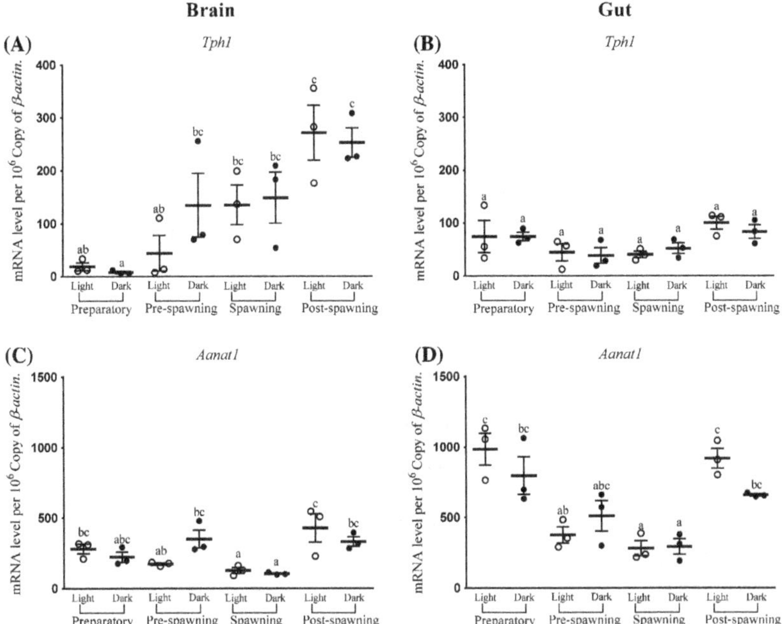

Figure 1. Seasonal and light–dark phase changes in the expression level of (A) *Tph1* in Brain, (B) *Tph1* in Gut and (C) *Aanat1* in Brain, (D) *Aanat1* in Gut were compared by taking the mean value of three time points during the light phase (represented by empty circle) and the dark phase (represented by filled circle).

Notes: Statistical analysis for brain *Tph1* expression was done by Kruskal–Wallis Test followed by Bonferroni. One-way ANOVA followed by Tukey's test were used to analyze gut *Tph1*, brain *Aanat1*, and gut *Aanat1*. Bar in the graph indicates mean ± S.E.M. Group sharing common letters has no statistical difference. All statistical analyses were conducted at 0.05 level of significance.

pre-spawning, post-spawning, and spawning phases (Figure 3). This change is inverse according to the changes in the water temperature (Tables 1 and 3).

The average *Tph1* transcripts in the gut in all four seasons were observed to be around 60 (Figure 1(B)). One-way ANOVA followed by Tukey HSD test and Cosinor analysis displayed insignificant changes in the expression of *Tph1* in the gut in all seasons as well as in the light–dark phases (*F*-ratio = 1.9, *p*-value = 0.14; Figure 1(B)). Like the brain, Cosinor analysis also portrayed insignificant diurnal rhythm of gut *Tph1* transcript in all the phases (Figure 3). Percentage of rhythm was lowest in spawning and then started to increase as it moved to post-spawning and became highest at preparatory; afterward, it started to fall in pre-spawning (Figure 3). This phenomenon was also negative with the water temperature or photic length (Table 1).

3.2. Expression of Aanat1 in the brain and gut

Average brain *Aanat1* transcripts were approximately 250 in all the seasons (Figure 1(C)). The highest expression level of brain *Aanat1* was during the day time of post-spawning and lowest in the spawning phase in the brain (Figure 1(C)). There was a significant change in the expression level of *Aanat1* in the brain between light and dark (*F*-ratio = 10.1, *p*-value < 0.01; Figure 1(C)) phases in all seasons, except spawning. Significant rhythm percentage was found in all the seasons, except post-spawning (the lowest) with the highest value in the preparatory phase (Figure 4). It is interesting that the acrophase of expression was shifted from late light phase to early light phase (spawning to preparatory phases) and finally moved to dark in the pre-spawning phase (Figure 4).

The average *Aanat1* transcript level detected in the gut was around 600 in all seasons (Figure 1(D)). The highest *Aanat1* expression in the gut was displayed during the daytime of preparatory and post-spawning phases and lowest during the spawning phase (*F*-ratio = 10.5, *p*-value < 0.01; Figure 1(D)). Except the spawning phase, a day–night variation was observed in the expression of gut *Aanat1* in all phases (Figure 1(D)). A significant rhythmic expression of gut *Aanat1* was observed in preparatory and post-spawning phases (Figure 4). There was an increase in the percentage of rhythm in the expression of gut *Aanat1* from spawning to preparatory phases and decreased at the pre-spawning phase (Figure 4).

3.3. Expression of Aanat2 in brain and gut

Brain *Aanat2* transcripts, on an average in four seasons, were about 300 (Figure 2(A)). There was no significant variation observed in the expression of brain *Aanat2* between light and dark in preparatory (highest) and spawning phases (lowest) (Figure 2(A)). The other two phases exhibited a day-night variation (F-ratio = 39.4, p-value < 0.01; Figure 2(A)). A significant rhythm during pre-spawning and post-spawning phases (highest) (Figure 5) was found. Except preparatory, the acrophase of brain *Aanat2* was found in dark (Figure 5).

Highest *Aanat2* expression in the gut was observed during the preparatory phase while lowest was during spawning (F-ratio = 13, p-value < 0.01; Figure 2(B)) and its overall average transcripts in four seasons were approximately 1,415 (Figure 2(B)). Except the preparatory phase, there was a significant difference in the expression of gut *Aanat2* in other three phases, amongst which pre-spawning and spawning phases were giving the highest value in dark (Figure 2(B)). The rhythm percentage of *Aanat2* in gut gradually increased from spawning to preparatory and ultimately became significantly rhythmic in pre-spawning phases (Figure 5). Expression of *Aanat1* and *Aanat2* in both organs was negatively related to the environmental photo-thermal conditions (Tables 1 and 3).

It has also been revealed that any one of the isoforms of *Aanat* was rhythmic in their expression in a particular season and in a particular organ, except in the spawning season in the gut (Figures 4 and 5)

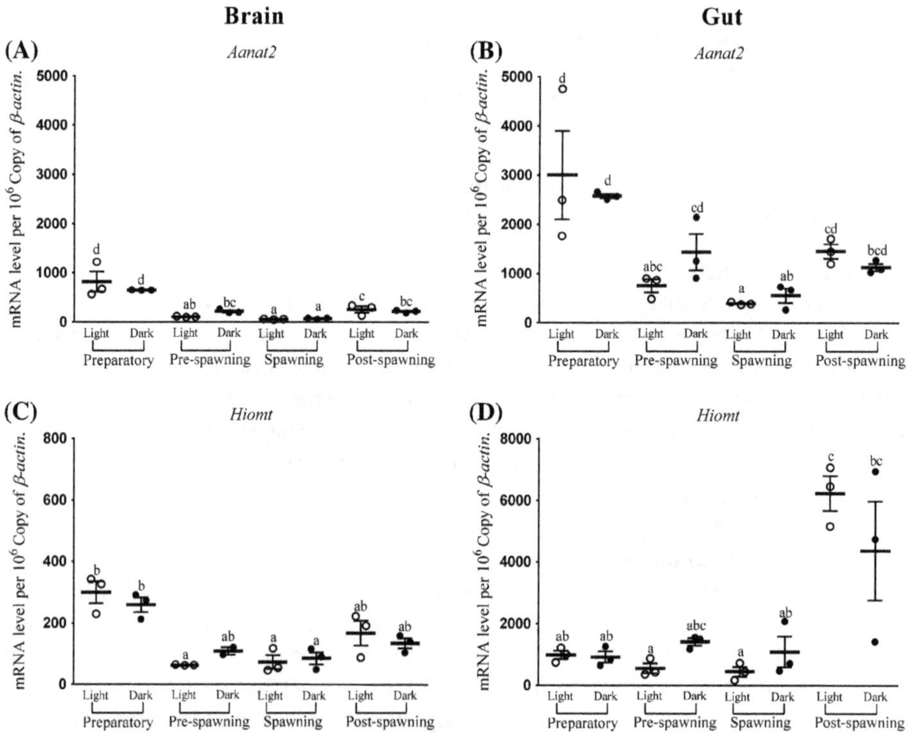

Figure 2. Seasonal and light–dark phase changes in the expression level of (A) *Aanat2* in Brain, (B) *Aanat2* in Gut and (C) *Hiomt* in Brain, (D) *Hiomt* in Gut were compared by taking the mean value of three time points during the light phase (represented by empty circle) and the dark phase (represented by filled circle).

Notes: Statistical analysis was done by one-way ANOVA followed by Tukey's test. Bar in the graph indicates mean ± S.E.M. Group sharing common letters has no statistical difference. All statistical analyses were conducted at 0.05 level of significance.

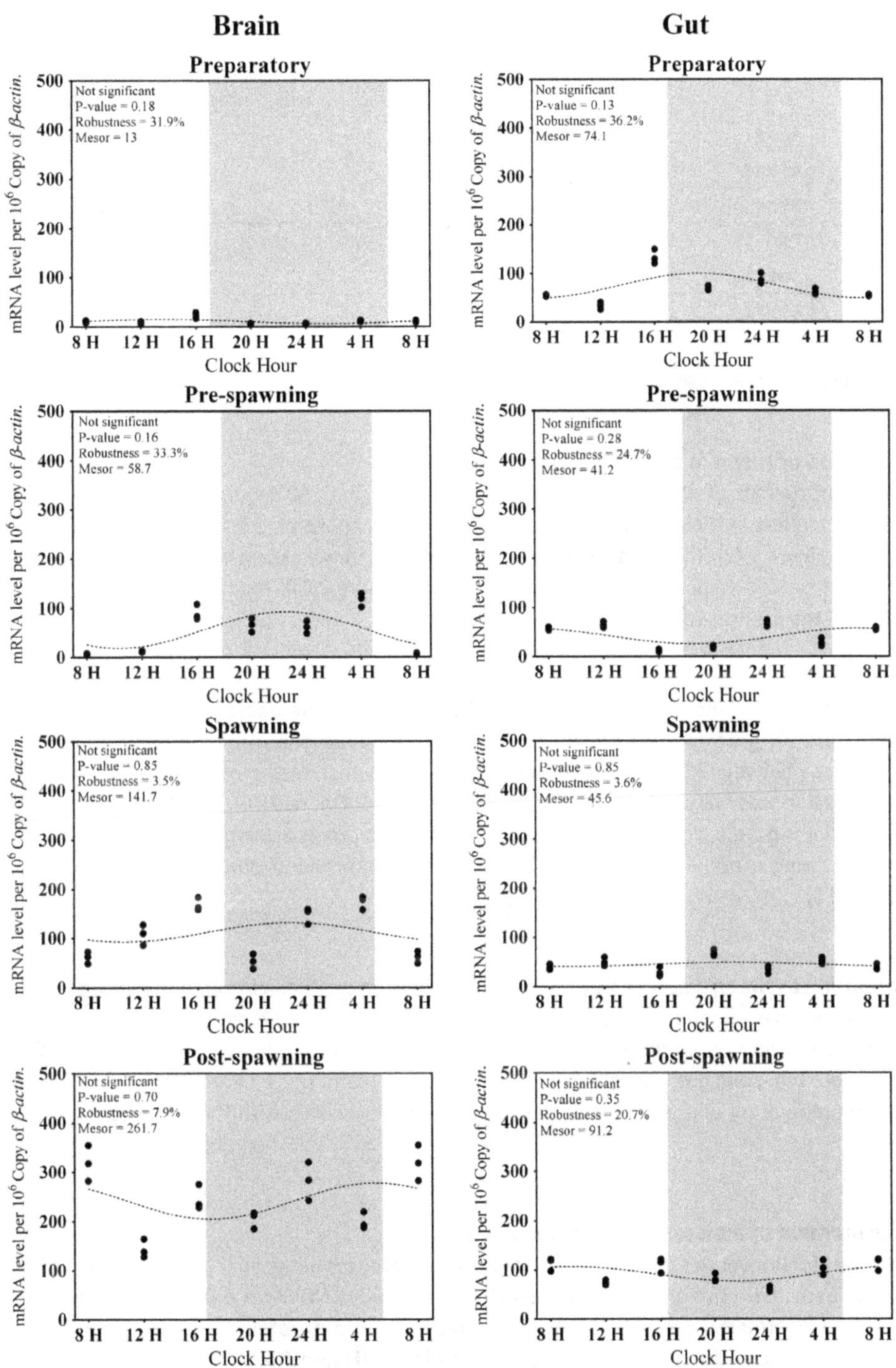

Figure 3. Changes in the expression of *Tph1* in six different time points (8, 12, 16, 20, 24, and 4 h) in different seasons ($n = 3$) were fitted to nonlinear regression curve using the formula "Y = Mesor + Amplitude Cos (Frequency X + Acrophase)".

Notes: Cosinor analysis was performed to obtain the rhythm parameters and the existence of rhythm. p-value < 0.05 is considered as significant.

where both isoforms were arrhythmic in their expression. Moreover, except the post-spawning phase, there was almost an identical acrophase of *Aanat1* and *Aanat2* in both brain and gut (Figures 4 and 5) in all the phases.

Table 3. Correlations study between water temperature and expression level

Tissue	Gene	Correlations coefficient	*p*-value
Brain	*Tph1*	−0.682*	0.005 (except preparatory phase)
	Aanat1	−0.315	0.130
	Aanat2	−0.943*	0.000
	Hiomt	−0.608*	0.004
Gut	*Tph1*	−0.533*	0.011
	Aanat1	−0.716*	0.001
	Aanat2	−0.633*	0.003
	Hiomt	−0.237	0.256

**p*-value < 0.05 is considered as significant.

3.4. Expression of Hiomt in the brain and gut

The average transcripts of *Hiomt* mRNA in the brain were around 160 (Figure 2(C)). The level of expression was highest in the preparatory phase and the light phase of pre-spawning was with the lowest value (Figure 2(C)). There was no significant difference in the expression of brain *Hiomt* in all phases except pre-spawning (*F*-ratio = 9.3, *p*-value < 0.01; Figure 2(C)). The highest rhythm was observed only in the pre-spawning phase (Figure 6).

In all four seasons, the average gut *Hiomt* mRNA transcripts were approximately 2,000 (Figure 2(D)). The expressivity of gut *Hiomt* was almost similar in all seasons with no or very less significant differences between light and dark phases, but at the post-spawning phase, suddenly, the overall expression was elevated (*F*-ratio = 9.4, *p*-value < 0.01; Figure 2(D)). Like the brain, the expression of *Hiomt* gave a significant rhythm only during the pre-spawning phase (Figure 6). This was also following a negative trend of relation with environmental photo-thermal conditions (Tables 1 and 3).

3.5. Correlation between water temperature and melatonin synthesizing enzyme genes

A significant negative correlation was found between the expression of melatonin synthesizing enzyme genes and water temperature from the correlation co-efficient analysis (Table 3) in all four seasons in both brain and gut, except for *Aanat1* (brain) and *Hiomt* (gut). In brain, the expression of *Tph1* in the preparatory phase did not reveal a negative relationship with the water temperature (Table 3).

3.6. Measurement of melatonin in vivo and in vitro

The brain melatonin level was found to have significant differences between the light and dark phases as well as between *in vivo* and *in vitro* as revealed by two-way ANOVA (*in vivo*/*in vitro* *F*-ratio = 8.6, *p*-value = 0.03; light/dark *F*-ratio = 16.5, *p*-value < 0.01). *In vitro* melatonin level was found to be more as compared to *in vivo* in both light and dark conditions (Figure 7(A)).

The gut melatonin level does not show any significant variation between light and dark phases in both *in vivo* and *in vitro* but the *in vitro* level was found to be higher than *in vivo* (*in vivo*/*in vitro* *F*-ratio = 41.5, *p*-value < 0.01; light/dark *F*-ratio = 0.1, *p*-value = 0.67; Figure 7(B)). Moreover, a considerable amount of melatonin was also detected during the light phase in both the tissues (Figures 7(A) and (B)).

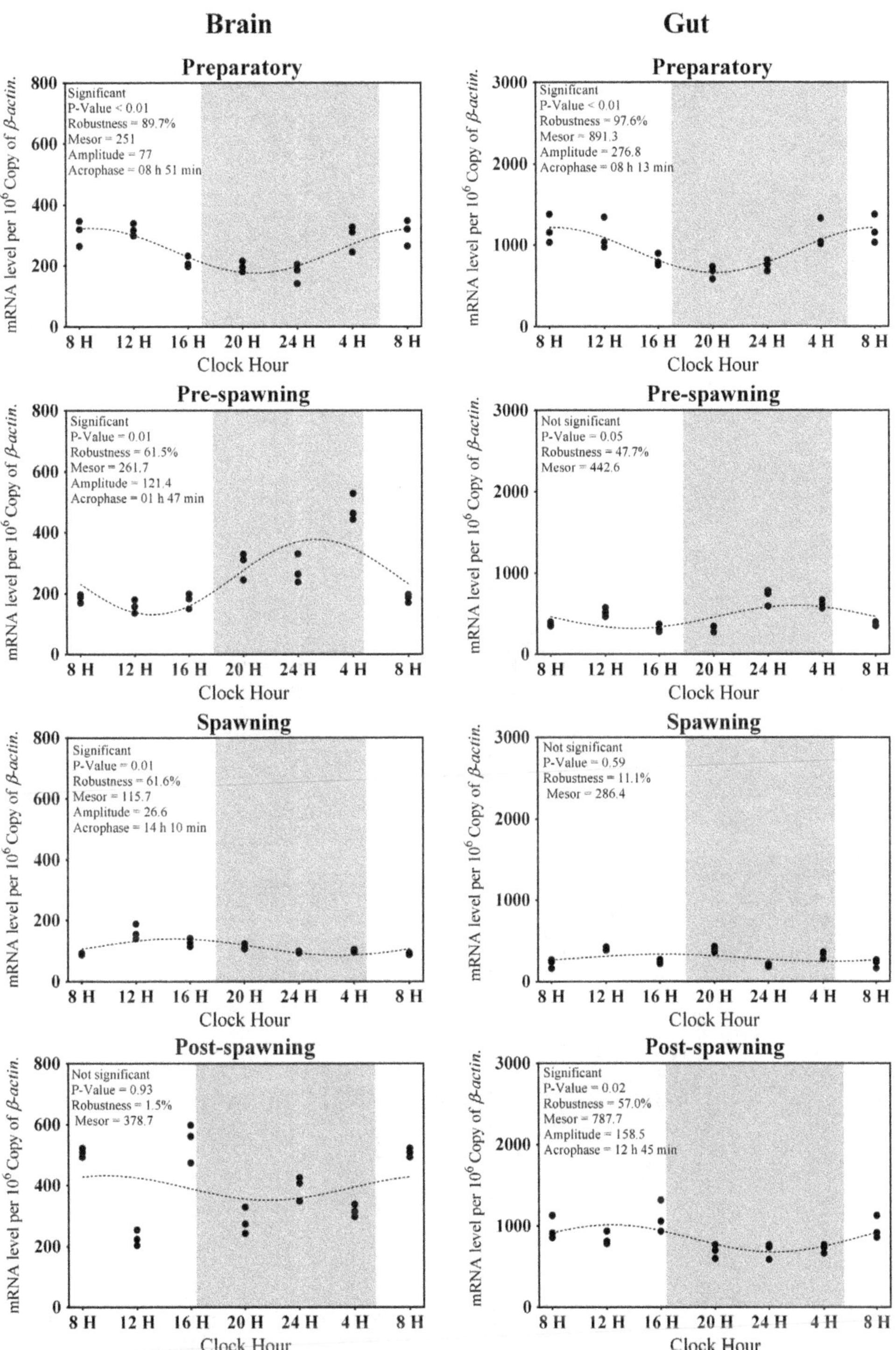

Figure 4. Changes in the expression of *Aanat1* in six different time points (8, 12, 16, 20, 24, and 4 h) in different seasons (n = 3) were fitted to nonlinear regression curve using the formula "Y = Mesor + Amplitude Cos (Frequency X + Acrophase)".

Notes: Cosinor analysis was performed to obtain the rhythm parameters and the existence of rhythm. p-value < 0.05 is considered as significant.

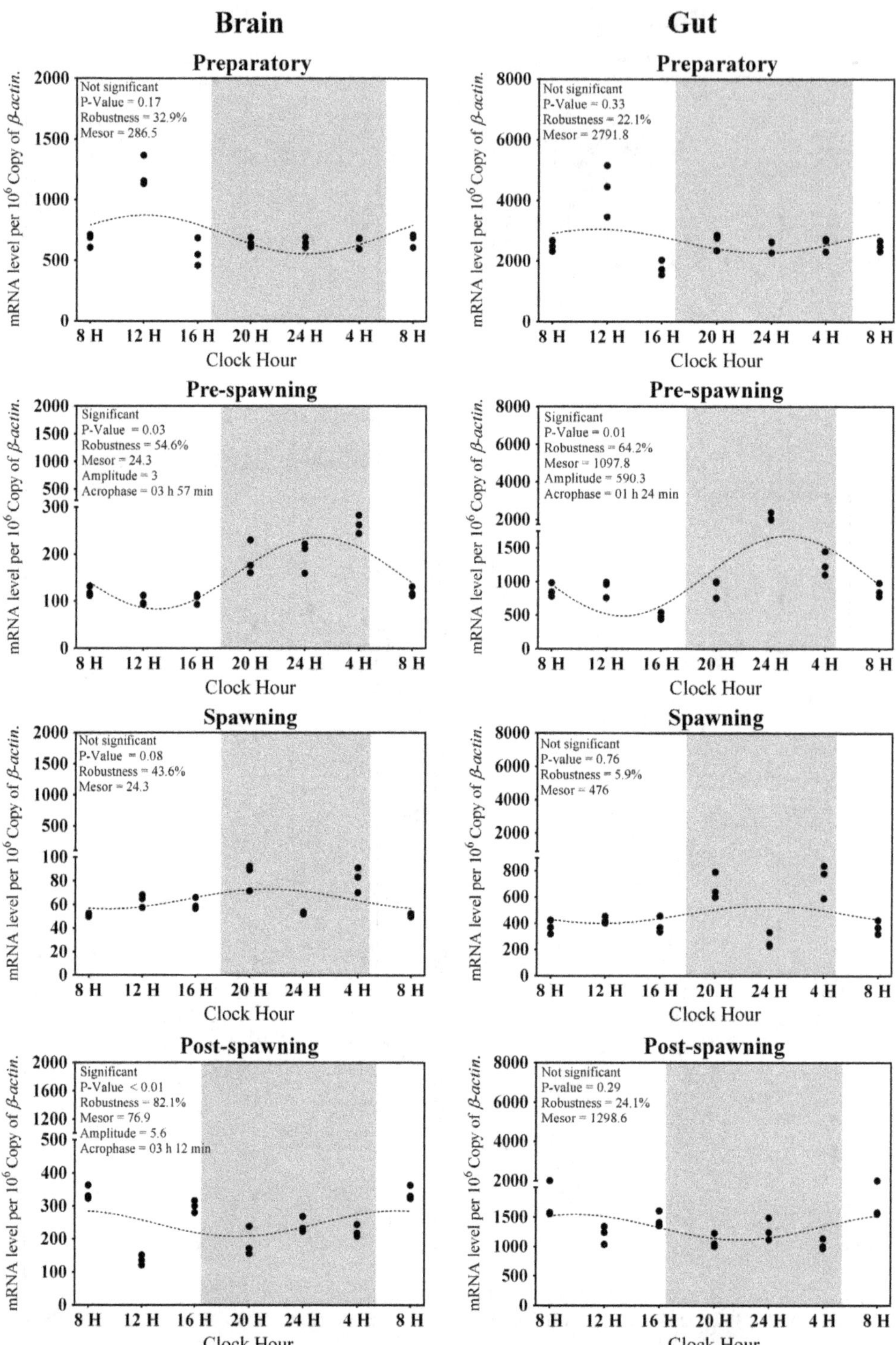

Figure 5. Changes in the expression of *Aanat2* in six different time points (8, 12, 16, 20, 24, and 4 h) in different seasons (n = 3) were fitted to nonlinear regression curve using the formula "Y = Mesor + Amplitude Cos (Frequency X + Acrophase)".

Notes: Cosinor analysis was performed to obtain the rhythm parameters and the existence of rhythm. p-value < 0.05 is considered as significant.

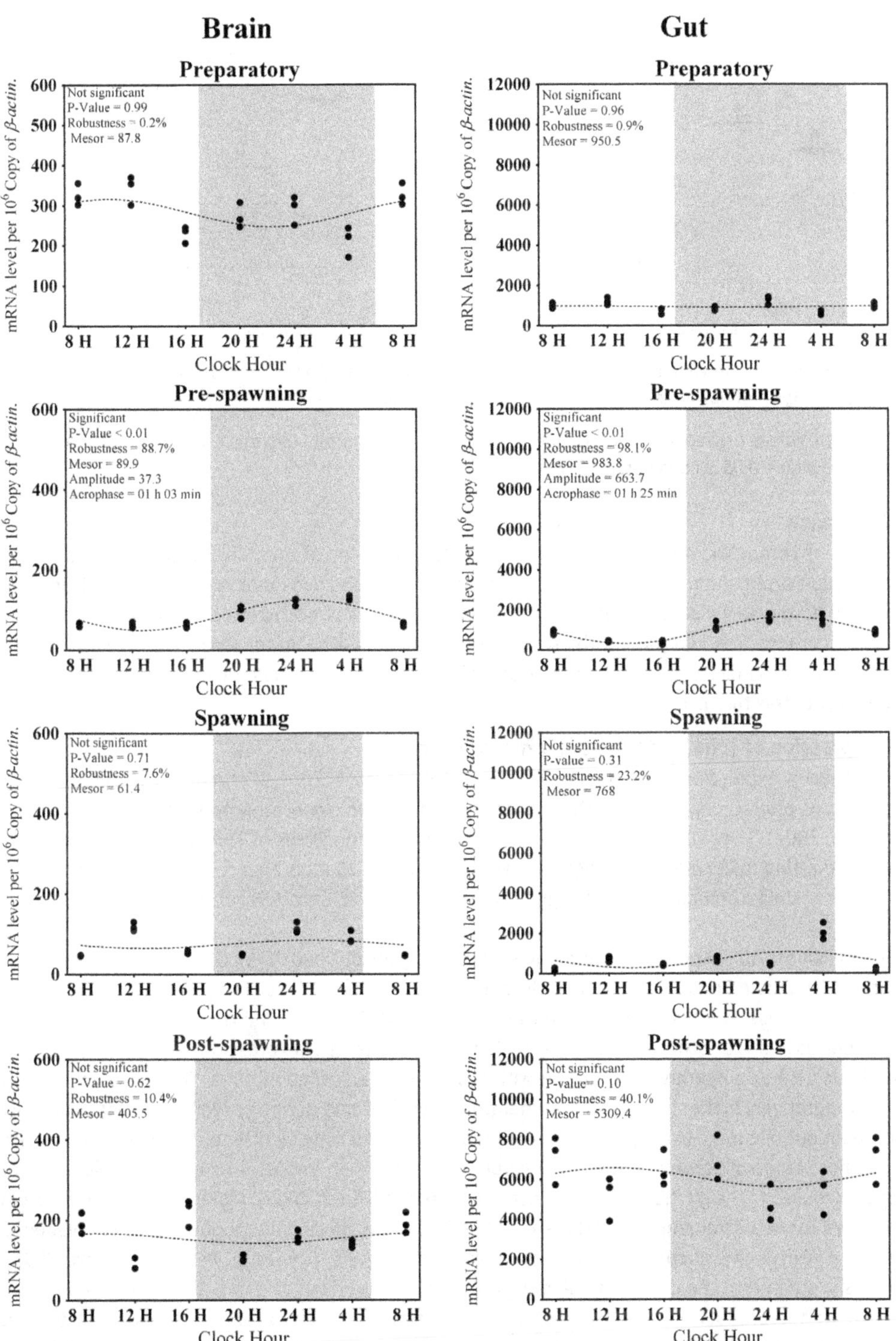

Figure 6. Changes in the expression of *Hiomt* in six different time points (8, 12, 16, 20, 24, and 4 h) in different seasons (n = 3) were fitted to nonlinear regression curve using the formula "Y = Mesor + Amplitude Cos (Frequency X + Acrophase)".

Notes: Cosinor analysis was performed to obtain the rhythm parameters and the existence of rhythm. p-value < 0.05 is considered as significant.

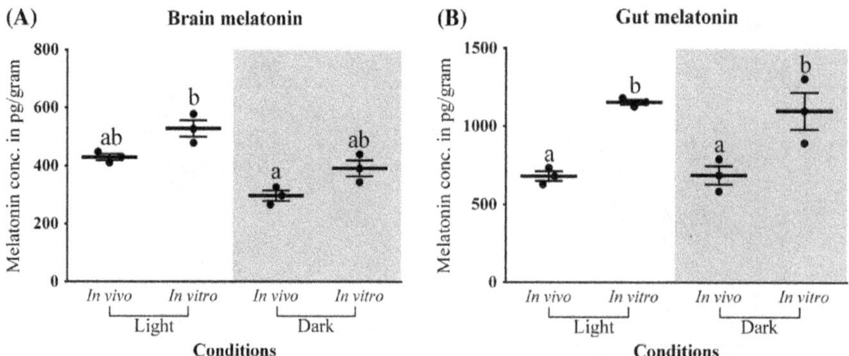

Figure 7. *In vivo* and *in vitro* melatonin synthesis in (A) Brain and (B) Gut tissues was analyzed by two-way analysis of variance followed by Student–Newman–Keuls.

Notes: The values are represented by circle and the bar in the graph indicates mean ± S.E.M with $n = 3$. p-value < 0.05 is considered as significant.

4. Discussion

The present communication demonstrates the presence of melatonin bio-synthesizing machinery both in adult brain and gut of any tropical carp with a daily and seasonal variation at the transcript level. Our data also validate the production of this indole amine both *in vivo* and *in vitro* cultures in these two organs. The variation in the mRNA of melatonin-synthesizing genes according to the environmental cues indicates the importance of brain and gut melatonin in the maintenance of annual physiology of this tropical carp.

4.1. Expression of Tph1 and Hiomt in brain and gut

Tissue-specific expression of *Tph* has been reported in zebrafish; *Tph1* (a and b) is involved mainly in the areas of photoreception, phototransduction, and production of melatonin (Bellipanni, Rink, & Bally-Cuif, 2002). The cloning, sequencing, and phylogenetic analyses of *Tph* and *Hiomt* from pineal gland and retina in *C. catla* (unpublished data from our group) identified them as *Tph1* and *Hiomt*, respectively, unlike goldfish *g-Hiomt-1* and *g-Hiomt-2* (Velarde, Cerda-Reverter et al., 2010).

The mRNA of *Tph1* in both brain and gut is giving a significant negative correlation with the water temperature (Table 3) and photoperiod (data not shown) in its expression. It is noteworthy that the preparatory phase expression of *Tph1* in the brain is not giving the same trend with photo-thermal conditions as it is with other genes. On the other hand, the level of expression of *Tph1* is very less in gut, though it has a negative correlation with the photo-thermal conditions. This low level of *Tph1* and comparatively higher levels of other melatonin synthesizing genes as well as higher melatonin content in gut (Figures 1(A), (B) and 7) in comparison to brain may be due to the possible existence of another source of serotonin in the gut as gut serotonin is an important signaling molecule that helps in regulating the gastrointestinal functions (Gershon & Tack, 2007). More serotonin in the brain is required for neurotransmission (Sugden, 2003). The tissue-specific distribution of *Tph1* revealed a weak and arrhythmic circadian expression in rat brain (Malek, Dardente, Pevet, & Raison, 2005), which is in support of the present finding (Figure 1(A)). This pattern of expression is also supporting the arrhythmic production of a trace amount of melatonin in the brain, as reported in the rat (Jimenez-Jorge et al., 2007). Further, the present study also reveals the presence of melatonin in both brain and gut at light phase, signifying a pineal-independent mechanism of expression of this gene in both brain and gut as suggested from the AANAT activity in rat brain (Jimenez-Jorge et al., 2007) and study on carp (Seth & Maitra, 2010).

On the other hand, the level of expression of *Hiomt* is very less in the brain as compared to gut (Figures 2(C) and (D)). The expression of *Hiomt* in gut portrays an insignificant negative trend of correlation with water temperature, though in the brain it is significant (Table 3). The rich melatonin content in fish gut found in this study (Figure 7) is also supportive of previous findings (Mukherjee & Maitra, 2015), and may be explained with the ample copy number of *Hiomt* mRNA, observed in this study. A recent study on rainbow trout is also showing a high transcription of *Hiomt* in gut

(Muñoz-Pérez et al., 2016). This particular observation may be linked to the notion that *Hiomt* acts as the key regulator in melatonin production (Liu & Borjigin, 2005). A lack of correlation of the expression of *Hiomt* in the gut with environmental factors (Table 3) also indicates a non-photo-thermal regulation of melatonin production (Mukherjee & Maitra, 2015; Velarde et al., 2009; Velarde, Delgado et al., 2010) in gut. Moreover, high expression of *Hiomt* in brain (Figures 2(C) and 6) during preparatory (phase of preparation for reproduction) is a sign of involvement of brain melatonin in the modulation of seasonal reproductive parameters through hypothalamo–pituitary–gonadal axis (Chattoraj et al., 2005; Falcón, Besseau, Sauzet, & Boeuf, 2007). Earlier studies on the same fish also demonstrated the role of environmental photo-thermal conditions with the development of gonads through melatonin (Bhattacharya et al., 2007; Dey et al., 2005; Maitra, Chattoraj, Mukherjee, & Moniruzzaman, 2013).

4.2. Expression of *Aanat1* and *Aanat2* in brain and gut

This is the first evidence for the presence of both isoforms of *Aanat* transcripts (*Aanat1* and *Aanat2*) in brain and gut in any tropical fish. Study on rainbow trout recently showed the transcription of *Aanat1* and *Aanat2* in gut (Muñoz-Pérez et al., 2016). The expression of *Aanat* mRNA has been demonstrated in various parts of the rat brain (Uz et al., 2002). The present study denotes an inverse relationship of the expression of *Aanat1* and *Aanat2* in both brain and gut with the environmental photo-thermal conditions in all phases, though in brain, *Aanat1* gave an insignificant value (Table 3). Involvement of environmental cues in the regulation of these enzymes is valid for both brain and gut in fish as evidenced from previous studies (Falcón et al., 2010). In both tissues, the expressivity of *Aanat2* is more than *Aanat1*, signifying the importance of *Aanat2* in the regulation of melatonin synthesis in carp (Figures 1(C) and (D), 2(A) and (B)). Presence of Val^{115} and Val^{161} is important for the temperature-dependent activity of AANAT2 (Cazamea-Catalan et al., 2013). This indicates AANAT2 might play a role in temperature regulation of melatonin bio-synthesis. Higher expression of *Aanat2* in the gut is also proportionate to the copy number of *Hiomt* and higher content of melatonin in the same tissue (Figures 2(B) and (D); Figure 7). Interestingly, in the brain, in every season, any one of the isoforms is rhythmic in their expression, but during the spawning phase, *Aanat* expression became arrhythmic (Figures 4 and 5). Similar kinds of observations are available in pike, where *Aanat1* is giving a peak at light in retina and *Aanat2* in dark in pineal (Coon, Begay, Deurloo, Falcon, & Klein, 1999), which may be explained through the proteasomal proteolysis of AANAT by light at the downstream of the clock (Gastel, Roseboom, Rinaldi, Weller, & Klein, 1998). The current finding demonstrates the alternation (season wise) of *Aanat1* and *Aanat2* rhythmicity and this further explains the probable reasons to have two isoforms of *Aanat* in fish. Both isoforms cannot remain rhythmic in same photo-thermal conditions in both brain and gut, except in the pre-spawning of the brain (Figures 4 and 5) which might be due to the difference in their optimum photo-thermal conditions. This mechanism may help the system react immediately to the length of photoperiod which cannot be anticipated by the endogenous clock (Coon et al., 1999). Insignificant variation of melatonin content between light and dark phases (Figure 7) suggests more involvement of endogenous cues in the regulation of melatonin synthesis in gut. The regulation of expression of *Aanat* through external cues (Table 3) supports the hypothesis of its chronobiological importance in the central nervous system (Uz et al., 2002). Study on the melatonin synthesizing machinery in the gut of rainbow trout also indicated a control by environmental light cycle and is likely to be under circadian regulation (Muñoz-Pérez et al., 2016).

Moreover, findings on goldfish (Velarde et al., 2009; Velarde, Cerda-Reverter et al., 2010) and mammals (Konturek, Brzozowski, & Konturek, 2011) suggest that the rhythm of *Aanat2* and *Hiomt* expression and melatonin production can be regulated by peripheral clocks entrained by non-photic cues (Liu & Borjigin, 2005). On the contrary, very recent studies indicate the synchronization of physiological rhythm in various parts of the body or even in the microbial population of the gut (Paulose, Wright, Patel, & Cassone, 2016) may be harmonized by melatonin. Further, these physiological processes of entrainment and/or synchronization are influenced by seasonal changes (Falcón et al., 2010; Maitra et al., 2013).

The recent study from our group has demonstrated that the synchronization of melatonin synthesizing genes and clock-associated genes in central (brain) and peripheral organs (ovary) of zebra fish is dependent on environmental photic conditions (Khan et al., 2016), supporting the "orchestra" model (Richards & Gumz, 2012). It was demonstrated earlier that the self-sustaining circadian oscillators exist in several peripheral organs in fish (Whitmore, Foulkes, Strahle, & Sassone-Corsi, 1998). The present work indicates a seasonal and diurnal variation of four major melatonin synthesizing enzyme genes in an annual cycle of a tropical carp in brain and gut. It is evident from the statistical analysis that the rhythm of transcription of all these genes is mostly insignificant in daily basis, but an overall variation can be seen depending on the photo-thermal conditions. Future study involving alteration of photo-thermal setting and pattern of expression of clock-associated genes will clarify the synchronization of melatonin synthesizing machinery in brain (central) and gut (peripheral) in this tropical carp.

Acknowledgments

Authors are grateful to the Director, IBSD India, We are thankful to Mr L. Umakanta Singh, Mr S. Surjit Singh, Mr W. Rahul, and Mr H. Bishorjit for their continuous support in the collection and maintenance of fish in laboratory conditions. Our sincere thanks to all of our colleagues of Fish Biology and Insect Biology Laboratories for their kind cooperations during the study. Our sincere gratitude to every Indian tax payer for their contribution to the funding for research. The authors are really grateful to three distinguished reviewers for their valuable comments and suggestions. The IBSD, Imphal manuscript number is 2015008.

Funding

The work was supported by IBSD (PhD JRF Programme); Department of Biotechnology, Ministry of Science and Technology [grant number BT/407/NE/U-Excel/2013]; CSIR-JRF Programme [grant number (09/1044(0002)/2012-EMR-I(14/11/2012)]; Science and Engineering Research Board [grant number SR/FT/LS-179/2010].

Competing Interests

The authors declare no competing interest.

Author details

Haobijam Sanjita Devi[1]
E-mail: sanjitahaobijam@gmail.com
Chongtham Rajiv[1]
E-mail: rajivchongtham@yahoo.com
Gopinath Mondal[1]
E-mail: gopinathmondal89@gmail.com
Zeeshan Ahmad Khan[1]
E-mail: acezeeshan@live.com
ORCID ID: http://orcid.org/0000-0002-0737-0253
Sijagurumayum Dharmajyoti Devi[1]
E-mail: shjyoti5@gmail.com
Thangal Yumnamcha[1]
E-mail: tyumnamcha@gmail.com
Rupjyoti Bharali[2]
E-mail: rupjyotibharali@gmail.com
Asamanja Chattoraj[1]
E-mails: achattoraj.ibsd@nic.in, asamanja.chattoraj@gmail.com
ORCID ID: http://orcid.org/0000-0002-8656-7459

[1] Biological Rhythm Laboratory, Animal Resources Programme, Department of Biotechnology, Institute of Bioresources and Sustainable Development, Government of India, Takyelpat, Imphal 795 001, Manipur, India.

[2] Department of Biotechnology, Gauhati University, Guwahati 781 014, Assam, India.

References

Appelbaum, L., & Gothilf, Y. (2006). Mechanism of pineal-specific gene expression: The role of E-box and photoreceptor conserved elements. *Molecular and Cellular Endocrinology, 252*, 27–33. doi:10.1016/j.mce.2006.03.021

Bellipanni, G., Rink, E., & Bally-Cuif, L. (2002). Cloning of two tryptophan hydroxylase genes expressed in the diencephalon of the developing zebrafish brain. *Mechanisms of Development, 119*, S215–S220. http://dx.doi.org/10.1016/S0925-4773(03)00119-9

Bertrand, P. P., Bertrand, R. L., Camello, P. J., & Pozo, M. J. (2010). Simultaneous measurement of serotonin and melatonin from the intestine of old mice: The effects of daily melatonin supplementation. *Journal of Pineal Research, 49*, 23–34. doi:10.1111/j.1600-079X.2010.00760.x

Bhattacharya, S., Chattoraj, A., & Maitra, S. K. (2007). Melatonin in the regulation of annual testicular events in carp *Catla catla* : Evidence from the studies on the effects of exogenous melatonin, continuous light, and continuous darkness. *Chronobiology International, 24*, 629–650. doi:10.1080/07420520701534665

Bubenik, G. A. (2002). Gastrointestinal melatonin: Localization, function, and clinical relevance. *Digestive Diseases and Sciences, 47*, 2336–2348. http://dx.doi.org/10.1023/A:1020107915919

Bubenik, G. A., & Brown, G. M. (1997). Pinealectomy reduces melatonin levels in the serum but not in the gastrointestinal tract of rats. *Neurosignals, 6*, 40–44. http://dx.doi.org/10.1159/000109107

Bubenik, G. A., & Pang, S. F. (1997). Melatonin levels in the gastrointestinal tissues of fish, amphibians, and a reptile. *General and Comparative Endocrinology, 106*, 415–419. doi:10.1006/gcen.1997.6889

Cazamea-Catalan, D., Magnanou, E., Helland, R., Besseau, L., Boeuf, G., Falcon, J., & Jorgensen, E. H. (2013). Unique arylalkylamine N-acetyltransferase-2 polymorphism in Salmonids and profound variations in thermal stability and catalytic efficiency conferred by two residues. *The Journal of experimental biology*. doi:10.1242/jeb.080960

Chattoraj, A., Bhattacharyya, S., Basu, D., Bhattacharya, S., Bhattacharya, S., & Maitra, S. K. (2005). Melatonin accelerates maturation inducing hormone (MIH): Induced oocyte maturation in carps. *General and Comparative Endocrinology, 140*, 145–155. doi:10.1016/j.ygcen.2004.10.013

Chattoraj, A., Bhattacharyya, S., Dey, R., Seth, M., & Maitra, S. K. (2006). The pineal organ and reproduction in fish. *Proceedings of the Zoological Society (Calcutta), 59*, 35–60.

Chattoraj, A., Seth, M., & Maitra, S. K. (2009). Localization and dynamics of Mel1a melatonin receptor in the ovary of carp *Catla catla* in relation to serum melatonin levels. *Comparative Biochemistry and Physiology Part A: Molecular & Integrative Physiology, 152*, 327–333. doi:10.1016/j.cbpa.2008.11.010

Chen, C. Q., Fichna, J., Bashashati, M., Li, Y. Y., & Storr, M. (2011). Distribution, function and physiological role of melatonin in the lower gut. *World Journal of Gastroenterology, 17*, 3888–3898. doi:10.3748/wjg.v17.i34.3888

Chowdhury, I., & Maitra, S. K. (2012). Melatonin in the promotion of health. In W. RR (Ed.), *Melatonin time line: From discovery to therapy* (pp. 1-60). Boca Raton, FL: Taylor and Francis.

Chowdhury, V. S., Yamamoto, K., Ubuka, T., Bentley, G. E., Hattori, A., & Tsutsui, K. (2010). Melatonin stimulates the release of gonadotropin-inhibitory hormone by the avian hypothalamus. *Endocrinology, 151*, 271–280. doi:10.1210/en.2009-0908

Coon, S. L., Begay, V., Deurloo, D., Falcon, J., & Klein, D. C. (1999). Two arylalkylamine N-acetyltransferase genes mediate melatonin synthesis in fish. *Journal of Biological Chemistry, 274*, 9076–9082. http://dx.doi.org/10.1074/jbc.274.13.9076

Dey, R., Bhattacharya, S., & Maitra, S. K. (2005). Importance of photoperiods in the regulation of ovarian activities in indian major carp *Catla catla* in an annual cycle. *Journal of Biological Rhythms, 20*, 145–158. doi:10.1177/0748730404272925

Dey, R., Bhattacharya, S., Maitra, S. K., & Banerji, T. K. (2003). The morpho-anatomy and histology of the pineal complex in a major indian carp, *Catla catla* : Identification of the pineal photoreceptor cells and their responsiveness to constant light and constant darkness during different phases of the annual reproductive cycle. *Endocrine Research, 29*, 429–443. http://dx.doi.org/10.1081/ERC-120026949

Falcón, J. (1999). Cellular circadian clocks in the pineal. *Progress in Neurobiology, 58*, 121–162. http://dx.doi.org/10.1016/S0301-0082(98)00078-1

Falcón, J., Besseau, L., Sauzet, S., & Boeuf, G. (2007). Melatonin effects on the hypothalamo-pituitary axis in fish. *Trends in Endocrinology & Metabolism, 18*, 81–88. doi:10.1016/j.tem.2007.01.002

Falcón, J., Migaud, H., Muñoz-Cueto, J. A., & Carrillo, M. (2010). Current knowledge on the melatonin system in teleost fish. *General and Comparative Endocrinology, 165*, 469–482. doi:10.1016/j.ygcen.2009.04.026

Fernández-Durán, B., Ruibal, C., Polakof, S., Ceinos, R. M., Soengas, J. L., & Míguez, J. M. (2007). Evidence for arylalkylamine N-acetyltransferase (AANAT2) expression in rainbow trout peripheral tissues with emphasis in the gastrointestinal tract. *General and Comparative Endocrinology, 152*, 289–294. doi:10.1016/j.ygcen.2006.12.008

Gastel, J. A., Roseboom, P. H., Rinaldi, P. A., Weller, J. L., & Klein, D. C. (1998). Melatonin production: Proteasomal proteolysis in serotonin N-acetyltransferase regulation. *Science, 279*, 1358–1360. http://dx.doi.org/10.1126/science.279.5355.1358

Gershon, M. D., & Tack, J. (2007). The serotonin signaling system: From basic understanding to drug development for functional GI disorders. *Gastroenterology, 132*, 397–414. doi:10.1053/j.gastro.2006.11.002

Hasan, K. N., Moniruzzaman, M., & Maitra, S. K. (2014). Melatonin concentrations in relation to oxidative status and oocyte dynamics in the ovary during different reproductive phases of an annual cycle in carp *Catla catla*. *Theriogenology, 82*, 1173–1185. doi:10.1016/j.theriogenology.2014.08.001

Iigo, M., Abe, T., Kambayashi, S., Oikawa, K., Masuda, T., Mizusawa, K., ... Yanagisawa, T. (2007). Lack of circadian regulation of *in vitro* melatonin release from the pineal organ of salmonid teleosts. *General and Comparative Endocrinology, 154*, 91–97. doi:10.1016/j.ygcen.2007.06.013

Isorna, E., Besseau, L., Boeuf, G., Desdevises, Y., Vuilleumier, R., Alonso-Gómez, A. L., ... Falcón, J. (2006). Retinal, pineal and diencephalic expression of frog arylalkylamine N-acetyltransferase-1. *Molecular and Cellular Endocrinology, 252*, 11–18. doi:10.1016/j.mce.2006.03.032

Jimenez-Jorge, S., Guerrero, J. M., Jimenez-Caliani, A. J., Naranjo, M. C., Lardone, P. J., Carrillo-Vico, A., ... Molinero, P. (2007). Evidence for melatonin synthesis in the rat brain during development. *Journal of Pineal Research, 42*, 240–246. doi:10.1111/j.1600-079X.2006.00411.x

Khan, Z. A., Yumnamcha, T., Rajiv, C., Devi, H. S., Mondal, G., Devi, S. D., & Chattoraj, .A. (2016). Melatonin biosynthesizing enzyme genes and clock genes in ovary and whole brain of zebrafish (Danio rerio): Differential expression and a possible interplay. *General and Comparative Endocrinology, 233*, 16–31. doi:10.1016/j.ygcen.2016.05.014

Konturek, P. C., Brzozowski, T., & Konturek, S. J. (2011). Gut clock: Implication of circadian rhythms in the gastrointestinal tract. *Journal of Physiology and Pharmacology : An Official Journal of the Polish Physiological Society, 62*, 139–150.

Korf, H. W., Schomerus, C., & Stehle, J. H. (1998). The pineal organ, its hormone melatonin, and the photoneuroendocrine system. *Advances in anatomy, embryology, and cell biology, 146*, 1–8. http://dx.doi.org/10.1007/978-3-642-58932-4

Kulczykowska, E., Kalamarz, H., Warne, J. M., & Balment, R. J. (2006). Day-night specific binding of 2-[125I] Iodomelatonin and melatonin content in gill, small intestine and kidney of three fish species. *Journal of Comparative Physiology B, 176*, 277–285. doi:10.1007/s00360-005-0049-4

Kumari, K., Pathakota, G.-B., Annam, P.-K., Kumar, S., & Krishna, G. (2015). Characterisation and validation of house keeping gene for expression analysis in *Catla catla* (Hamilton). *Proceedings of the National Academy of Sciences, India Section B: Biological Sciences, 85*, 993–1000. doi:10.1007/s40011-014-0482-9

Lillesaar, C. (2011). The serotonergic system in fish. *Journal of Chemical Neuroanatomy, 41*, 294–308. doi:10.1016/j.jchemneu.2011.05.009

Liu, T., & Borjigin, J. (2005). N-acetyltransferase is not the rate-limiting enzyme of melatonin synthesis at night. *Journal of Pineal Research, 39*, 91–96. doi:10.1111/j.1600-079X.2005.00223.x

Maitra, S., & Chattoraj, A. (2007). Role of photoperiod and melatonin in the regulation of ovarian functions in Indian carp *Catla catla*: Basic information for future application. *Fish Physiology and Biochemistry, 33*, 367–382. http://dx.doi.org/10.1007/s10695-007-9174-1

Maitra, S. K., Chattoraj, A., Mukherjee, S., & Moniruzzaman, M. (2013). Melatonin: A potent candidate in the regulation of fish oocyte growth and maturation. *General and Comparative Endocrinology, 181*, 215–222. doi:10.1016/j.ygcen.2012.09.015

Malek, Z. S., Dardente, H., Pevet, P., & Raison, S. (2005). Tissue-specific expression of tryptophan hydroxylase mRNAs in the rat midbrain: Anatomical evidence and daily profiles. *European Journal of Neuroscience, 22*, 895-901. doi:10.1111/j.1460-9568.2005.04264.x

Messner, M., Huether, G., Lorf, T., Ramadori, G., & Schwörer, H. (2001). Presence of melatonin in the human hepatobiliary-gastrointestinal tract. *Life Sciences, 69*, 543–551. http://dx.doi.org/10.1016/S0024-3205(01)01143-2

Mukherjee, S., & Maitra, S. K. (2015). Gut melatonin in vertebrates: Chronobiology and physiology. *Frontiers in endocrinology, 6*, 112. doi:10.3389/fendo.2015.00112

Mukherjee, S., Moniruzzaman, M., & Maitra, S. K. (2014). Daily and seasonal profiles of gut melatonin and their temporal relationship with pineal and serum melatonin in carp *Catla catla* under natural photo-thermal conditions. *Biological Rhythm Research, 45*, 301–315. doi:10.1080/09291016.2013.817139

Muñoz, J. L., Ceinos, R. M., Soengas, J. L., & Míguez, J. M. (2009). A simple and sensitive method for determination of melatonin in plasma, bile and intestinal tissues by high performance liquid chromatography with fluorescence detection. *Journal of Chromatography B, 877*, 2173–2177. doi:10.1016/j.jchromb.2009.06.001

Muñoz-Pérez, J. L., López-Patiño, M. A., Álvarez-Otero, R., Gesto, M., Soengas, J. L., & Míguez, J. M. (2016). Characterization of melatonin synthesis in the gastrointestinal tract of rainbow trout (Oncorhynchus mykiss): Distribution, relation with serotonin, daily rhythms and photoperiod regulation. *Journal of Comparative Physiology B, 186*, 471–484. doi:10.1007/s00360-016-0966-4

Nelson, W., Tong, Y. L., Lee, J. K., & Halberg, F. (1979). Methods for cosinor-rhythmometry. *Chronobiologia, 6*, 305–323.

Nisembaum, L. G., Tinoco, A. B., Moure, A. L., Alonso Gómez, A. L., Delgado, M. J., & Valenciano, A. I. (2013). The arylalkylamine-N-acetyltransferase (AANAT) acetylates dopamine in the digestive tract of goldfish: A role in intestinal motility. *Neurochemistry International, 62*, 873–880. doi:10.1016/j.neuint.2013.02.023

Noche, R. R., Lu, P. N., Goldstein-Kral, L., Glasgow, E., & Liang, J. O. (2011). Circadian rhythms in the pineal organ persist in zebrafish larvae that lack ventral brain. *BMC Neuroscience, 12*, 7. doi:10.1186/1471-2202-12-7

Paulin, C. H., Cazamea-Catalan, D., Zilberman-Peled, B., Herrera-Perez, P., Sauzet, S., Magnanou, E., & Besseau, L. (2015). Subfunctionalization of arylalkylamine N-acetyltransferases in the sea bass Dicentrarchus labrax: Two-ones for one two. *Journal of Pineal Research, 59*, 354–364. doi:10.1111/jpi.12266

Paulose, J. K., Wright, J. M., Patel, A. G., & Cassone, V. M. (2016). Human gut bacteria are sensitive to melatonin and express endogenous circadian rhythmicity. *Plos One, 11*, e0146643. doi:10.1371/journal.pone.0146643

Piesiewicz, A., Kedzierska, U., Turkowska, E., Adamska, I., & Majewski, P. M. (2015). Seasonal postembryonic maturation of the diurnal rhythm of serotonin in the chicken pineal gland. *Chronobiology International, 32*, 59–70. doi:10.3109/07420528.2014.955185

Portaluppi, F., Smolensky, M. H., & Touitou, Y. (2010). Ethics and methods for biological rhythm research on animals and human beings. *Chronobiology International, 27*, 1911–1929. doi:10.3109/07420528.2010.516381

Refinetti, R., Cornélissen, G. C., & Halberg, F. (2007). Procedures for numerical analysis of circadian rhythms. *Biological Rhythm Research, 38*, 275–325. doi:10.1080/09291010600903692

Reiter, R. J. (1991a). Melatonin synthesis: Multiplicity of regulation. *Advances in Experimental Medicine and Biology, 294*, 149–158. http://dx.doi.org/10.1007/978-1-4684-5952-4

Reiter, R. J. (1991b). Melatonin: The chemical expression of darkness. *Molecular and Cellular Endocrinology, 79*, C153–C158. http://dx.doi.org/10.1016/0303-7207(91)90087-9

Reiter, R. J., Tan, D.-X., Manchester, L. C., & Gitto, E. (2010). Gastrointestinal tract and melatonin: Reducing pathophysiology. *Polish Gastroenterology, 17*, 213–218.

Richards, J., & Gumz, M. L. (2012). Advances in understanding the peripheral circadian clocks. *The FASEB Journal, 26*, 3602–3613. doi:10.1096/fj.12-203554

Seth, M., & Maitra, S. K. (2010). Importance of light in temporal organization of photoreceptor proteins and melatonin-producing system in the pineal of carp *Catla catla*. *Chronobiology International, 27*, 463–486. doi:10.3109/07420521003666416

Seth, M., & Maitra, S. K. (2011). Neural regulation of dark-induced abundance of arylalkylamine N -acetyltransferase (AANAT) and melatonin in the carp (*Catla catla*) pineal: An in vitro study. *Chronobiology International, 28*, 572–585. doi:10.3109/07420528.2011.590913

Stefulj, J., Hortner, M., Ghosh, M., Schauenstein, K., Rinner, I., Wolfler, A., & Liebmann, P. M. (2001). Gene expression of the key enzymes of melatonin synthesis in extrapineal tissues of the rat. *Journal of Pineal Research, 30*, 243–247. http://dx.doi.org/10.1034/j.1600-079X.2001.300408.x

Sugden, D. (2003). Comparison of circadian expression of tryptophan hydroxylase isoform mRNAs in the rat pineal gland using real-time PCR. *Journal of Neurochemistry, 86*, 1308–1311. http://dx.doi.org/10.1046/j.1471-4159.2003.01959.x

Uz, T., Qu, T., Sugaya, K., & Manev, H. (2002). Neuronal expression of arylalkylamine N-acetyltransferase (AANAT) mRNA in the rat brain. *Neuroscience Research, 42*, 309–316. http://dx.doi.org/10.1016/S0168-0102(02)00011-1

Velarde, E., Cerdá-Reverter, J. M., Alonso-Gómez, A. L., Sánchez, E., Isorna, E., & Delgado, M. J. (2010). Melatonin-synthesizing enzymes in pineal, retina, liver, and gut of the goldfish (Carassius): mRNA expression pattern and regulation of daily rhythms by lighting conditions. *Chronobiology International, 27*, 1178–1201. doi:10.3109/07420528.2010.496911

Velarde, E., Delgado, M. J., & Alonso-Gómez, A. L. (2010). Serotonin-induced contraction in isolated intestine from a teleost fish (Carassius auratus): Characterization and interactions with melatonin. *Neurogastroenterology & motility, 22*, e364–e373. doi:10.1111/j.1365-2982.2010.01605.x

Velarde, E., Haque, R., Iuvone, P. M., Azpeleta, C., Alonso-Gomez, A. L., & Delgado, M. J. (2009). Circadian clock genes of goldfish, Carassius auratus: cDNA Cloning and rhythmic expression of Period and Cryptochrome transcripts in retina, liver, and gut. *Journal of Biological Rhythms, 24*, 104–113. doi:10.1177/0748730408329901.

Whelan, J. A., Russell, N. B., & Whelan, M. A. (2003). A method for the absolute quantification of cDNA using real-time PCR. *Journal of Immunological Methods, 278*, 261–269. doi:10.1016/S0022-1759(03)00223-0

Whitmore, D., Foulkes, N. S., Strahle, U., & Sassone-Corsi, P. (1998). Zebrafish clock rhythmic expression reveals independent peripheral circadian oscillators. *Nature neuroscience, 1*, 701–707. doi:10.1038/3703

Permissions

All chapters in this book were first published in CB, by Cogent OA; hereby published with permission under the Creative Commons Attribution License or equivalent. Every chapter published in this book has been scrutinized by our experts. Their significance has been extensively debated. The topics covered herein carry significant findings which will fuel the growth of the discipline. They may even be implemented as practical applications or may be referred to as a beginning point for another development.

The contributors of this book come from diverse backgrounds, making this book a truly international effort. This book will bring forth new frontiers with its revolutionizing research information and detailed analysis of the nascent developments around the world.

We would like to thank all the contributing authors for lending their expertise to make the book truly unique. They have played a crucial role in the development of this book. Without their invaluable contributions this book wouldn't have been possible. They have made vital efforts to compile up to date information on the varied aspects of this subject to make this book a valuable addition to the collection of many professionals and students.

This book was conceptualized with the vision of imparting up-to-date information and advanced data in this field. To ensure the same, a matchless editorial board was set up. Every individual on the board went through rigorous rounds of assessment to prove their worth. After which they invested a large part of their time researching and compiling the most relevant data for our readers.

The editorial board has been involved in producing this book since its inception. They have spent rigorous hours researching and exploring the diverse topics which have resulted in the successful publishing of this book. They have passed on their knowledge of decades through this book. To expedite this challenging task, the publisher supported the team at every step. A small team of assistant editors was also appointed to further simplify the editing procedure and attain best results for the readers.

Apart from the editorial board, the designing team has also invested a significant amount of their time in understanding the subject and creating the most relevant covers. They scrutinized every image to scout for the most suitable representation of the subject and create an appropriate cover for the book.

The publishing team has been an ardent support to the editorial, designing and production team. Their endless efforts to recruit the best for this project, has resulted in the accomplishment of this book. They are a veteran in the field of academics and their pool of knowledge is as vast as their experience in printing. Their expertise and guidance has proved useful at every step. Their uncompromising quality standards have made this book an exceptional effort. Their encouragement from time to time has been an inspiration for everyone.

The publisher and the editorial board hope that this book will prove to be a valuable piece of knowledge for researchers, students, practitioners and scholars across the globe.

List of Contributors

Aakanksha Pant and Prem Prakash
Department of Botany, Government P.G. College, Kumaun University, Dwarahat, Almora 263653, India

Rakesh Pandey
Department of Microbial Technology and Nematology, Central Institute of Medicinal and Aromatic Plants, P.O. CIMAP, Almora, Lucknow 226 015, India

Rishendra Kumar
Department of Biotechnology, Kumaun University, Nainital 263136, India

Randal S. Stahl, Richard M. Engeman and Richard E. Mauldin
United States Department of Agriculture/Animal and Plant Health Inspection Service/Wildlife Services/National Wildlife Research Center, 4101 LaPorte Ave., Fort Collins, CO, USA

Michael L. Avery
United States Department of Agriculture/Animal and Plant Health Inspection Service/Wildlife Services/National Wildlife Research Center Florida Field Station, 2820 East University Ave., Gainesville, FL, USA

Malcolm E. Connor and Thomas R. King
Department of Biomolecular Sciences, Central Connecticut State University, 1615 Stanley Street, New Britain, CT 06053, USA

Jessica Benkaroun, Ashley N.K. Kroyer and Andrew S. Lang
Department of Biology, Memorial University of Newfoundland, 232 Elizabeth Ave., St. John's, NL, Canada A1B 3X9

Dany Shoham
Begin-Sadat Center for Strategic Studies, Bar Ilan University, Building 109, Ramat Gan 5290002, Israel

Hugh Whitney
Newfoundland and Labrador Forestry and Agrifoods Agency, St. John's, NL A1E 3Y5, Canada

Lidianne Salvatierra
Instituto Nacional de Pesquisas da Amazônia, Laboratório de Ecologia e Sistemática de Invertebrados do Solo, Av. André Araújo, 2.936, CEP 69067-375, Manaus, Amazonas, Brazil

Marlus Q. Almeida
Laboratório de Ecologia Terrestre, Universidade Federal do Amazonas, Av. Gal. Rodrigo Otávio Jordão Ramos 3000, CEP 69067-000, Manaus, Amazonas, Brazil

S. Roy, B. Ukil and L.M. Lyndem
Parasitology Research Laboratory, Department of Zoology, Visva-Bharati University, Santiniketan 731235, West Bengal, India

Nuno M. Félix, Hugo Pissara and Maria M.R.E. Niza
Faculty of Veterinary Medicine, CIISA, ULisboa, Avenida da Universidade Técnica, 1300-477 Lisboa, Portugal

Isabelle Goy-Thollot
VetAgro Sup - Hémostase, Inflammation et Sepsis, SIAMU, Université de Lyon, F-69280 Marcy L'Etoile, France

Ronald S. Walton
Speciality and Emergency of Tacoma, 5608 S. Durango Street, Tacoma, Washington 98409, USA

Pedro M. Borralho and Cecília M.P. Rodrigues
Faculty of Pharmacy, Research Institute for Medicines (iMed. ULisboa), Universidade de Lisboa, Avenida Professor Gama Pinto, 1649-003 Lisboa, Portugal

Ana S. Matos
Faculdade de Ciências e Tecnologia, Departamento de Engenharia Mecânica e Industrial, Universidade Nova de Lisboa (UNIDEMI), 2829-516 Caparica, Portugal

Hong Ngoc Pham and Le Son Hoang
Department of Biochemistry, School of Biotechnology, International University - Vietnam National University, HCMC, Vietnam

Van Trung Phung
Institute of Chemical Technology - Vietnam Academy of Science and Technology, HCMC, Vietnam

Colm O'Reilly and Naomi Harte
School of Engineering, Sigmedia, ADAPT Centre, Trinity College Dublin, Dublin, Ireland

Olufemi Samuel Bamidele, Joshua Oluwafemi Ajele and Folasade Mayowa Olajuyigbe
Enzymology Research Unit, Department of Biochemistry, The Federal University of Technology, P. M. B 704, Akure, Nigeria

Y.X. Wang, T. Engelmann and W. Schwarz
Shanghai Research Center for Acupuncture and Meridians, Shanghai, China
Shanghai Key Laboratory of Acupuncture Mechanism and Acupoint Function, Fudan University, Shanghai, China
Institute for Biophysics, Goethe-University Frankfurt, Frankfurt am Main, Germany

Y.F. Xu
Shanghai Research Center for Acupuncture and Meridians, Shanghai, China
Shanghai Key Laboratory of Acupuncture Mechanism and Acupoint Function, Fudan University, Shanghai, China

Kenneth M. Palanza, Legairre A. Radden, Mohammed A. Rabah, Tu V. Nguyen, Audra C. Kohm, Malcolm E. Connor, Morgan M. Ricci, Jachius J. Stewart, Sidney Eragene and Thomas R. King
Department of Biomolecular Sciences, Central Connecticut State University, 1615 Stanley Street, New Britain, CT 06053, USA

Zongqiang Lian and Xudong Wu
Ningxia Fisheries Research Institute, Yinchuan 750001, China
Ningxia Engineering Research Center for Fisheries, Yinchuan 750001, China

Brian Walcott and Mahipal Singh
Animal Biotechnology Program, Agricultural Research Station, Fort Valley State University, Fort Valley, GA 31088, USA

Christian R. Moreno
Faculty of Biology, Department of Human and Animal Biology, Havana University, Havana, Cuba

Lida Sanchez
Faculty of Biology, Department of Human and Animal Biology, Havana University, Havana, Cuba
Division of Ecology and Evolutionary Biology, Graduate School of Life Sciences, Tohoku University, Sendai 980-8578, Japan

Emanuel C. Mora
Faculty of Biology, Department of Human and Animal Biology, Havana University, Havana, Cuba
Instituto de Ciencias Biomédicas, Universidad Autónoma de Chile, El Llano Subercaseaux 2801, Santiago de Chile, Chile

Haobijam Sanjita Devi, Chongtham Rajiv, Gopinath Mondal, Zeeshan Ahmad Khan, Sijagurumayum Dharmajyoti Devi, Thangal Yumnamcha and Asamanja Chattoraj
Biological Rhythm Laboratory, Animal Resources Programme, Department of Biotechnology, Institute of Bioresources and Sustainable Development, Government of India, Takyelpat, Imphal 795 001, Manipur, India

Rupjyoti Bharali
Department of Biotechnology, Gauhati University, Guwahati 781 014, Assam, India

Index

www.ingramcontent.com/pod-product-compliance
Lightning Source LLC
Chambersburg PA
CBHW082018190326
41458CB00010B/3224